D0782418

Chance, Strategy, and Choice

Games and elections offer familiar, current, and lively subjects for a course in mathematics. With applications in a range of areas such as economics, political science, and sociology, these topics are of broad interest. This undergraduate textbook, primarily intended for a general-education course in game theory at the freshman or sophomore level, provides an elementary treatment of games and elections. Starting with basics such as gambling games, Nash equilibria, zero-sum games, social dilemmas, combinatorial games, and fairness and impossibility theorems for elections, the text then goes further into the theory with accessible proofs of advanced topics such as the Sprague–Grundy Theorem and Arrow's Impossibility Theorem. The book

- uses an integrative approach to probability theory, game theory, and social choice theory by highlighting the mix of ideas occurring in seminal results on games and elections such as the Minimax Theorem, allowing students to develop intuition in all areas while delving deeper into the theory;
- provides a gentle introduction to the logic of mathematical proof, thus equipping readers with the necessary tools for further mathematical studies, a feature not shared by most game theory texts;
- contains numerous exercises and examples of varying levels of difficulty to help students learn and retain the material;
- requires only a high school mathematical background, thus making the text accessible to a large range of students.

Samuel Bruce Smith is Professor and Chair of Mathematics at Saint Joseph's University and a past director of the University Honors Program. In 2012, he won the Tengelmann Award for Distinguished Teaching and Research.

CAMBRIDGE MATHEMATICAL TEXTBOOKS

Cambridge Mathematical Textbooks is a program of undergraduate and beginning graduate level textbooks for core courses, new courses, and interdisciplinary courses in pure and applied mathematics. These texts provide motivation with plenty of exercises of varying difficulty, interesting examples, modern applications, and unique approaches to the material.

Chance, Strategy, and Choice

An Introduction to the Mathematics of Games and Elections

Samuel Bruce Smith
Saint Joseph's University, Philadelphia, PA, USA

CAMBRIDGE
UNIVERSITY PRESS

32 Avenue of the Americas, New York, NY 10013-2473, USA

Cambridge University Press is part of the University of Cambridge.

It furthers the University's mission by disseminating knowledge in the pursuit of education, learning, and research at the highest international levels of excellence.

www.cambridge.org
Information on this title: www.cambridge.org/9781107084520

First published 2015

Printed in the United States of America

A catalog record for this publication is available from the British Library.

Library of Congress Cataloging in Publication Data
Smith, Samuel B., 1966–
Chance, strategy, and choice : an introduction to the mathematics of games and elections / Samuel Bruce Smith, Saint Joseph's University, Philadelphia, PA, USA.
pages cm
Includes bibliographical references and index.
ISBN 978-1-107-08452-0 (hardback)
1. Game theory–Textbooks. 2. Games of chance (Mathematics)–Textbooks. I. Title.
QA269.S587 2015
519.3–dc23 2015003107

ISBN 978-1-107-08452-0 Hardback

To Patty, David, and Ned

Contents

Preface *page* ix

Part I First Notions

1 Introduction 3
2 Games and Elections 11
3 Chance 23
4 Strategy 37
5 Choice 57
6 Strategy and Choice 77
7 Choice and Chance 89
8 Chance and Strategy 105
9 Nash Equilibria 119
Web Resources 135

Part II Basic Theory

10 Proofs and Counterexamples 139
11 Laws of Probability 157
12 Fairness in Elections 175
13 Weighted Voting 187
14 Gambling Games 203
15 Zero-Sum Games 217
16 Partial-Conflict Games 233
17 Takeaway Games 251
18 Fairness and Impossibility 263
Suggestions for Further Reading 277

Part III Special Topics

19 Paradoxes and Puzzles in Probability 281
20 Combinatorial Games 299
21 Borda versus Condorcet 315
22 The Sprague–Grundy Theorem 335
23 Arrow's Impossibility Theorem 349
Suggestions for Further Reading 371

Bibliography **373**
Index **377**

Preface

This book was written to teach an accessible and authentic mathematics course to a general audience. Games and elections are ideal topics for such a course. The motivating examples are familiar and engaging: from the 2000 U.S. presidential election, to economic pricing wars, to gambling games such as Poker and Blackjack. The elementary theory of these topics involves manageable computations and emphasizes analytic reasoning skills. Most important, the study of games and elections has produced a wealth of beautiful theorems, many within the last century. Games and elections offer a wide selection of elegant results that can be approached without extensive technical background. A course on these topics provides a view of mathematics as a powerful and thriving modern discipline.

There are excellent texts covering aspects of the theory of games and elections, such as those of Straffin [48] and Taylor and Pacelli [49]. The topics are also frequently covered as chapters in liberal arts mathematics texts. This text is pitched at a level between these two types of treatments. The book is designed to be more mathematically ambitious than the general textbook but to allow for a less demanding course than one offered using a more specialized text.

A novelty of this book is the integrative approach taken to the three subjects: probability theory, game theory, and social choice theory. This approach highlights the mix of ideas occurring in seminal results on games and elections such as the Jury Theorem, the Minimax Theorem, and the Gibbard-Satterthwaite Theorem. On a practical level, the integrative approach allows for a more gradual development of material. Rather than taking a steep descent in one area, the chapters follow a spiraling path from chance to strategy to social choice and back again, exploring examples and developing techniques and intuitions in all areas while delving deeper into the theory of games and elections.

Structure of the Book

This book is divided into three parts. Part I introduces the main devices used in the text. These are the probability tree, the game tree, the payoff matrix, and the preference table. Expected values, dominated strategies, backward induction, and rational play are then defined for the various types of games. The idea of preference ballots is motivated with examples from sports and politics, and the basic voting methods are introduced.

Part I includes three chapters exploring two of the three notions of chance, strategy, and choice together. The paradox of the chair, the Gibbard-Satterthwaite Theorem, swing states in the Electoral College, the Jury Theorem, and a simplified version of Poker are introduced in these chapters. The final chapter of Part I presents Nash equilibria as a unifying concept, tying together several examples in the categories of zero-sum, partial-conflict, mixed-strategy, and electoral games.

Part II focuses more directly on the development of mathematical theory. A chapter on logic prepares the way, with topics from the first part of the book serving as representative examples of conditional and universal statements. The method of proof by induction for game trees is illustrated with a version of Zermelo's Theorem. Conditionals are explored further in the probability setting, while fairness criteria for voting methods provide exercises with universal statements and an exposure to the dichotomy between proofs and counterexamples in mathematics. Counting techniques are developed for computing the power index of a weighted voting system. The problem of determining weights for a yes-no system and the notion of a mathematical invariant are introduced, leading up to a statement of the Taylor-Zwicker Theorem.

Part II also includes a series of chapters exploring the various types of games. Counting techniques are applied again to compute probabilities with card hands in a chapter on gambling games. The game Craps shows the power of introducing a conditional, whereas a problem in Poker motivates Bayes' formula. A chapter on zero-sum games contains the proof of the 2×2 Minimax Theorem and introduces techniques for the $2 \times n$ case. A chapter on partial-conflict games explores backward induction through a variety of examples, including the Iterated Prisoner's Dilemma, the Chain Store Game, the Dollar Auction, and Bram's Theory of Moves. Next, a chapter on takeaway games introduces the Sprague–Grundy numbers and includes a proof of Bouton's beautiful solution of the game Nim. The final chapter in Part II is a return to the question of fairness for voting methods. Fairness criteria for social welfare methods including Arrow's famous Independence criteria are introduced. May's Theorem, a simplified version of Arrow's Theorem, and Sen's Impossibility Theorem are proved in this chapter.

Part III consists of five independent chapters offering possible enhancements to a basic course. A chapter on probability explores famous paradoxes ranging from the Monte Hall Problem to Bertrand's Paradox in geometry. Nim Misère, Chomp, and Hex are analyzed in a chapter on combinatorial games. The debate over fairness between Borda and Condorcet is explored, with the arguments for both sides interlaced with three theorems: a version of the Jury Theorem, a modern result of McGarvey's on the realization of social preference graphs, and Borda's Theorem from his original paper introducing the Borda count. The final two chapters focus on proofs of two beautiful results on games and elections. The Sprague–Grundy Theorem for combinatorial games is proved, introducing the notion of Nim sums. The game Hackenbush is analyzed as an application. The final chapter is devoted to a proof of Arrow's Impossibility Theorem. This chapter includes a brief discussion of the modern era of social choice theory with the proof of Sen's Coherence Theorem and the introduction of nonpreferential voting methods.

Designing a Course

The text was written to allow for flexibility in designing a course. I have taught a general-education course from this book for several years. I typically cover the first two parts of the text treading lightly on the discussion of induction in Chapter 10 and subsequent proofs by induction in Part II. On occasion, I have had time to add one of the chapters in Part III as a last topic. I have also taught an honors version of this course for a number of semesters. This course moved quickly through Part I and included one or two chapters in Part III according to student interest.

It is easy to design a course emphasizing one topic or another from the various areas. For students who have had a course in probability, Chapters 3 and 11 can be skipped. A course emphasizing social choice theory might omit the developments on combinatorial games in Chapter 17. Induction proofs beginning with Zermelo's Theorem and the Law of the Probability Tree in Chapters 10 and 11 can be omitted depending on the audience. The book can also be used for an elective course for sophomore-level mathematics majors. Such a course would emphasize the proof techniques including induction. The final chapters on the Sprague–Grundy Theorem and Arrow's Impossibility Theorem are worthy end goals for such a course.

Acknowledgments

I am very grateful to Kaitlin Leach, editor at Cambridge University Press, for her expert help and enthusiastic support for this project. I thank the anonymous reviewers of the book for their insightful comments, critiques, and suggestions and James Madru for his careful editing. I am indebted to Professor Gregory Manco for many helpful discussions about teaching these topics. Finally, it is a pleasure to acknowledge the students from the course "Math of Games and Politics" at Saint Joseph's University, who have inspired me to write this book.

Part I

First Notions

1 Introduction

In their great variety, from contests of global significance such as a championship match or the election of a president down to a coin flip or a show of hands, games and elections share one common feature: each game or election offers the possibility of a final, decisive result obtained according to well-established rules, a public **outcome**. Mathematics offers a similar possibility. In mathematics, the rules are founded in the laws of logic and represent a formalization of our basic common sense. An outcome in mathematics, to pursue the analogy, is a **theorem**, a statement that can be proven true. The outcome of a game or election may be surprising or expected. Similarly, a theorem can either defy intuition or confirm a well-evidenced conjecture. Just as a game or an election separates winners from losers, a proven theorem distinguishes true statements from false ones, creating a new fact-of-the-matter about mathematics.

This book is an introduction to the mathematical theory of games and elections. We pursue the analogy between analyzing a game or election and developing a mathematical theory somewhat further before turning to our main topics. First, just as a game or an election creates a new language of specialized terminology, a mathematical theory begins with the formulation of *definitions*. A mathematical **definition** is a precise and verifiable description of an object of study. We adopt the following convention for definitions: when we define a new term, we use **boldface**. Thus the terms "outcome," "theorem," and "definition" were defined earlier. (Of course, subsequent definitions will be more technical than these.) We use *italics* to indicate we are mentioning a term that has not been defined yet but will be defined later. For example, we mentioned the term "definition" before subsequently defining it earlier.

The best way to learn to play a new game is often to just give it a try. In mathematics, the corresponding hands-on method of learning is the study of *examples*. By an **example**, we mean a specific instance of a definition or, alternately, a particular consequence of a theorem. A great advantage of our chosen topics is the wealth of examples. Not only do we have many familiar games and election methods available, creating new examples is as easy as making up rules. When referring to games or election methods, we will use capital letters.

We organize our study around the roles of *chance, strategy, and choice*. We define these terms loosely here. By an act of **chance**, we mean an event whose outcome is not predetermined or controlled by human agency. By an act of **strategy**, we mean a situation in which a person chooses among possible moves. Finally, by a **social choice**, we mean a decision by a group of individuals (the society) to select one of several competing alternatives. We briefly discuss the history of the study of these topics. As we do so, we highlight three questions arising from representative examples in each area. These questions will motivate more advanced developments in the second part of the book.

The study of pure chance belongs to the field of **probability theory**, the science of odds. The modern theory has roots in gambling games in Paris casinos. In the 1650s, a correspondence between two luminaries of the time, Blaise Pascal and Pierre de Fermat, led to a solution of a basic problem on iterated rolls of the dice, laying the groundwork for the modern mathematical theory. The twentieth century saw deep entrenching of the probabilistic viewpoint with the discovery of quantum mechanics and the development of mathematical statistics. Probability theory is now part of the common vernacular as we talk about sports, the stock market, and the weather.

Among the simplest versions of Poker is the game called **Five-Card Stud**. Five-Card Stud can be played between two or among several players. Each player is first dealt two **down** cards (cards not revealed to other players). The remaining three cards for each player are dealt **up** (revealed to all players) one by one. After each player receives a new up card, the players bet on their evolving hands. When a player makes a bet, each subsequent player may either **fold** (resign and give up any chance at the pot), **call** (match the bet), or **raise** (match the bet and make a new, added bet). When all the cards are turned and all players have either folded or called the last bet, the eligible players show their down cards. The player with the best Poker hand takes the pot. Here is an example.

Example 1.1 A Hand of Five-Card Stud

We consider the following scenario in which only two players survive to the final round of betting. The cards are as shown. The pot is \$100,000. Our opponent has bet \$50,000 more. We are faced with two choices: call the \$50,000 bet or fold and lose our chance at the pot.

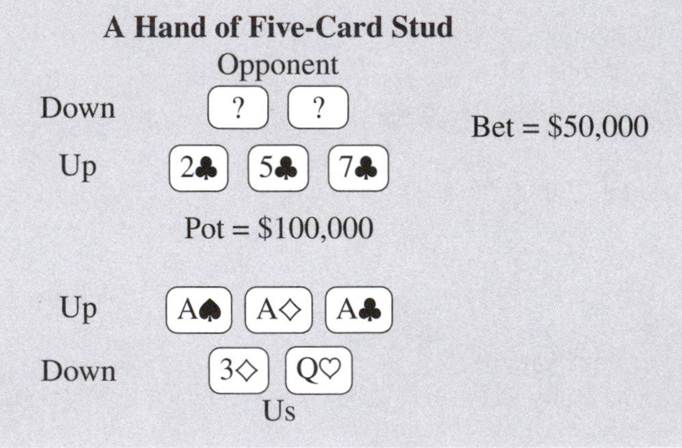

We will introduce the rankings of Poker hands in Chapter 14. Here we simply observe that our hand is **Three of a Kind** (three cards with the same value, in this case aces). The only possible hand our opponent can have that beats ours is a **Flush**, five cards of the same suit. Specifically, our opponent must be holding two clubs to beat us.

Poker is a game involving both chance and strategy. The strategy in Five-Card Stud arises solely in the betting, the decision to call or fold or how much to bet to open and whether to raise. To keep this particular example in the realm of pure chance, we assume that we know that our opponent bets on a losing hand only 3 percent of the time. A bet on a losing hand made with the intention of inducing an opponent to fold is called a **bluff**. Our first question is the following:

Question 1.2 *Should we call or fold in A Hand of Five-Card Stud?*

If we call, we risk $50,000 more but have a chance at what would be a $200,000 pot! If we fold, the game is over. Of course, the decision to call or fold in a game of Poker is a personal and financial one. We will propose an answer to Question 1.2 in the form of an *expected value* in Chapter 14. The analysis of A Hand of Five-Card Stud will lead us to discover a powerful method in probability theory called *Bayes' Law*.

Poker was a central example in the development of **game theory**, the study of games of strategy and strategic conflict. Émile Borel began the mathematical study of strategy in the 1920s. His work included an analysis of a simplified version of Poker. Borel conjectured the existence of an *equilibrium* for games in which the players are in total conflict. In 1928, John von Neumann proved Borel's conjecture true in what is now known as the *Fundamental Theorem of Game Theory*. We will prove a special case of this theorem in Chapter 15.

John Nash analyzed a three-player Poker game in his seminal 1950s Ph.D. thesis. The analysis was a first application of his spectacular generalization of von Neumann's Theorem. The celebrated Nash Equilibrium Theorem is now the centerpiece of modern game theory, with applications in fields ranging from economics to business to evolutionary biology. Nash was awarded the Nobel Prize in Economics in 1994 for his work.

Perhaps the most purely mathematical games are those with no role for chance. Games such as Checkers and Chess present contests of pure strategy and high complexity, rewarding superior cleverness and the ability to foresee future contingencies. We introduce a simple game of this class that can be played with just a handful of Poker chips, 11 chips in this case. Our example is a variation on the game called *Nim*.

Example 1.3 A Game of Nim

The game we propose is played as follows: the first player, referred to as Player 1, is handed 11 Poker chips. She divides the chips into any number of piles. The second player, Player 2, then takes any number of chips (but at least one) from one of the piles. Player 2 can only take chips from one pile but can take as many as desired, including the whole pile. Player 1 goes next, taking any number of chips (but at least one) from one of the remaining piles. The game continues in this way, with each player taking chips from one pile on his or her turn. The game is over when all the chips are taken. The last player to take chips is the winner. Here is a sample round:

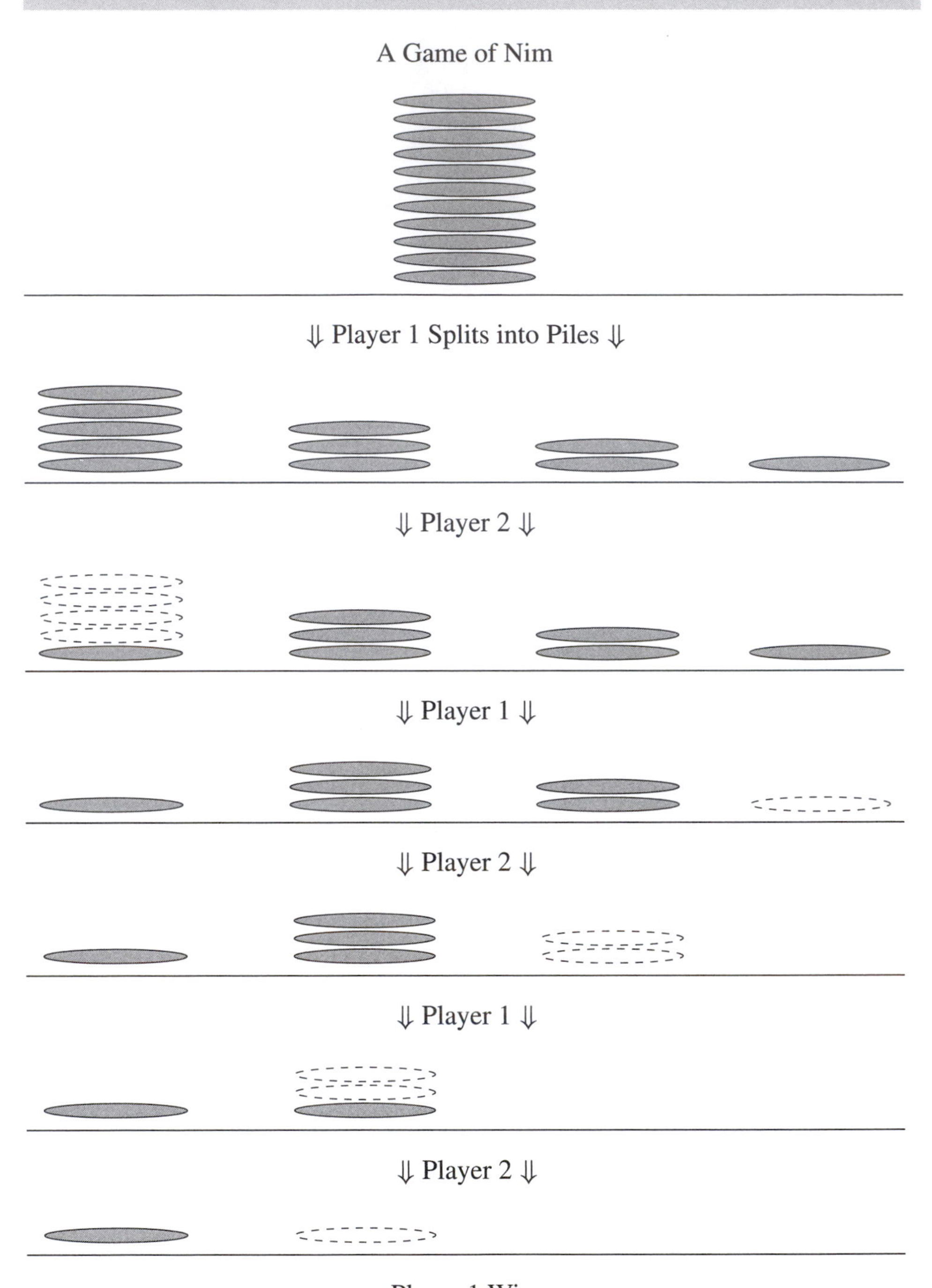

Of course, the preceeding scenario is only one of the many ways the game can be played. We pose the following question:

Question 1.4 *In A Game of Nim, which player has the advantage, Player 1 or Player 2?*

We address Question 1.4 in Chapter 17. The answer comes as a consequence of a remarkable theorem published by Charles Bouton in 1901. Bouton gave a complete mathematical solution to the game of Nim, setting the stage for extensive twentieth-century research on the mathematics of games of pure strategy.

The study of elections, or **social choice theory**, is an interdisciplinary field with branches in political science and philosophy as well as mathematics. The question of how to conduct elections is one faced by every political entity. The answers often reveal the character of the government. Even among democracies, the ideals of fairness and universal representation are pursued in various ways. Here we may compare, for example, the majority runoff methods for presidential elections of Ireland and France with the method of delegates used in the U.S. Electoral College.

The mathematical theory of elections traces back to a debate carried out in academic journals in late eighteenth-century France. The Chevalier Jean-Charles de Borda introduced the rank-ordering method now bearing his name for elections involving multiple candidates. Borda argued for the suitability of his method by introducing what is now known as a *fairness criterion*. During this same period, the Marquis Nicolas de Condorcet uncovered a basic obstacle to fairness for elections called the *Condorcet Paradox*. Condorcet critiqued Borda's method for ignoring the majority opinion in certain cases. He argued for a voting method featuring head-to-head elections. Condorcet based his arguments for his majoritarian voting method on his proof of the Jury Theorem. We prove a version of this theorem in Chapter 21.

The foundation for the modern theory of social choice is the 1950s' doctoral thesis of Kenneth Arrow. Arrow developed a mathematical language for the field, giving precise definitions for election mechanisms and their fairness. Within this framework, he proved his celebrated Impossibility Theorem, establishing that all democratic election mechanisms allowing more than two candidates violate particular principles of fairness. His work opened the door to wide-ranging research on the mathematics of voting and elections. Arrow was awarded the Nobel Prize in 1972. We prove Arrow's Impossibility Theorem in Chapter 23.

Social choice theory offers a wealth of problems concerning the distribution of power in legislative systems. We pose a simple question about one of the most elegant legislative bodies, the U.S. Senate.

Example 1.5 The U.S. Senate

The U.S. Senate is comprised of 100 senators: two senators for each of the 50 states. A bill passes the Senate with a majority vote. However, in the event of a 50–50 tie, the vice president, who is known as the President of the Senate, casts the tie-breaking vote.

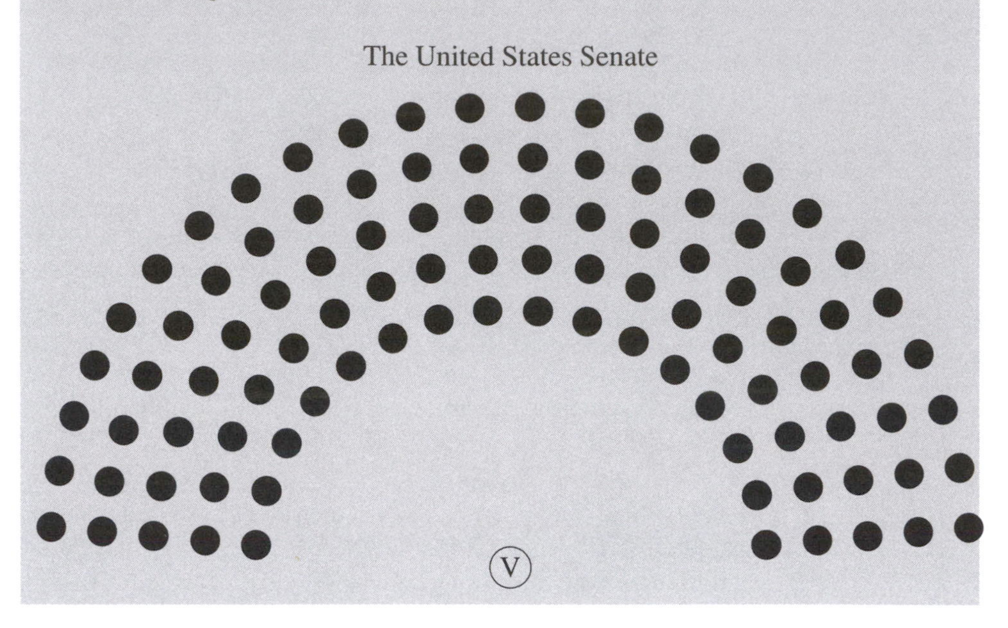

The question we pose here is the following:

Question 1.6 *Who has more power in the U.S. Senate, an individual senator or the vice president?*

We note that the U.S. Senate presents many fascinating problems of voting strategy, with roll-call votes, amendment riders, and filibusters available among the various methods. Our question concerns only the basic power distribution of the voting system. We answer Question 1.6 in Chapter 13 using the idea of a *power index*, a notion that connects the study of elections with probability theory. We will see that the answer to Question 1.6 is essentially the proof of a famous *combinatorial identity*.

We conclude with some general remarks about the text. As the preceding discussion suggests, we emphasize the interplay between the ideas of chance, strategy, and choice in our study of these three topics. The integration will be especially apparent in the beginning chapters. As we delve deeper into the separate fields of probability, game, and social choice theory, we will focus more closely on the area at hand.

In addition to the topics of games and elections, the text is intended to give an introduction to mathematics. As discussed earlier, we will begin with precise definitions and make use of examples to illustrate ideas and build intuition. Although our approach will be elementary, we will move steadily in our analysis so that by the end of the text we will be in a position to prove some important theorems about games and elections.

2 Games and Elections

In this chapter we formulate mathematical definitions of a *game* and an *election* and give examples. The examples of games will indicate some of the categories we will study and will also serve to introduce the two main devices we will use to represent games: the *game tree* and the *payoff matrix*. For elections, we introduce the notions of *true preferences* and *insincere voting*. We consider the famous example of the 2000 presidential election in the state of Florida as an illustration of the connection between games and elections.

We begin by thinking a bit about how to define the notion of a *game*. The term has broad meanings. What properties should we use to limit the scope? We can probably agree that a game should involve an act of competition with roles for skill, ingenuity, and maybe luck. Should a game be fun? Should it be fair? Fun is, of course, a matter of taste. Fairness also seems difficult to ensure. A grandmaster can play a game of Chess against a novice, but surely this this is not a "fair" game. However, allowing a weaker player to have a handicap destroys the intrinsic fairness of a game but may be the only way to create a good match.

We define a game in a manner that is quite inclusive, making no commitment either way as to the questions of fun, fairness, or other such qualities. We focus instead on what a game produces. This is the **outcome**, by which we mean some particular, well-defined final status such as "win," "loss," or "tie." Whatever the method or means of play, when the dust settles and the game is over, we expect to have a well-defined final result.

Further, we often want to know the score of the game or the winnings for each player. We will refer to these as the *payoffs*. Precisely, the **payoffs** are numeric quantities assigned to each player for each particular outcome. We can think of the payoff to a player as either points or a dollar amount. Payoffs will be either zero (no gain), positive (a gain), or negative (a loss). When appropriate, we will express payoffs as dollar amounts. We can now formulate a mathematical definition of a game.

Definition 2.1 A **game** is an event or mechanism involving two or more players that produces a well-defined outcome resulting in a payoff to each player.

We can picture the playing of a game, here, a four-player game, as a process:

A Four-Player Game

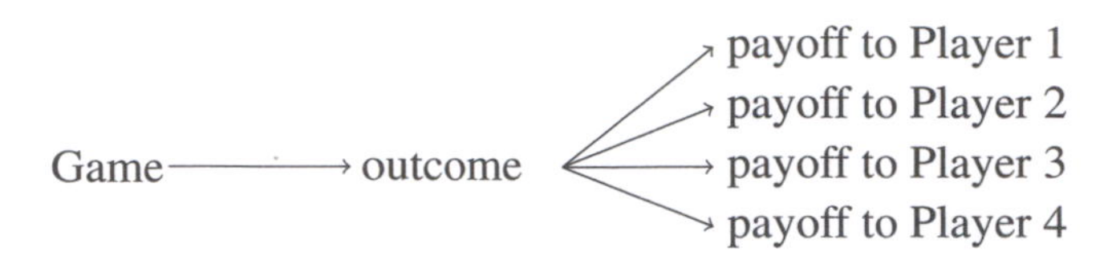

We often focus on two-player games. In this setting, we have the special class of **win-lose games,** two-player games in which the only outcomes are win and lose. If ties are also possible, then we say this is a **win-lose-tie game.** The payoffs for such games may be taken, implicitly, to be win $= +1$, lose $= -1$, and tie $= 0$.

We can classify games according to the method by which the outcome is achieved. By a **game of pure chance** we mean a game in which the outcome is produced by an act of chance. Examples include the children's game Chutes and Ladders and the casino game Roulette. A **game of pure strategy** is a game in which there is no role for chance. The strategic decisions of the players lead directly to the outcome of the game. Examples here include Tic-Tac-Toe, Checkers, and Chess. Perhaps the simplest example of a game is the following:

Example 2.2 Heads or Tails

This is the classic method of making a decision between two competing options. One player calls "Heads" or "Tails." The other player flips a coin. The calling player wins if he or she calls the flip correctly. Otherwise, the flipping player wins.

What kind of game is Heads or Tails? Note that the outcome of Heads or Tails is achieved in two steps. The first step in the game involves a strategic decision, albeit a rather trivial one. The second step is an act of chance. We refer to games involving both chance and strategy as **games mixing chance and strategy**. These include many of the most entertaining games, such as Monopoly, Scrabble, Poker, and Bridge.

The game Heads or Tails serves to introduce an important device called the *game tree*. A game tree is used to represent games in which players take turns, one by one, until the final outcome is produced. Such a game is called a **sequential-move game**. Most board games are sequential-move games. The game tree shows the progress of such a game as it moves turn by turn to the outcome. The last stage is the outcome of the game and so, in our case, produces a winner and a loser. We label accordingly:

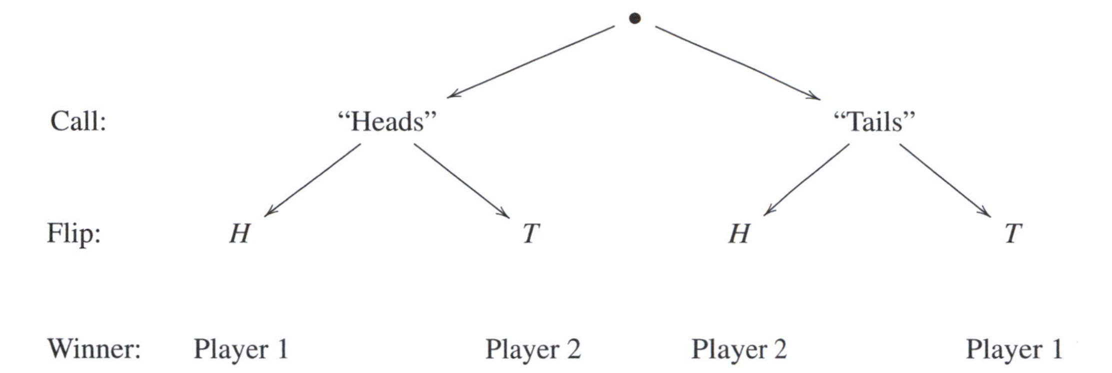

In general, we have the following definition.

Definition 2.3 The **game tree** for a sequential-move game is a system of **branches** (arrows) and **nodes** (points) showing all the steps of the game from start to finish. The uppermost node (usually denoted •) is called the **root node** and represents the start of the game. The other nodes are labeled positions in the game. The branches emanating from each node show all the possible positions that can be reached from the current position in the next step of the game. The final nodes (with no outgoing branches) are called the **end nodes**. These correspond to the outcomes of the game and will often be labeled accordingly.

Even for relatively simple games, the game tree will be unwieldy. For instance, in Tic-Tac-Toe, the first player has nine possible moves corresponding to the nine squares in the grid. From each of these positions, there are eight possible moves for the second player. We could take advantage of symmetry to reduce the possibilities somewhat, but the full game tree will still be quite large. However, at least Tic-Tac-Toe ends in a fixed number of steps. Imagine the game tree for Checkers or Chess!

The game Heads or Tails offers a fair method for choosing between two options. Can this be achieved without a coin? The following game does the trick. It also serves to introduce a useful device called a *payoff matrix*

Example 2.4 Odds and Evens

This is a two-player game giving the strategic equivalent of a Coin Toss. The players decide beforehand who is "Odd" and who is "Even." The players then simultaneously produce either a closed fist (0) or a single finger (1). The "Even" player wins if the sum of their choices is even, and "Odd" player wins if the sum is odd.

Odds and Evens is not a sequential-move game. If it were, Player 2 could always win! We refer to a game of pure strategy in which both players make their strategic decisions simultaneously as a **simultaneous-move game**. The game Odds and Evens is summarized by the following table:

Win Loss Matrix for Odds and Evens

"Even" player

"Odd" player

Strategy	0	1
0	"Even" player	"Odd" player
1	"Odd" player	"Even" player

Here the innermost entries indicate the winner.

As with Heads or Tails, the game Odds and Evens involves an initial step that is rather arbitrary, namely, the assigning of "Odd" and "Even" to the players. To obtain a more standard form for our device, we assume that Player 1 is "Odd" and Player 2 is "Even." Then, following our convention for assigning payoffs in win-lose games, (win = +1, lose = −1), we indicate the payoffs for each player using an ordered pair

(Player 1's payoff, Player 2's payoff)

The result is the following:

Payoff Matrix for Odds and Evens

Player 2

Player 1

Strategy	0	1
0	$(-1, 1)$	$(1, -1)$
1	$(1, -1)$	$(-1, 1)$

The *payoff matrix* will be our main device for representing simultaneous-move games. We make an official definition:

Definition 2.5 The **payoff matrix** for a two-player simultaneous-move game is the array whose rows and columns are labeled with the strategy choices for Player 1 and Player 2, respectively. The entries of the matrix then correspond to the outcomes of the game and indicate the payoffs in an ordered pair

(Player 1's payoff, Player 2's payoff)

While Odds and Evens is the strategic equivalent of a Coin Toss, simultaneous-move games can be used to model social interaction and conflict. Here is a famous example:

Example 2.6 Battle of the Sexes

A husband and wife decide between attending a game and going to a movie. The husband would prefer to go to the game, and the wife would prefer to go to the movie. If they do not agree on where to go, they will stay home. Both would prefer going out to staying home regardless of the destination.

We imagine this as a simultaneous-move game of strategy. Each spouse chooses between "Game" or "Movie" without discussion. There are then four possible outcomes for the game with two corresponding to the worst-case scenario for both. Because there are no monetary payoffs, we simply rank the outcomes with 2 for the best outcome and 0 for worst. Since the wife prefers to go to the movie first, the game second, and doing nothing last, we have assigned her the payoffs 2, 1, and 0 for these outcomes. The payoffs for the husband are the same. We have the following payoff matrix:

Payoff Matrix for Battle of the Sexes

Husband

	Strategy	Game	Movie
Wife	Game	(1, 2)	(0, 0)
	Movie	(0, 0)	(2, 1)

Let us emphasize an important difference between the two proceeding simultaneous-move games. In Odds and Evens, a win for Player 1 is a loss for Player 2 and vice versa. The payoff of $+1$ for Player 1 means a payoff of -1 for Player 2 and vice versa. A two-player game of pure strategy such that, for each outcome of the game,

$$\text{payoff from outcome for Player 1} + \text{payoff from outcome for Player 2} = 0$$

is called a **zero-sum game.** Zero-sum games were among the first games studied in modern game theory and the topic of the first great theorem, the *Minimax Theorem*, discussed in Chapter 15.

The game Battle of the Sexes is not a zero-sum game. Indeed, the husband and wife are not in direct conflict. Both can be happy (although not equally so) with the same outcome. We will call such games *partial-conflict game*. Examples include the famous *Prisoner's Dilemma* introduced in Chapter 4, as well as many other models of social conflict discussed in Chapter 16.

One way to appreciate the difference between these two games is to imagine playing them both, say, ten times with the same opponent. In Odds and Evens, it would be a dubious strategy for Player 1 to simply declare his or her 10 choices all at once. Player 2 could win every round. In Battle of the Sexes, however, there might be an advantage to this approach. Can you see why?

Finally, we analyze an election as a particular type of game. We will assume that we are given a group $\mathcal{V}$ of voters and a list A, B, C, D, and so on of candidates. The first question we must address is who should be the players in the game? There are two choices: the voters or the candidates. Suppose that we treat the candidates as the players in the game. We then have a political contest to win an election. Although the analysis of strategy for political candidates is certainly a rich topic, it is not mathematics.

We suggest that it makes more sense to take the voters to be the players in the game of an election. After all, the voters are the true stakeholders in the outcome. This point of view has immediate implications, however. The ideals of democracy depend on elections being faithful accountings of social preferences not strategic contests between voters. As we will see, exploring the game-theoretic properties of elections provides us with a powerful lens for viewing the problem presented by the need to make a social choice.

Having settled on the voters as the players in our "game," we now identify the strategies available to each voter. If the candidates are A, B, C, D, and X, say, then a strategy for a voter is often a choice of the preferred candidate. We will expand on this idea and require that voters rank all candidates from most preferred to least preferred. For example, a voter P might choose the following ranking:

***P*'s Chosen Ranking**

Rank	Candidate
1st	A
2nd	X
3rd	B
4th	C
5th	D

For each voter P, a choice of rankings such as this will constitute a choice of strategy. When each voter has chosen his or her strategy (i.e., his or her ranking of the candidates), we are ready for the election. All that is needed now is a *voting method.*

Definition 2.7 A **voting method** is a mechanism that takes the chosen rankings of the candidates submitted by a group of voters and produces a winning candidate called the **social choice**.

Remark 2.8 On Ties in the Social Choice We emphasize one aspect of the preceding definition that will become a relevant issue later. We have defined the social choice to be an individual candidate, not a set of candidates. In other words, we do not allow ties as outcomes of elections. We impose this restriction both as a convenience and to emphasize that, in political elections, ties ultimately must be broken.

We will introduce a variety of voting methods in Chapter 5. Notice that as described so far, an election is a simultaneous-move game of pure strategy – a game with many players, of course! Once all the voters (players) have chosen their strategies, as P did earlier, the voting method produces the outcome.

It remains to tackle the problem of assigning payoffs. A peculiar feature of the "election as game" now emerges. In order to assess the value of the outcome of an election for an individual voter, we will need to know the voter's *true preferences* for the candidates. By the **true preferences** for a voter P, we mean a fixed ranking of the candidates corresponding to P's honest preferences for the candidates. For example, P might have the following true preferences:

***P*'s True Preferences**

Rank	Candidate
1st	X
2nd	A
3rd	B
4th	C
5th	D

However P votes, we know he or she likes candidate X the most, A second, and so on. Now we can assign voter P a payoff based on his or her true preferences for the candidates versus the social choice. In the absences of ties, this is easy. We simply assign the payoff = 4 to voter P when X is the social choice (because X is P's true favorite candidate). We assign payoff = 3 to P when A is the social choice, payoff = 2 when B, payoff = 1 when C, and payoff = 0 to P when his or her least-favorite candidate D is the social choice. Of course, if there are more candidates, then these payoffs should be adjusted accordingly. We emphasize that P does not need to submit his or her true preferences to the voting method. Voter P is free to vote however he or she likes. Voter P's true preferences determine his or her payoff from the game. Summarizing, we have

Definition 2.9 Suppose that we are given a group of voters, a slate of candidates, and a voting method as in Definition 2.7. The **election** is then the simultaneous-move game of pure strategy in which the voters are the players and the strategies are the rankings the voters have submitted of the candidates. The outcome of the election is the social choice produced by the voting method. We assume that each voter has true preferences for the candidates. The payoffs to the voters for a given outcome are determined by these true preferences, as described in the preceding paragraph.

The preceding definition may seem excessively formal. After all, elections are held every day without preference rankings and payoffs. The extra details here allow for a mathematical analysis of elections. Real-world examples will often require some adaptation to use this framework. We illustrate this process with a famous example:

Example 2.10 The 2000 Florida Presidential Election

The 2000 U.S. presidential election stands as one of the great spectacles of modern politics. The saga played out for over a month, from Election Day, November 7, 2000, until the controversial Supreme Court decision *Bush v. Gore* on December 12, 2000. This decision ended the recounting of votes in the pivotal state of Florida. The final tallies for votes in Florida are summarized in the following table:

Final Vote Count: 2000 Florida Presidential Election

Candidate	Party	Votes	Percent
George Bush	Republican	2,912,790	48.85
Al Gore	Democratic	2,912,253	48.84
Ralph Nader	Green	97,488	1.63
Other	—	40,579	0.68

While there are many issues one could consider regarding this example, we focus on placing the election within the framework of Definition 2.9. First, we must come to terms with the large number of voters. A natural solution to this problem of size is the idea of a *voting bloc*.

Definition 2.11 A **voting bloc** in an election is a group of voters who have the same preferences for the candidates and who submit the same ranking (which might be different from their true ranking) as their vote.

For the purposes of this analysis, we ignore the 40,579 voters who voted for "other" candidates. We introduce voting blocs by party: the Republican bloc with 2,912,790 voters, the Democratic bloc with 2,912,253 voters, and the Green Party voting bloc with 97,488 voters. We can now speculate on the true preferences for these voting blocs based on the political positions of the blocs and the candidates.

The Green Party candidate Ralph Nader ran on an extremely liberal platform. We can thus surmise that Nader was the last choice for the Republican bloc and the second choice for the Democratic bloc. Similarly, Gore would be the natural second choice for the Green Party bloc. We arrive at the following true preferences for our voting blocs:

True Preferences for Blocs in Florida 2000 Election

Republican Bloc		Democratic Bloc		Green Party Bloc	
Rank	**Candidate**	**Rank**	**Candidate**	**Rank**	**Candidate**
1st	Bush	1st	Gore	1st	Nader
2nd	Gore	2nd	Nader	2nd	Gore
3rd	Nader	3rd	Bush	3rd	Bush

To simplify matters even further, let us suppose that the Republican bloc votes sincerely. That is, all the voters in this bloc vote for George Bush. We allow the Democratic and Green Party blocs to decide between voting for their first- or second-choice candidate. Using our convention for payoffs in elections, we obtain a simultaneous-move game of strategy summarized as follows:

Payoff Matrix when the Republican Bloc votes for Bush

		Democratic Bloc	
	Strategy	Gore	Nader
Green Party Bloc	Gore	(1, 2)	(0, 0)
	Nader	(0, 0)	(2, 1)

We have recovered the Battle of the Sexes game (Example 2.6). This example gives a glimpse of the important connection between the analysis of games and the analysis of elections.

Exercises

2.1 The game **Rock-Paper-Scissors** is a two-player simultaneous-move game of strategy. Each player chooses from one of three options: “Rock” (a closed fist), “Paper” (an open hand), or “Scissors” (two fingers). The outcome is determined by the rules: rock smashes scissors, scissors cuts paper, and paper covers rock. If the players choose the same option, the game is a tie. Write down the payoff matrix for Rock-Paper-Scissors.

2.2 Suppose that Rock-Paper-Scissors is played with the same rules but that Player 1 is not allowed to play “Rock.” Write down the payoff matrix. Does the game favor either player? Explain.

2.3 Play the following zero-sum game 20 times with an opponent, 10 times as Player 1 and 10 times as Player 2. Pen and paper are useful to play. Who has the advantage in this game? What is a good total score for Player 1? For Player 2? Explain your reasoning.

		Player 2	
	Strategy	X	Y
Player 1	A	(1, −1)	(−5, 5)
	B	(−1, 1)	(1, −1)

2.4 Suppose that Odds and Evens (Example 2.4) is played as a sequential-move game. That is, Player 1 (the "Odd" player) first chooses 0 or 1 and then Player 2 (the "Even" player) chooses 0 or 1. Write down the game tree for this game. Explain how Player 2 should play this game. Why is this game fundamentally different from the original simultaneous-move version of the game?

2.5 Suppose that the Battle of the Sexes (Example 2.6) is played as a sequential-move game. That is, the wife first chooses "Game" or "Movie" and then the Husband chooses "Game" or "Movie." Write down the game tree for this game. How would you play this game as the wife? How would you play as the husband?

2.6 Define a simultaneous-move zero-sum game as follows. Player 1 has one $10 bill and two $1 bills. Player 2 has two $5 bills and one $1 bill. Each player orders his or her bills first, second, and third, simultaneously. The player with the larger bill in each position takes the other player's bill and keeps his or her own. If the bills are the same in a position, each player recovers his or her bill. The payoff to each player is the profit made. Write down the payoff matrix for this game. Notice that each player has three strategies determined by where he or she plays the odd bill. Which player, if any, has the advantage?

2.7 Define a simultaneous-move game extension of Odds and Evens. Each player simultaneously lists his or her strategy for two rounds of play. That is, each player selects from the four choices: 00 (play 0 twice), 01 (play 0 first and 1 second), 10, or 11. The game Odds and Evens is then played with these strategy choices. The payoff to the players is the sum of the payoffs for the two rounds. Write down the payoff matrix for this game. Compare the strategy choices for either player.

2.8 Repeat Exercise 2.7 for three rounds of Odds and Evens. Notice that in this case each player chooses from among eight strategies.

2.9 Consider the zero-sum game with the following payoff matrix:

		Player 2	
	Strategy	X	Y
Player 1	A	(10, −10)	(−7, 7)
	B	(12, −12)	(14, −14)
	C	(11, −11)	(−8, 8)

Which strategy would you choose if you were Player 1? Based on this answer, what strategy should you choose if you are Player 2.

2.10 Define a sequential-move game of strategy using the payoff matrix in Exercise 2.9. Player 1 goes first, choosing A, B or C. Player 2 then goes, choosing X or Y. The payoffs to the players are given by the corresponding entry in the payoff matrix. Write down the game tree for this game. How would you play the game as Player 1?

2.11 The three leading candidates in the 1992 U.S. presidential election were Bill Clinton, George H. W. Bush, and Ross Perot. Here are the final vote totals in the state of Georgia:

Final Vote Count: 1992 Georgia Presidential Election

Candidate	Party	Votes	Percent
Bill Clinton	Democratic	1,008,966	43.47
George H. W. Bush	Republican	995,252	42.88
Ross Perot	Independent	309,657	13.34
Other	—	7,258	0.01

Introduce voting blocs and decide on their preference rankings for the three candidates based on political platforms. Ignore the "other" voters as we did in Example 2.10. Assuming that the Democratic bloc votes sincerely, create a two-player simultaneous-move game of strategy between two of the voting blocs. Did you obtain a game equivalent to Battle of the Sexes (Example 2.6)? Do you obtain the same game if the Republican bloc votes sincerely and the two-player game is between the Democratic and Independent blocs? Explain.

2.12 Define a **third-party candidate** to be a candidate in an election with less support than the two leading candidates. Following Example 2.10 and Exercise 2.11, we say a third-party candidate is a **spoiler** in an election under the following circumstances: when the voting bloc supporting the leading candidate votes sincerely and the other two voting blocs choose from among their top two candidates, the resulting simultaneous-move game is equivalent to Battle of the Sexes. Find three further examples of real-world elections (in addition to Florida 2000 and Georgia 1992) with spoiler candidates. Give precise vote counts, and explain your preference rankings.

2.13 Explain why, in a three-candidate election, a third-party candidate can only be a spoiler for one other candidate.

2.14 A selection committee of four has three members and a chair. We assume that there are three candidates A, B, and C running for an open position. The committee votes as follows: first, the chair names his or her favorite candidate. The members then vote on the other two candidates. The winner of the member election runs against the chair's candidate, but again, only the members vote. The winner of the second member election is the social choice. Suppose that the members' true preferences are as follows:

True Preferences for Committee Members

Member 1		Member 2		Member 3	
Rank	**Candidate**	**Rank**	**Candidate**	**Rank**	**Candidate**
1st	A	1st	B	1st	C
2nd	B	2nd	C	2nd	A
3rd	C	3rd	A	3rd	B

(a) Suppose that the chair chooses candidate A. Who wins the election?
(b) Suppose that the chair chooses candidate B. Who wins the election?
(c) Suppose that the chair chooses candidate A again. Show that member 2 can vote insincerely and improve the outcome of the election according to his or her true preferences.
(d) Suppose that the chair chooses candidate C. Which member can vote insincerely and improve his or her outcome in this case?
(e) Explain why the chair cannot improve his or her outcome by voting insincerely when all members vote their true preferences.

3 Chance

With no role for cleverness or skill, games of pure chance represent the essence of gambling. Examples in this category include casino games such as Roulette and Craps, games designed to attract large bets on an act of chance whether it be the spin of a wheel or the roll of the dice. As in the case of Heads or Tails (Example 2.2), these games do involve strategy in the sense that the players choose how to place their bets. We assume here that the bets of the players are preassigned as part of the game so as to focus on the role of chance.

Mathematically, the analysis of a win-lose game of chance boils down to the calculation of a *probability*. We introduce probability informally in this chapter using concrete examples with coins, dice, and cards. We then construct sequential-move games from these elements and define the *probability tree*. We state the *Law of the Probability Tree* (Theorem 3.6), a result we will use throughout this text, and gives some simple examples.

By a **gambling game**, we mean a game of pure chance with payoffs. The analysis of a gambling game serves to introduce the central notion of an *expected value*. For sequential-move gambling games, the Law of the Probability Tree extends to a powerful principle in probability theory called the *Linearity of Expectation*. We state a simple version of this general principle here (Theorem 3.13) and apply the result to compute the expected value of a sequential-move gambling game involving coins, dice, and cards. We conclude with some examples from the game *Blackjack*.

Probability

Games of chance are created using various physical devices. When we flip a coin, roll dice, or draw cards from a deck, we perform what is called a **probability experiment**, an act of pure chance. An **outcome** of a probability experiment is one particular occurrence of the experiment among the many possible. An **event** is a collection of outcomes that can occur in this experiment. We note that the outcome of a game of chance usually corresponds to an event in a probability experiment. We will often express events as quoted phrases, such as $A =$ "Player 1 wins."

Given an event A, the **probability** of A is a number p in the range $0 \leq p \leq 1$ whose relative size measures the likelihood that event A occurs. We will use the following notations:

$$p = P(A) = \text{probability } A \text{ occurs}$$

If all outcomes are equally likely, then

$$p = \frac{\text{number of outcomes in } A}{\text{total number of possible outcomes}}$$

The probability that A does not occur is then $1 - p$.

We introduce three familiar probability experiments used in games.

Example 3.1 Coin Flip

This is the simplest example of a probability experiment. A coin flip presents two outcomes: "Heads" and "Tails." These events are equally likely:

$$P(\text{"Heads"}) = P(\text{"Tails"}) = \frac{1}{2}$$

More interesting probabilities arise from dice and cards.

Example 3.2 Rolling Dice

A die is a cube with sides labeled

The Faces of a Die

⚀, ⚁, ⚂, ⚃, ⚄, ⚅

Rolling a single die presents six possible outcomes corresponding to the sides of the cube. We assume these rolls are equally likely so that each has probability 1/6.

Many games of chance involve a pair of dice and focus on events defined in terms of the sum of the two dice. Notice that these sums range from 2 (⚀⚀) through 12 (⚅⚅). The various sums are not all equally likely. For example, a sum of 4 can only occur three ways, whereas there are six different ways to roll a sum of 7. To assign probabilities to the various sums, it is helpful to imagine that the dice are of different colors, say, red and blue. We can then summarize all 36 possible outcomes using a **dice table**, the 6×6 array of possible rolls. Here is the picture with the sums 4 and 7 indicated:

Two Sums on a Dice Table

		Blue die					
		⚀	⚁	⚂	⚃	⚄	⚅
	⚀			X			0
	⚁		X			0	
Red die	⚂	X			0		
	⚃			0			
	⚄		0				
	⚅	0					

X = "sum of 4," 0 = "sum of 7"

Because there are 36 equally likely outcomes of a roll of the dice, we have

$$P(\text{"sum} = 4\text{"}) = \frac{3}{36} \qquad \text{and} \qquad P(\text{"sum} = 7\text{"}) = \frac{6}{36}$$

Example 3.3 Drawing a Card

A deck of cards offers a rich variety of possible games. The standard deck consists of 52 cards. Each card has a **value** and a **suit** defined by

Card values: 2, 3, 4, 5, 6, 7, 8, 9, 10, J (Jack), Q (Queen), K (King), A (Ace)
Card suits: ♣ (clubs), ♢ (diamonds), ♡ (hearts), ♠ (spades)

Notice that there are 13 values and 4 suits and so each of the $52 = 13 \cdot 4$ cards is uniquely distinguished by its value and suit. Some sample events for a single card drawn from a deck include "Jack" and "Diamond." We see

$$P(\text{"Jack"}) = \frac{4}{52} \quad \text{and} \quad P(\text{"Diamond"}) = \frac{13}{52}$$

Of course, most card games involve events with multiple cards. We will develop counting techniques for computing probabilities with multiple cards in Chapter 14.

Probability Trees

As we imagine combining these simple experiments to create more interesting games of chance, we arrive at one of the main computational devices in this book, the *probability tree*. To begin, consider a simple win-lose game of chance. Suppose that the probability that Player 1 wins is p. We picture this game tree with one branching:

A Win-Lose Game of Chance

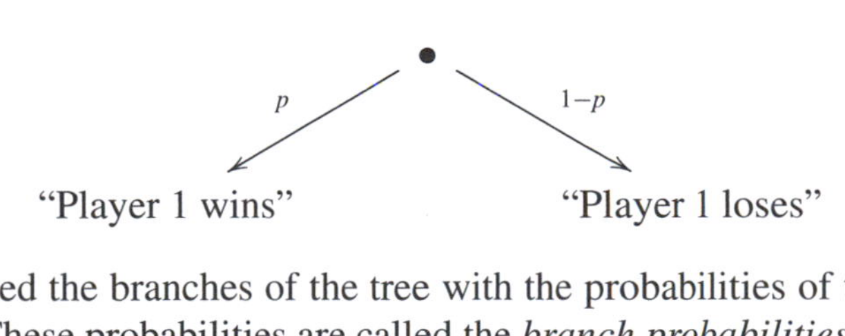

We have labeled the branches of the tree with the probabilities of taking this step in the game. These probabilities are called the *branch probabilities*.

Next, consider the game tree for a sequential-move game of chance with multiple stages. Each stage of the game corresponds to a probability experiment. The nodes of the game tree correspond to certain events for this experiment. Labeling the branches of the tree with the probabilities as earlier yields our basic device:

Definition 3.4 The **probability tree** for a sequential-move game of chance is the game tree in which each branch connecting one node to another is labeled with the probability of taking this step in the game. The labels on the branches of the probability tree are called the **branch probabilities**.

We make an observation that reveals the central role of the probability tree. A game of pure chance may always be taken to be a sequential-move game. Any simultaneous random events that occur in the course of the game can always be viewed as occurring in sequence. Thus every game of chance can be described by a probability tree. Here is a first example.

Example 3.5 Flipping Coins

Two players flip coins. Player 1 wins if the coins match (both heads or both tails). Otherwise, Player 2 wins. In this case, the branch probabilities are all 1/2. Writing H for the event "Heads" and T for "Tails," we have

Probability Tree for Flipping Coins

	1/2		1/2	
Flip 1:	H		T	
	1/2	1/2	1/2	1/2
Flip 2:	H	T	H	T
Winning player:	1	2	2	1

The power of the probability tree for calculation derives from the following important theorem:

Theorem 3.6 The Law of the Probability Tree *Suppose that we are given a probability tree for a game of chance with branches labeled with branch probabilities. Then the probability of reaching any node in the tree is obtained by multiplying together all the branch probabilities arising in the unique path from the root node to the node in question. Further, the probability of reaching any one of several chosen nodes is the sum of the probabilities of reaching each node individually.*

We will prove this theorem in the second part of this book. For the present, we focus on understanding and using the result. Regarding the end nodes, the outcomes of the game, Theorem 3.6 tells us that to compute the probability, we simply multiply together all the branch probabilities that arise in the path to this end node from the root. If we want to combine end nodes, say, compute the probability of a win for Player 1, we simply add the probabilities of reaching each selected end node, in this case those labeled with a 1. In the probability tree for Flipping Coins earlier, the probability of reaching each of the four outcomes is $p = (1/2) \cdot (1/2) = 1/4$. The probability of Player 1 winning is $1/2 = 1/4 + 1/4$. From here on, we will label the end nodes with these probabilities when writing a probability tree. We consider a simple example with cards.

Example 3.7 Drawing for Diamonds

Player 1 draws one card from a full deck. If it is a Diamond, the game is over and Player 1 wins. If the first card is not a Diamond, Player 2 draws a second card from the remaining 51 cards. If this card is a Diamond, then Player 1 wins. Otherwise, Player 2 wins.

Here is the probability tree for the game:

Probability Tree for Drawing for Diamonds

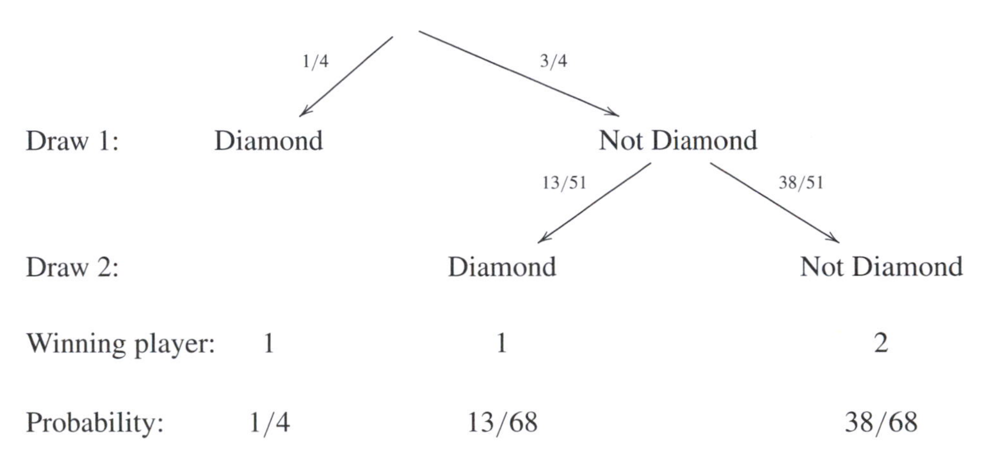

The probabilities for the first draw are clear. The interesting node occurs when Player 1 does not draw a Diamond. In this case, Player 2 draws a card from the remaining 51 cards. Notice that there are still 13 Diamonds because Player 1 did not draw one. The end-node probabilities were obtained by multiplying along the branches: for example, $13/68 = (3/4) \cdot (13/51)$. We can now compute that the probability that Player 1 wins the game is $13/68 + 1/4 = 30/68$. The probability that Player 2 wins is $1 - 30/68 = 38/68$.

It is worth noting a difference between Flipping Coins and Drawing for Diamonds. In the latter game, the outcome of the first card drawn affects the probabilities governing the second card drawn. In other words, Drawing for Diamonds cannot be played as a simultaneous-move game. The probability of drawing a diamond on the second draw is called a *conditional probability*. Conditional probabilities are, in fact, a key ingredient in the proof Theorem 3.6.

Payoffs and Expected Value

For gambling games, the probabilities of winning and losing do not give a complete analysis. Before deciding to place a bet, a gambler weighs the possible gains and losses together with the probabilities. To analyze a gambling game, we must take into account the payoffs resulting from the various outcomes. This analysis leads to the notion of the *expected value* of a game. The following example will serve to illustrate the idea.

Example 3.8 Hoping for Heads

Players 1 and 2 each flip a coin. If the coins both come up heads, Player 2 pays \$10 to Player 1. If both come up tails, then Player 1 pays \$5 to Player 2. If both heads and tails appear, Player 1 pays \$1 to Player 2.

We take Player 1's perspective and so insert the payoffs to Player 1 in our probability tree.

Probability Tree for Hoping for Heads

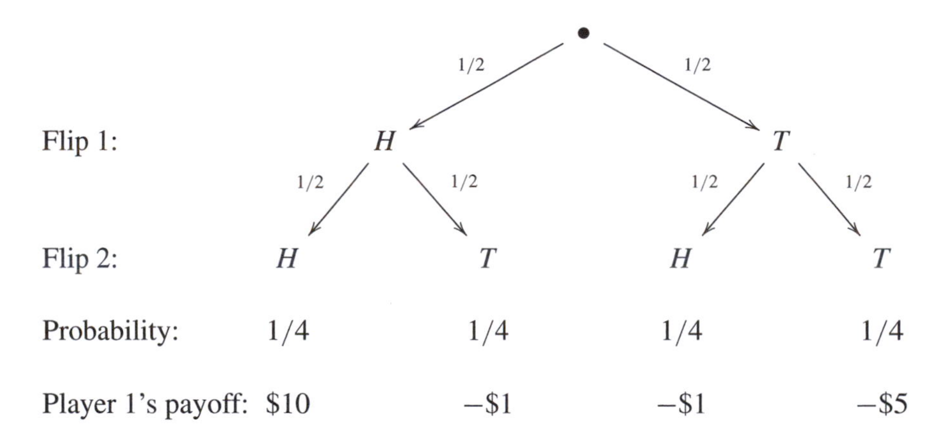

Player 1 is no doubt attracted to the game by the \$10 payoff. We denote this outcome by HH. Suppose that Player 1 plans to play 20 rounds of the game. He or she can expect HH to occur $(20)(1/4) = 5$ times, resulting in a profit of \$50 dollars from HH. Of course, this is just an average; HH could come up more or less than 5 times out of 20. We have observed

$$\begin{aligned}&\text{Average payoff from } HH \text{ for 20 rounds}\\&= (\text{payoff from } HH)\cdot(20 \text{ rounds})\cdot\left(\frac{1}{4}\right) = \$50\end{aligned}$$

Because the number 20 was rather arbitrary, it makes sense to consider the average amount Player 1 can expect to win *per round* of the game. That is, we replace 20 with 1 and find

$$\text{Average payoff from } HH \text{ per round } = \$10\left(\frac{1}{4}\right) = \$2.50$$

We can make the same calculations for the outcomes HT, TH, and TT at which Player 1 loses money. These are negative payoffs for Player 1:

$$\text{Average payoff from } HT \text{ per round } = (-\$1)\left(\frac{1}{4}\right) = -\$0.50$$

$$\text{Average payoff from } TH \text{ per round } = (-\$1)\left(\frac{1}{4}\right) = -\$0.50$$

$$\text{Average payoff from } TT \text{ per round } = (-\$5)\left(\frac{1}{4}\right) = -\$1.25$$

The individual outcomes each contribute an average payoff per round. To obtain the overall average, we simply add these values to obtain

$$E = \$10\left(\frac{1}{4}\right) - \$1\left(\frac{1}{4}\right) - \$1\left(\frac{1}{4}\right) - \$5\left(\frac{1}{4}\right) = +\$0.75 \text{ per round of the game}$$

The value $E = \$0.75$ is called the *expected value* of Hoping for Heads for Player 1. The expected value E is the average payoff for a round of the game. The game Hoping for Heads favors Player 1. Over time, he or she will average 75 cents in profit per round. We define this important notion generally.

Definition 3.9 Suppose that a game of pure chance has n possible game outcomes having probabilities $p_1, p_2, \ldots, p_n$ and payoffs for Player 1 given as $\$D_1, \$D_2, \ldots, \$D_n$. Here the $\$D_i$ could be positive (a profit for Player 1), zero (no gain), or negative (a loss for Player 1). The **expected value** of the game for Player 1 is then

$$E = \$D_1 \cdot p_1 + \$D_2 \cdot p_2 + \cdots + \$D_n \cdot p_n$$

The expected value is a remarkably powerful tool in probability theory. For a gambling game, E is a theoretical price tag telling a player the cost or the profit of a round, on average. Of course, anything is possible on a single round of the game regardless of the odds. It is this disconnect between theory and practice that makes gambling entertaining.

We give another illustration of calculating the expected value, this time for a game with cards.

Example 3.10 Hoping for Jacks

Player 1 draws a card from deck of 52, and Player 2 draws a second card from the remaining 51 cards. Player 2 pays Player 1 \$5 per Jack drawn. If no Jacks are drawn, Player 1 pays \$2 to Player 2.

Here is the probability tree for the game:

Probability Tree for Hoping for Jacks

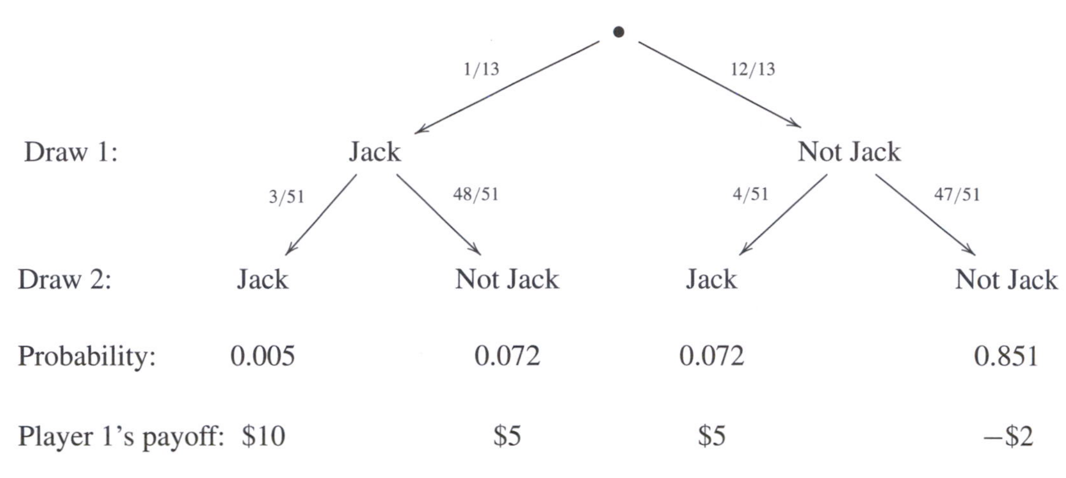

Computing the expected value, we find that Player 1 can expect to lose money on this game:

$$E = \$10 \cdot (0.005) + \$5 \cdot (0.072) + \$5 \cdot (0.072) - \$2 \cdot (0.851) = -\$0.93$$

Combining Games

Games of chance can be combined to make more complicated games. We illustrate the idea using our two gambling games: Hoping for Heads and Hoping for Jacks. We add a dice roll to the beginning to create a game involving all three elements.

Example 3.11 Dice, Coins, and Cards

Player 1 rolls a pair of dice. If the sum $= 7$, then the game Hoping for Heads (Example 3.8) is played. If the sum $\neq 7$, then the game Hoping for Jacks (Example 3.10) is played.

The game Dice, Coins, and Cards serves to introduce an important fact about expected values. We can use our previous calculations of the expected values of these games to compute the expected value of the new game. To explain, we introduce some terminology: The games Hoping for Heads and Hoping for Jacks represent games within the game of Dice, Coins, and Cards. These are called *subgames*.

Definition 3.12 A **subgame** of a sequential-move game is the game that begins at a particular event occurring in the original game.

The subgame Hoping for Heads occurs at the event "sum = 7" in Dice, Coins, and Cards. The subgame Hoping for Jacks occurs at the event "sum $\neq$ 7." We write the probability tree for Dice, Cards, and Coins in an abbreviated form that treats these subgames as end nodes. Generally, we refer to a game tree that summarizes certain steps of the game as an **abridged game tree**. For games of chance, we have an **abridged probability tree**, which is the abridged game tree with probabilities labeling the shown edges. For our game, we have

Abridged Probability Tree for Dice, Coins, and Cards

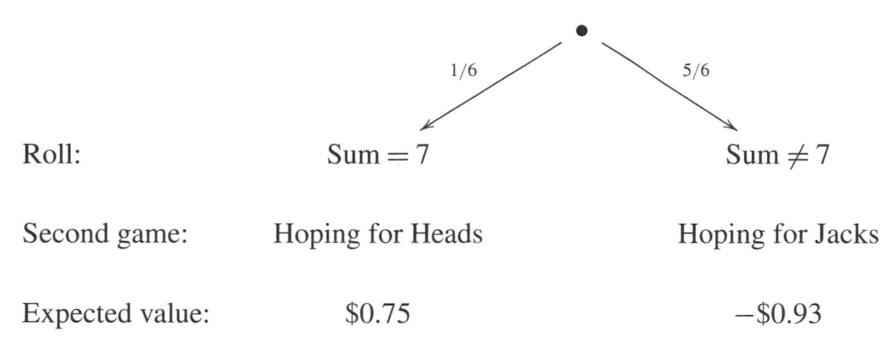

The expected value E for Player 1 in Dice, Coins, and Cards can be computed directly in terms of the expected values of the subgames. Observe that E will be the sum of two terms, one corresponding to the event "sum = 7" and the other to "sum $\neq$ 7." In the first case, the term is the same as that used to compute the expected value for Hoping for Heads except that each probability is multiplied by $p = 1/6$ to account for the first step of the new game. Thus the first summand corresponding to "sum = 7" is just $1/6 \cdot (\$0.75)$. Similarly, the contribution to E when "sum $\neq$ 7" is just the expected value for Hoping for Jacks multiplied by the probability $1 - p = 5/6$. We conclude that

$$E = \left(\frac{1}{6}\right)(\$0.75) + \left(\frac{5}{6}\right)(-\$0.93) = -\$0.65$$

The argument for Dice, Coins, and Cards directly extends to prove the following important result:

Theorem 3.13 Linearity of Expectation *Suppose that a sequential-move game of chance is constructed as follows: at the first stage, a probability experiment is performed with an event A. If the event A occurs, Game 1 is played. If A does not occur, Game 2 is played. Let p denote the probability that A occurs. Let E_1 be the expected value for Player 1 for Game 1 and E_2 be the expected value for Game 2. Then the expected value E of the main game for Player 1 is*

$$E = pE_1 + (1 - p)E_2$$

■

We conclude by briefly considering a familiar gambling game.

Example 3.14 Blackjack

The casino version of the game **Blackjack** pits all bettors against the dealer (the **house**), who plays according to fixed rules. The object of the game is to obtain a higher sum of cards than the dealer without going over 21. Cards are counted as follows: aces can be worth either 1 point or 11 points as desired; face cards (K, Q, J) are each worth 10 points; and numbered cards (2, 3, …, 10) are worth the value of the card. Each player starts with two cards and can take as many cards, called **hits**, one by one from the deck to build his or her hand until he or she is satisfied and **stands**. If a player goes over 21, he or she **busts** and loses his or her bet. When all the players have either busted or are standing (on numbers ≤ 21), it is the dealer's turn. The dealer must hit on 16 and must stand on 17. If the dealer busts, all standing players win. Otherwise, all standing players win who have hands higher than the dealer's hand. Ties are called **pushes** and result in no loss or gain.

Blackjack mixes chance and strategy. The strategy for the players lies in the decision to hit or stand. In the casino, the dealer's first card is shown to the players. We consider an unrealistic scenario in which both the dealer's initial cards are known

(3.1) **A Hand of Blackjack**

We compute the expected value of winning when we stand on 14. Notice that the dealer must take a hit because he or she holds 16. We win if the dealer draws any card other than an Ace, 2, 3, 4, or 5. We take into account the cards showing as we count. There are 48 cards left in the deck. Notice that because the four of hearts is showing, there are 19 cards (Ace, 2, 3, 4, 5) that beat us. The probability that we lose this hand is $p = 19/48$. The expected value of standing is thus

$$E = \$10,000\left(\frac{29}{48}\right) - \$10,000\left(\frac{19}{48}\right) = \$2083.33$$

Although this calculation suggests that we should stand on 14, in order to make a truly informed decision, we should also compute the expected value of hitting and compare. The latter calculation involves too many contingencies to attempt here. We can imagine that it is not as safe a bet as standing.

Games of pure chance are arguably the best suited among the various types of games we consider for pure quantitative description. While the calculation of probabilities and expected values can be challenging, understanding the game at the level of expected outcomes is a matter of pure computation. As a case in point, we formulate a precise definition of *fairness* for two-player games of chance.

efinition 3.15 We say that a two-player game of pure chance with payoffs is **fair** if the expected value for each player is zero.

Defining fairness for games with strategy and, especially, elections will be a much richer question, as we will see.

Exercises

3.1 Devise a two-player game of pure chance with three flips of a coin such that the probability of Player 1 winning is $p = 3/8$. Remember, you decide who wins and who loses based on the outcome of the three flips.

3.2 Consider the following variation on Drawing for Diamonds (Example 3.7): Player 1 draws a card. If the card is red (diamond or heart), the game is over, and Player 1 wins. If the first card is not red, Player 2 draws a second card. If this card is a diamond, then Player 1 wins. Otherwise, Player 2 wins. Compute the probability of winning this game for Player 1.

3.3 Consider the following version of Drawing for Diamonds (Example 3.7) with payoffs: Player 1 draws a card. If the card is red, then Player 1 pays \$10 to Player 2. Otherwise, Player 2 draws a second card. If the second card is red, Player 1 pays \$5 to Player 2. If the second card is not red, Player 2 pays \$15 to Player 1. Compute the expected value of this game for Player 1.

3.4 The game **Roulette** is played with a large spinning wheel and a ball. The wheel is divided into 38 pockets each equally likely to capture the ball when the wheel stops spinning. The pockets are numbered $0, 00, 1, 2, 3, \ldots, 36$ and colored as follows:

Green	0, 00
Red	1, 3, 5, 7, 9, 12, 14, 16, 18, 19, 21, 23, 25, 27, 30, 32, 34, 36
Black	2, 4, 6, 8, 10, 11, 13, 15, 17, 20, 22, 24, 26, 28, 29, 31, 33, 35

(a) A bet on "Red" pays even money. That is, if a gambler bets \$D on "Red," then he or she wins $\$D$ if the ball lands in a red pocket and loses $\$D$ if not. Compute the expected value of a bet on "Red."

(b) A bet on the "First Dozen" pays 2-1 odds. That is, a bet of $\$D$ wins $\$2D$ if the ball lands in a pocket numbered 1–12 and loses $\$D$ if not. Compute the expected value of a bet on the "First Dozen."

(c) A general bet in Roulette is made by choosing n pockets on the wheel. The bet wins if the ball lands in one of the chosen pockets. The payoff for a bet of D dollars is given by the formula

$$\text{Payoff} = \$D\left(\frac{36-n}{n}\right)$$

Prove that the expected value for all general bets in Roullete are equal.

3.5 Consider the following variation on Hoping for Heads (Example 3.8): flip a coin three times. Player 1 wins \$5 for each "Heads" flipped. Player 2 wins \$20 if all three flips are "Tails." Write down the probability tree including the payoffs for Player 1, and compute the expected value of this game for Player 1.

3.6 Write down the full game tree for Dice, Coins, and Cards (Example 3.11), and compute the expected value directly.

3.7 Consider the following two games:

Game 1: Draw a card from a deck. If the card is a Jack, Player 2 pays \$20 to Player 1. If the card is not a Jack, flip a coin. Player 2 pays \$5 to Player 1 if the coin is "Heads," and Player 1 pays \$10 to Player 2 if "Tails."

Game 2: Flip a coin. If "Heads," draw two cards from a deck. If "Tails," draw one card. Player 2 pays \$10 to Player 1 per Jack drawn. If no Jacks are drawn, Player 1 pays \$10 to Player 2.

Use Theorem 3.13 to answer the following:

(a) Compute the expected value of the game in which the first stage is flipping a coin. If the coin is "Heads," then Game 1 is played. If the coin is "Tails," then Game 2 is played.

(b) Compute the expected value of the game in which the first stage is rolling a pair of dice. If "sum = 2," then Game 1 is played. Otherwise, Game 2 is played.

(c) Choose an event A for the experiment of rolling a pair of dice such that the game whose first stage is rolling the dice and whose second stage is Game 1 when A occurs and Game 2 when A does not occur has expected value E for Player 1 with $E < -\$3$.

3.8 A number of gambling games can be created with a pair of dice. We assume here in all cases that the winning player is payed the amount of the dice. Rather than write down the full probability tree, it is easier to use the following table:

Sum of dice	Probability	Payoff
2	1/36	$\pm\$2$
3	2/36	$\pm\$3$
4	3/36	$\pm\$4$
5	4/36	$\pm\$5$
6	5/36	$\pm\$6$
7	6/36	$\pm\$7$
8	5/36	$\pm\$8$
9	4/36	$\pm\$9$
10	3/36	$\pm\$10$
11	2/36	$\pm\$11$
12	1/36	$\pm\$12$

The sign $\pm$ on the payoffs will be determined by the rules of the game. For each of the following examples, compute the expected value of the game for Player 1:

(a) Player 1 wins with a sum of 3, 6, 9, or 12; Player 2 wins otherwise.
(b) Player 1 wins with a sum of 10, 11, or 12; Player 2 wins otherwise.
(c) Player 1 wins with a sum of 2, 3, 7, or 11; Player 2 wins otherwise.
(d) Player 1 wins in all cases. In this case, the expected value E is "average roll" of the dice. Can you guess the answer? Check to see if your guess is correct.

3.9 Use Theorem 3.13 and your results in Exercise 3.8 to compute the expected value of the given games.

(a) A pair of dice are rolled. If the sum of the dice is 7, then the game described in Exercise 3.8(a) is played. If the sum $\neq 7$, then the game from Exercise 3.8(b) is played.
(b) A pair of dice are rolled. If the sum $= 6$, then the game described in Exercise 3.8(b) is played. If sum $\neq 6$, then the game from Exercise 3.8(c) is played.
(c) Draw a card from a deck. If it is a Jack, Player 1 pays \$10 to Player 2. Otherwise, the game from Exercise 3.8(b) is played.

3.10 Consider the following scenarios in Blackjack (Example 3.14)

(a) Compute the expected value when taking a hit on 17.

(b) Compute the expected value for taking exactly one hit on 15. Remember, the payoff is zero in the event of a tie.

(c) Compute the expected value of standing with 17.

3.11 Investigate the odds of winning Blackjack at a casino. Can you explain why the dealer has the advantage? How do the odds of winning Blackjack compare with the odds of playing "Red" in Roulette (Exercise 3.4)?

4 Strategy

The class of strategic games presents examples ranging from simple decision games such as Rock-Paper-Scissors, to models for economic and political interactions, to parlor games such as Checkers and Chess. Mathematically, the focus on strategy as opposed to chance represents a paradigm shift. Whereas games of pure chance are described by numbers (probabilities and expected values), the analysis of strategic games involves various notions of "solution." We introduce three related solution concepts in this chapter.

Given the large variety of examples of games of strategy, it is natural to introduce some groupings. Focusing on the payoffs, we obtain one useful division. A **total-conflict game** is a two-player game in which the sum of the payoffs of the players is always the same amount at every outcome. In a total-conflict game, one player's gain is another's loss. Examples include win-lose games and zero-sum games, as introduced in Chapter 2. A **partial-conflict game** is any game that is not a total-conflict game. We have seen an example of a two-player partial conflict game with Battle of the Sexes (Example 2.6).

We begin with simultaneous-move partial-conflict games. We introduce the famous *Prisoner's Dilemma* as a representative example, giving both the standard presentation and a version arising in economics. We also analyze an example of an *auction* that serves to motivate our first solution concept, a *dominated strategy solution*. We next consider an example of a sequential-move partial-conflict game and introduce the notion of a *backward induction solution*.

Turning to games of total conflict, we apply our solution concepts to the case of zero-sum games and identify the *maximin* strategy. Finally, we consider the class of win-lose sequential-move games called *combinatorial games*. We introduce the *Binary Labeling Rule* to solve certain examples of games in this class. Our partial results here motivate *Zermelo's Theorem* (Theorem 4.17), which asserts the existence of a solution to all finite combinatorial games.

Partial-Conflict Games

Games in which the players are not in total conflict with one another have proven extremely useful for modeling social and political interactions. Perhaps the most famous example is the following:

Example 4.1 Prisoner's Dilemma

Two men are arrested, accused of a common crime. They are held in separate prison cells. Each prisoner is given the same choice between two options: to rat on his partner or to remain silent. If both prisoners rat, then both get 10-year sentences. If both prisoners remain silent, then both receive 1-year sentences. Finally, if one prisoner rats and the other stays silent, the prisoner who rats on the other goes free, whereas the prisoner who stays silent gets 20 years.

The simultaneous-move aspect of this game is a key feature. The prisoners must make their choices without communication, and yet the correct choice seems to depend on how both prisoners behave. Or does it? Let us imagine the game from Prisoner 1's point of view. If Prisoner 2 is planning to rat, then Prisoner 1 should rat too. What if Prisoner 2 is loyal and stays silent? In this case, both prisoners can do minimal time if Prisoner 1 also stays silent. However, if Prisoner 1 rats on his loyal partner, Prisoner 1 goes free. By this reasoning, Prisoner 1 should rat on Prisoner 2 in either case.

Prisoner 2 can follow the exact same logic, of course, to convince himself to rat on Prisoner 1. When both follow this logic, they end up serving a long term together, knowing that the extra years are due directly to their disloyalty to each other.

Let us treat a prison sentence as a negative payoff. This maintains our convention that the higher the payoff, the better is the outcome. We then have

Payoff Matrix for Prisoner's Dilemma

Prisoner 2

	Strategy	Stay silent	Rat
Prisoner 1	Stay silent	$(-1,-1)$	$(-20,0)$
	Rat	$(0,-20)$	$(-10,-10)$

The Prisoner's Dilemma highlights a basic obstruction to *cooperation* in personal or political interactions. A player in a partial-conflict game is said to **cooperate** if he or she chooses a strategy resulting in less payoff for himself or herself but more for other players. A player who does not cooperate is said to **defect**. In the presence of a Prisoner's Dilemma, the collective ideal of mutual cooperation, be it between individuals, companies, or nations, is threatened by the temptation of the higher payoffs from defection. Here is an example from the realm of economics.

Example 4.2 Pricing Game

In 2009, Amazon and Walmart engaged in a highly publicized competition for online book sales. The conflict played out day by day on the Web, with slashed prices on best-sellers appearing on one company's website matched or lowered still further as soon as the next day on the other company's site. Strictly speaking, this competition corresponds to a sequential-move game. We imagine a simplified version as a simultaneous-move game.

We focus on one best-seller from 2009 and suppose that each company chooses between charging $9 or $10 for the book. We further imagine that 1,000 customers buy the book. If both websites offer the book for the same price, the customers split evenly between the websites. If one website lists the book at a cheaper price than the other, all the customers buy the book for the cheaper price.

The question of payoffs for this game is somewhat delicate. If the price is too low, the companies lose money on sales. In the actual case there was much more at stake than short-term profits. We simplify matters and define

$$\text{Payoff} = \text{revenue from sales of the book}$$

We then have the following payoff matrix:

Payoff Matrix for Pricing Game

Walmart

	Strategy	Charge $9	Charge $10
Amazon	Charge $9	($4,500, $4,500)	($9,000, 0)
	Charge $10	(0, $9,000)	($5,000, $5,000)

The Pricing Game is a Prisoner's Dilemma with cooperation here corresponding to charging the larger amount, namely, $10 for the book. If the companies call a truce on the price war and both charge $10, they will split the customers and the revenue. A collective strategy to raise prices by players in economic competition is called **collusion**. The dilemma lies in the fragility of such an agreement to collude. At the last moment, either company can renege on the agreement and charge $9, gaining all the customers. Notice that the temptation to defect by the companies has the effect of lowering prices and so, in principle, represents a benefit to customers.

We consider another example arising in economics in order to introduce our first solution concept. Generally, an **auction** involves a collection of bidders competing for a desired object. The object may be appraised differently by the individual bidders. Representing this as a game, we treat the bidders as players and the strategies as the bids. To be concrete, we introduce a familiar method for conducting an auction defined as a strategic game:

Definition 4.3 A **sealed-bid auction** is a simultaneous-move game of pure strategy. We assume that each player has a **true valuation** of the object up for bid. Each player simultaneously submits a bid for the object. The highest bidder wins the object and pays his or her bid. The other players win nothing and pay nothing and so have zero payoff. The payoff for the winning bidder is given by

$$\text{Payoff} = \text{amount bid} - \text{true valuation}$$

We emphasize the similarity to our definition of an election (Definition 2.9). As in that definition, the payoffs depend on the "true" feelings of the players. We also note that the definition makes no allowances for a tie in the highest bids. This can be a tricky problem in practice. We will arrange things so that there are no such ties in the examples we consider. Here is an example.

Example 4.4 Baseball Card Auction

A baseball card is to be auctioned off by sealed bid between two collectors. The first collector values the card at $6, whereas the second collector values the card at $7. Collector 1 decides between submitting three bids: either $1, $4, or $5 Collector 2 decides between bidding $2, $3, or $6.

Here is the payoff matrix for the game:

Payoff Matrix for Baseball Card Auction

Collector 2

Collector 1

Strategy	Bid $2	Bid $3	Bid $6
Bid $1	(0, $5)	(0, $4)	(0, $1)
Bid $4	($2, 0)	($2, 0)	(0, $1)
Bid $5	($1, 0)	($1, 0)	(0, $1)

The intrigue of an auction derives from the uncertainty about the various players' valuations of the object and the bids they consider making. In order to analyze the Baseball Card Auction, we eliminate this uncertainty and assume that both players have knowledge of the other's valuations and possible bids. We make this key assumption official with the following definition:

Definition 4.5 A player in a simultaneous-move game of pure strategy has **perfect information** if he or she has full knowledge of the payoff matrix for the game.

Now let us consider the game Baseball Card Auction from Collector 1's point of view. He or she must choose between three strategies. It is helpful to compare two at a time. We first look at the payoffs to Collector 1 from bidding $1 versus bidding $4 against all possible bids by Collector 2. We obtain the following table:

Comparison of Bidding $1 versus $4 for Collector 1

Collector 1's bid	Collector 2's bid	Payoff to Collector 1
$1	$2	0
$4	$2	$2 ✓
$1	$3	0
$4	$3	$2 ✓
$1	$6	0
$4	$6	0

Notice that in the two cases with checkmarks, the payoff from bidding $4 is better for Collector 1. In the third case, when Collector 2 bids $6, the payoff to Collector 1 is zero either way. We conclude that the payoff from bidding $4 is never worse and sometimes better than bidding $1 for Collector 1. We define this important relationship between strategies.

Definition 4.6 Suppose that a player in a simultaneous-move game of pure strategy has two strategies S and T. We say that S **weakly dominates** T for this player if the payoff to this player from playing S is always at least as large and, in at least one case, strictly larger than the payoff from playing T for every possible choice of strategy by the other players.

The term "weakly" refers to the fact that strategy S does not always produce a better outcome for the player than T does; in some cases, S and T produce the same payoff for this player. The relation of weak dominance is in fact a *strong* condition! Notice that if strategy S weakly dominates strategy T for Player 1, then Player 1 can eliminate strategy T from consideration because he or she is paid more by choosing S in all cases.

Most pairs of strategies will not satisfy a relation of weak dominance. When it is neither the case that S weakly dominates T nor that T weakly dominates S, we say S and T have **no relation of weak dominance**. For instance, we can see that for Collector 2, the strategies of bidding $2 and $6 have no relation of weak dominance.

No Relation of Weak Dominance Bidding $2 versus $6 for Collector 2

Collector 2's bid	Collector 1's bid	Payoff to Collector 2
$2	$1	$5 ✓
$6	$1	$1
$2	$4	0
$6	$4	$1 ✓
$2	$5	0
$6	$5	$1 ✓

Returning to Collector 1, we next check that bidding \$4 weakly dominates bidding \$5 for this player.

Bidding \$4 Weakly Dominates Bidding \$5 for Collector 1

Collector 1's bid	Collector 2's bid	Payoff to Collector 1
\$4	\$2	\$2 ✓
\$5	\$2	\$1
\$4	\$3	\$2 ✓
\$5	\$3	\$1
\$4	\$6	0
\$5	\$6	0

We have seen, then, that bidding \$4 weakly dominates both bidding \$1 and bidding \$5 for Collector 1. It seems clear that Collector 1 should bid \$4! We say that a player in a simultaneous-move game of pure strategy has a **weakly dominant strategy** S if strategy S weakly dominates every other strategy choice for this player.

We have settled on Collector 1's best option. Let us look at Collector 2's options next. We can quickly see that Collector 2 does not have a weakly dominant strategy. However, his or her knowledge of the payoff matrix allows him or her to conclude that Collector 1 will bid \$4. Collector 2's best move, then, is to bid \$6 and win the object for a payoff of \$1.

The Baseball Card Auction illustrates our first solution concept.

Definition 4.7 Suppose that in a two-player simultaneous-move game of pure strategy, one player has a weakly dominant strategy S. Let T be the strategy for the other player that produces the best payoff for this player when his or her opponent plays S. The outcome corresponding to the strategy choices S and T is called the **dominated-strategy solution** of the game.

As argued earlier, the dominated-strategy solution to the Baseball Card Auction occurs when Collector 1 bids \$4 and Collector 2 bids \$6.

Some remarks are in order regarding this solution concept. Many simultaneous-move games do not have dominated-strategy solutions. For example, in Odds and Evens (Example 2.4), there is no relation of weak dominance between playing 0 or 1. Second, we emphasize that without perfect information, the dominated-strategy solution may exist in theory without being evident in practice. Finally, consider the Prisoner's Dilemma. The dominated-strategy solution occurs when both prisoners rat because ratting weakly dominates staying silent (see Exercise 4.1). The dominated-strategy solution may lead to a less than ideal outcome.

We next consider an example of a sequential-move game.

Example 4.8 Splitting the Allowance

A father pays his two sons their combined $20 weekly allowance by playing the following game: the father gives four $5 bills to his older son (who does most of the actual work) and tells him that he can offer any number of these $5 bills to his brother as his share and keep the rest for himself. The younger son can either accept the offer or reject it. If the younger son accepts the offer, then both keep their shares, but if he rejects the offer, then the father takes back the $20 and gives each son $1 for their weekly allowance.

For simplicity, we assume that the older son chooses between two options: "Offer $5" or "Offer $10." The game tree is then as shown:

Game Tree for Splitting the Allowance

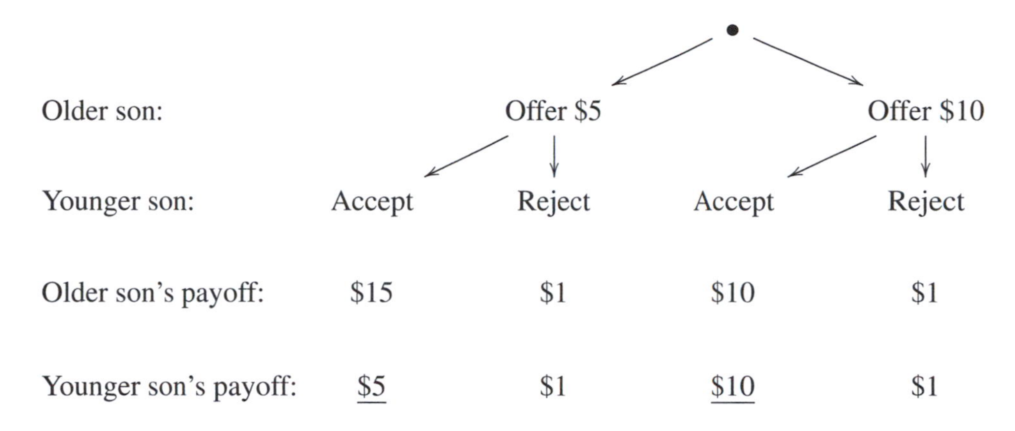

The older son can analyze the consequences of his two strategies and determine what his younger brother is likely to do in each case. We make this analysis now. Our analysis again depends on an assumption of *perfect information*, in this case a version appropriate for a sequential-move game.

Definition 4.9 A player in a sequential-move game of pure strategy is said to have **perfect information** if he or she has full knowledge of the complete game tree for the game.

We assume that the older son has perfect information in Splitting the Allowance. Now suppose that the older son offers $5. The younger son is then faced with a choice. Accept the offer and receive $5 or reject the offer and receive $1. Putting aside hurt feelings, the choice that pays the most for the younger son when offered $5 is to take it. Similarly, when offered $10 by the older son, the younger son should accept the offer. We have underlined the younger son's maximal payoff in each case in the tree. Notice that we are simply comparing the payoffs available to the younger son from each second-stage node in the tree and picking the larger one. Summarizing, we have the the following tree:

Younger Son Maximizes Payoff in Splitting the Allowance

Older son:	Offer $5		Offer $10	
Younger son:	Accept ✓	Reject	Accept ✓	Reject
Older son's payoff:	$15	$1	$10	$1
Younger son's payoff:	$5	$1	$10	$1

The older son can now compare the payoffs for himself corresponding to his two strategy choices, assuming that his brother will use the checked branches on his turn. We conclude that the older son should offer $5.

Backward Induction Solution for Splitting the Allowance

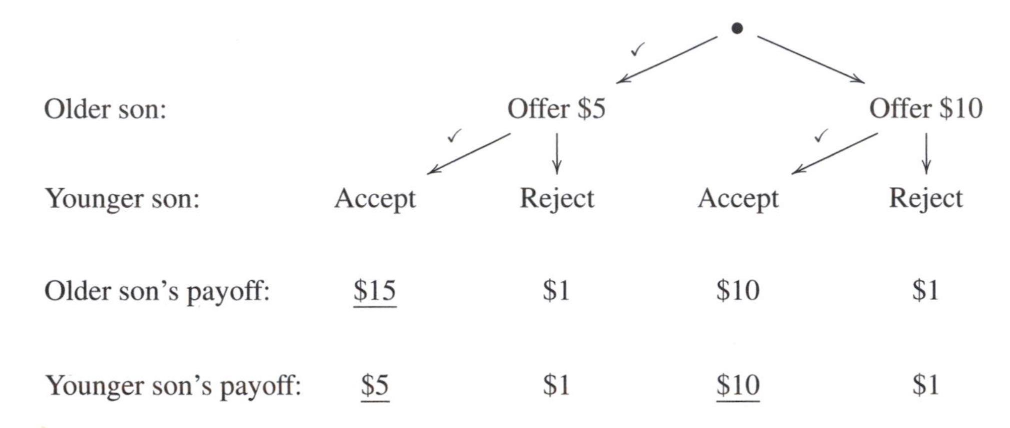

This reasoning is an example of our a second solution concept, this one for sequential-move games.

Definition 4.10 A **backward induction solution** to a sequential-move strategic game is the sequence of strategy choices at each stage for both players arrived at by first analyzing the last stage and then the next-to-last stage and so on until the first stage of the game is reached.

Games with more than two stages may require more complicated analyses. As we will see shortly, backward induction is particularly effective for solving win-lose sequential-move games.

Total-Conflict Games

Games of total conflict arise from interactions in which the players are in direct opposition. Whether it be splitting a sum of money or playing a win-lose game, there is no room for compromise in these games as each player pursues his or her own interests.

Simultaneous-move games of total conflict correspond to zero-sum games as introduced in Chapter 2, as we now explain. Recall that in a zero-sum game we have $\text{Payoff to Player 1} + \text{Payoff to Player 2} = 0$ for all outcomes. For a total-conflict game, the assumption is that there is some fixed amount, say, A such that

$$\text{Payoff to Player 1} + \text{Payoff to Player 2} = A$$

for all outcomes. For the purposes of analysis, in the latter case we can subtract $A/2$ from all payoffs for both players. The resulting game will be a zero-sum game with the same strategic properties as the original. We analyze a zero-sum game using the ideas just developed.

Example 4.11 A Zero-Sum Game

We define a simultaneous-move game by giving the payoff matrix:

A Zero-Sum Game

Player 2

Player 1

Strategy	X	Y	Z
A	(3, −3)	(1,−1)	(5, −5)
B	(−2, 2)	(0, 0)	(−5, 5)
C	(1, −1)	(4, −4)	(−3, 3)

First, observe that strategies A and C both weakly dominate strategy B for Player 1. We can thus eliminate the row, giving the following:

Zero-Sum Game with a Dominated Strategy Removed

Player 2

Player 1

Strategy	X	Y	Z
A	(3, −3)	(1,−1)	(5, −5)
C	(1, −1)	(4, −4)	(−3, 3)

Now there are no further relations of weak dominance between strategies for either player. The game does not have a dominated-strategy solution as in Definition 4.7.

How should Player 1 play this game? We have ruled out strategy B, but how does the player choose between strategies A and C?

We use the idea of backward induction to discover a conservative strategy for Player 1 called the *maximin strategy*. We imagine that the simultaneous-move game is actually played sequentially with Player 1 first and Player 2 second. Here is the game tree with the backward induction solution shown:

Backward Induction Solution for A Zero-Sum Game Played Sequentially

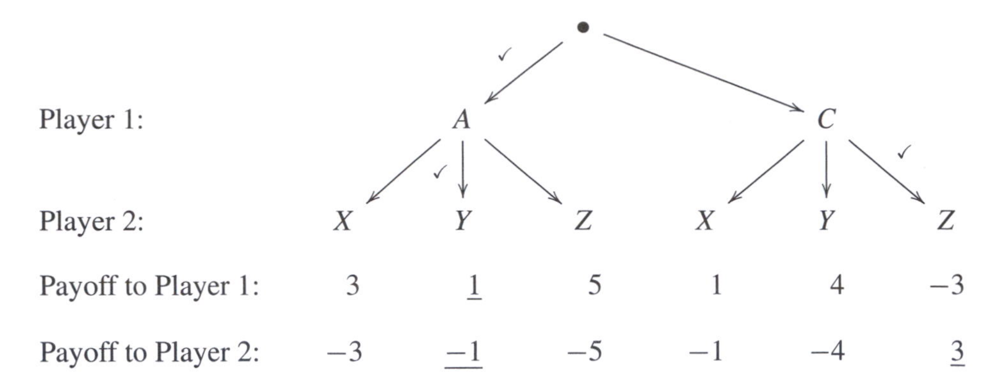

The backward induction strategy for Player 1 is strategy A. Returning to the simultaneous-move game, we see that Player 1 can tip his or her hand and commit to strategy A ahead of time. He or she is assured of at least the payoff 1 no matter what Player 2 does. Strategy A maximizes the worst-case (minimum) payoff possible for Player 1. This feature of the strategy choice explains the name.

Definition 4.12 A strategy S in a zero-sum game is a **maximin strategy** for a player if the minimum possible payoff from strategy S is at least as large as the minimum possible payoff from strategy T for all other strategy choices T for this player.

Some remarks are in order regarding this definition. First, the definition applies to either player. For zero-sum games, the maximin strategy for Player 2 is called the *minimax strategy* for technical reasons that we will see in Chapter 9. Second, we observe that backward induction is certainly not required to find the maximin strategy. Just list the minimum possible payoffs for each strategy choice for a given player, and the strategy producing the largest minimum is the maximin strategy. The backward induction argument highlights the fact that the maximin strategy produces the best payoff if the player has to reveal his or her strategy ahead of time. We will revisit maximin strategies in the context of *Nash equilibria* in Chapter 9.

Combinatorial Games

We turn now to a class of strategic games without payoffs or hidden cards, just two players matching wits turn by turn until a winner is determined.

efinition 4.13

A **combinatorial game** is a finite two-player win-lose sequential-move game of pure strategy with perfect information.

This definition rules out famous games such as Checkers and Chess, which can go on indefinitely, and also finite games such as Tic-Tac-Toe, which may end in a tie. These restrictions will allow for a more complete mathematical theory.

We focus on a class of combinatorial games that we call *takeaway games*. An example of such a game is A Game of Nim (Example 1.3) from Chapter 1. Here is a whole family of games that like the latter require only a handful of chips to play.

Example 4.14 Taking Chips

Let k and n be positive integers with $k < n$. Start with a pile of n chips. On each turn, a player must take at least one chip but no more than k chips from the pile. The last player to take a chip or chips wins the game.

The game tree in the case $n = 5$ and $k = 3$ follows. The node labels indicate the number of chips left at each stage.

Game Tree for Taking Chips with $n = 5$ and $k = 2$

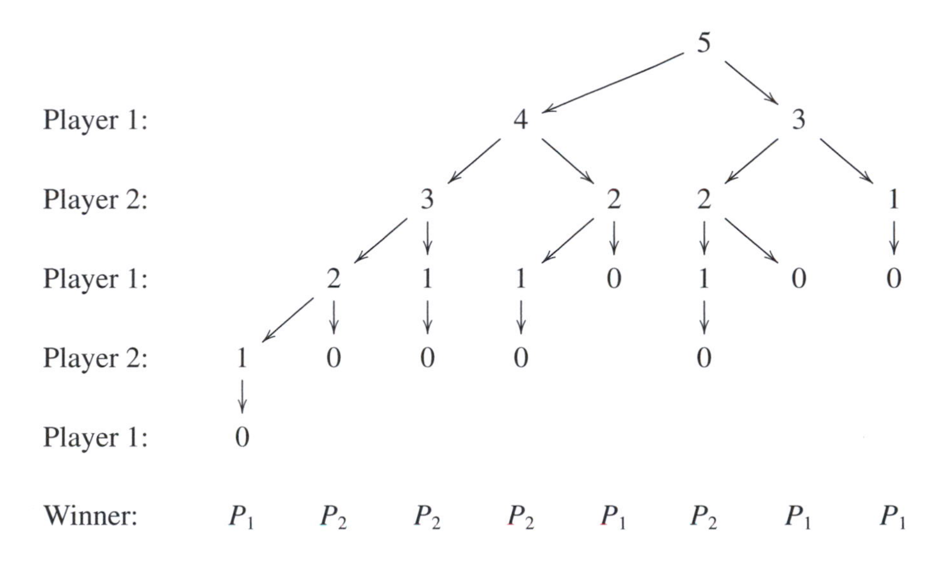

Taking Chips has the property that the last player to move wins the game. We single out these combinatorial games with the following definition:

finition 4.15

A **takeaway game** is a combinatorial game in which the last player to move wins.

We introduce a simple but powerful method for finding the backward induction solution of a takeaway game. We call this the *Binary Labeling Rule* because we use the numbers 0 and 1.

Definition 4.16 Suppose that we are given the game tree of a takeaway game. The **Binary Labeling Rule** is the following procedure for labeling the nodes of the tree with either a 0 or 1: working from the end nodes back up the tree toward the root, assign labels to the nodes of the tree as follows:

0 nodes: The end nodes are all labeled 0. Any node that does not directly branch to at least one node labeled 0 is labeled 0.

1 nodes: Any node directly branching to at least one node labeled 0 is labeled 1.

The resulting labeled tree is called the **binary labeled tree**.

For the game Taking Chips with $n = 5$ and $k = 2$, we have the following binary labels, shown as superscripts on the node labels:

Binary Labeling of Game Tree for Taking Chips with $n = 5$ and $k = 2$

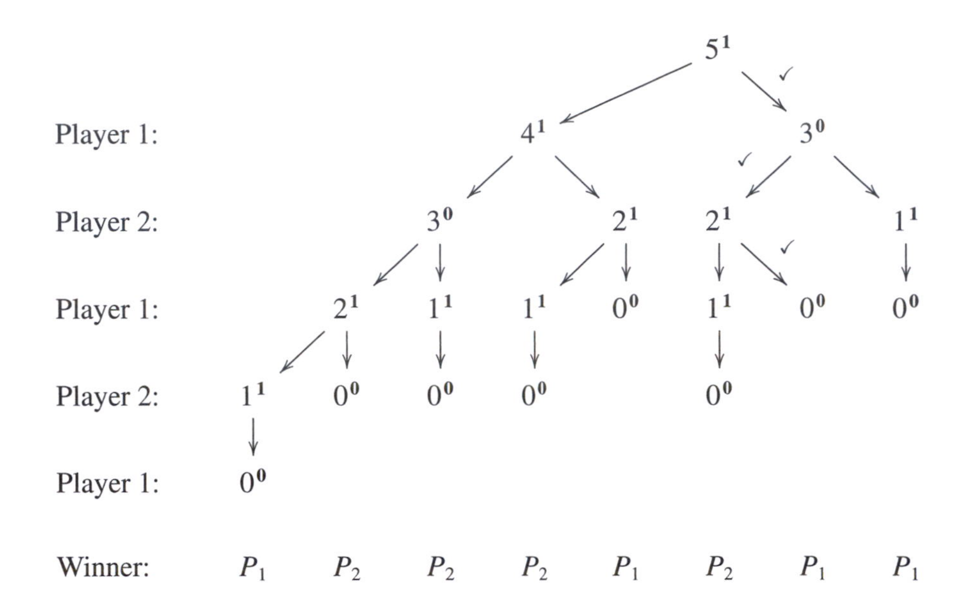

The binary labels determine the backward induction solution of the game. The root node is labeled with a 1, which means that Player 1 can guarantee victory in this game. The winning strategy for Player 1 is to always move to a node labeled 0. Notice that Player 2 will always begin his or her turn on a node labeled 0 and so, by the definition of the binary labels, will only be able to move to a node labeled 1. One particular version of the play of this game is shown. Note that Player 2 has two choices on his or her first turn, but either way, he or she loses on the next turn.

It is perhaps not surprising to learn that one player, either Player 1 or Player 2, is in a position to guarantee victory in any version of Taking Chips. The player who has this power, however, depends on n and k. In fact, we have the following result, proved by Ernst Zermelo in 1913, regarding all combinatorial games:

Theorem 4.17 Zermelo's Theorem *In every combinatorial game, one player has a strategy that will guarantee victory.* ■

We prove Zermelo's Theorem for takeaway games in Chapter 10 using the Binary Labeling Rule. Zermelo's Theorem implies that combinatorial games are hardly games at all because one player can, at least in principle, steer the result to inevitable victory. However, when the game tree is too large to write down completely, the Binary Labeling Rule is not much use. Finding the winning strategy often requires real cleverness. As a case in point, we define perhaps the most important game in combinatorial game theory.

efinition 4.18

The game **Nim** is a takeaway game played starting with several piles of chips. On his or her turn, Player 1 selects one of the piles and removes as many chips as he or she likes (but at least one) from this pile. Player 1 can take the entire pile if desired, but he or she must take chips from this pile exclusively. Player 2 then chooses any pile and takes as many chips as desired (but at least one) from that pile alone. The players continue in this way, with each player taking chips from one of the remaining piles on his or her turn, until all the chips are taken. The player who takes the last chip(s) wins.

We have seen a variation on this game in Example 1.3 in which Player 1 determines the starting position for a game of Nim with 11 total chips. Even with a relatively small number of chips, the game tree for Nim is too large to completely write down. One of the key features of Nim, however, is the fact that each subgame of Nim is itself another game of Nim. Recall that a subgame of a sequential-move game is a game that begins at some position in the main game. For Nim, the subgames begin with smaller and possibly fewer piles than the original game. Analyzing versions of Nim with few chips can lead to discoveries about the general game. We illustrate this idea focusing on the following special case of the general game:

Example 4.19 Two-Pile Nim

Let m and n be two positive integers. The game is then Nim starting with two piles: pile 1 with m chips and pile 2 with n chips.

It is helpful to experiment and play Two-Pile Nim for various choices of m and n. For our analysis, we introduce some notation. Write (a,b) for the position in the game in which pile 1 has a chips and pile 2 has b chips. Of course, we must have $a \leq m$ and $b \leq n$. Write $\text{Nim}(a,b)$ for the game of Nim that begins at node (a,b). Our game is thus $\text{Nim}(m,n)$.

We now make a key observation: the binary label on node (a,b) is the same whether this node is a part of the game tree for $\text{Nim}(m,n)$ or whether (a,b) is the root node for a new game $\text{Nim}(a,b)$. This fact makes the Binary Labeling Rule effective for searching for patterns in Nim. For instance, we can write down the full game tree for the game Nim(2, 1) as follows:

Binary Labels for Nim(2, 1)

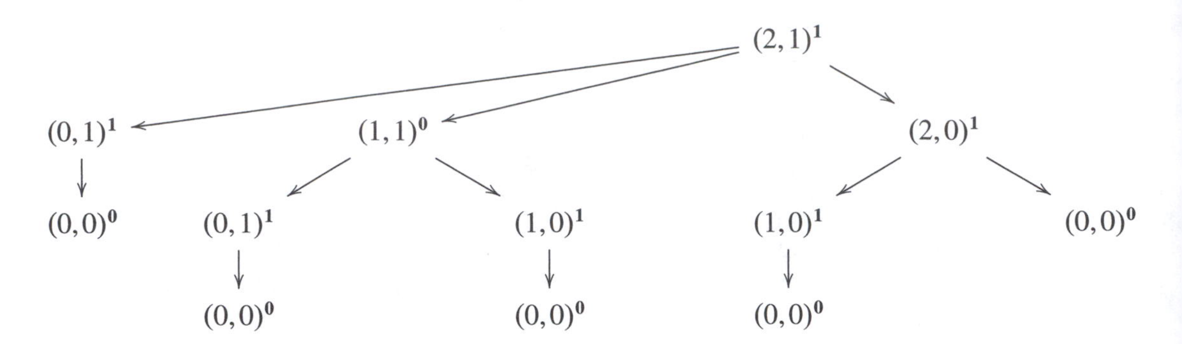

We now know the binary label for node $(2,1)$ in any game of Nim. We can then progressively analyze more complicated versions of Nim. Rather than write down the full game tree, in each case we stop when we reach a node whose binary label is known. Recall that we refer to a game tree that summarizes the play of a game without including all nodes as an *abridged* game tree.

Note that nodes of the form $(0,b)$ and $(a,0)$ have binary label 1. If one pile has no chips, then the player whose turn it is can take the remaining pile and win. Here are some further cases:

Binary Labeling for Abridged Game Trees for Nim

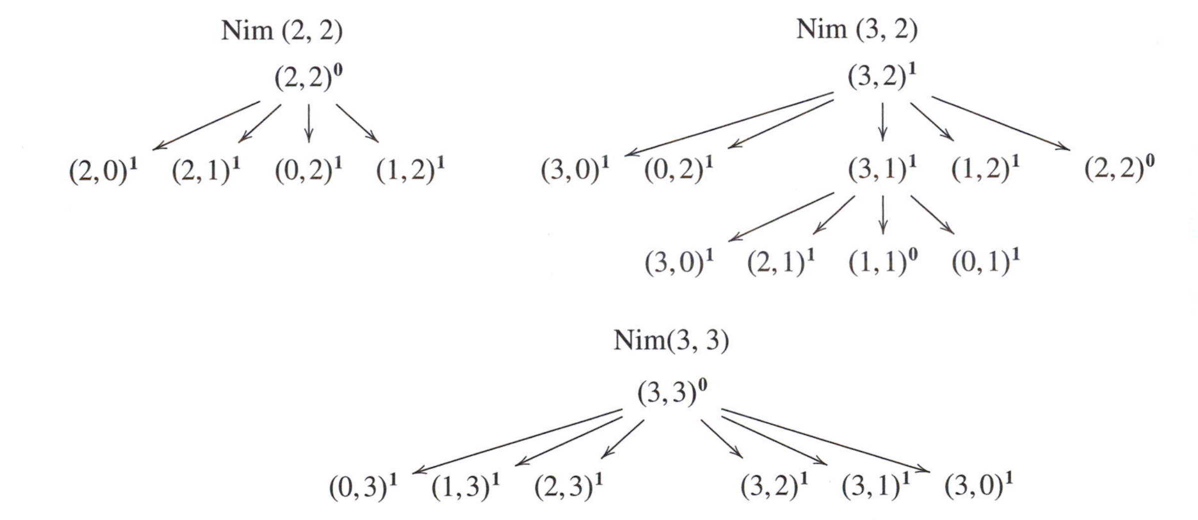

Since this process is far from efficient, it is time to look for a pattern. So far we have three examples of nodes with binary label 0. These are $(3,3),(2,2)$, $(1,1)$, and of course, $(0,0)$. All the other nodes we have encountered are labeled with a 1. If this pattern continues, Player 1 should win the game $\text{Nim}(m,n)$ in rational play when $m \neq n$ and lose when $m = n$. Looking closer at our examples, we can glean the winning strategy: when facing unbalanced piles, the winning strategy is to balance them. In fact, these observations hold in general, and we may state the following complete result for any game of Two-Pile Nim.

Theorem 4.20 Two-Pile Nim *Player 1 can guarantee victory in Nim(m,n) if and only if $m \neq n$. When $m \neq n$, Player 1's winning strategy is to balance the piles on each of his or her turns. Otherwise, when $m = n$, Player 2 can guarantee victory with this strategy of balancing.* ■

While the result represents a complete solution for the case of two piles, there are many more versions of Nim to consider, namely, those involving more than two piles. In particular, Theorem 4.20 does not resolve Question 1.4, although it does tell us that Player 1 should not split the 11 chips into two piles! We return to this question and the general game of Nim in Chapter 17.

For combinatorial games, the guaranteed winning strategy is a true solution to the game. We provide the following definition:

Definition 4.21 A player in a combinatorial game following a strategy guaranteed to win the game is said to exhibit **rational play**.

Of course, only one player in a combinatorial game can play rationally, the player whose good fortune it is to be in the winning position. The other player will lose regardless of his or her strategy, provided that the opponent plays rationally. For this unlucky player, the only hope is that the opponent makes a misstep in the execution of the winning strategy.

Exercises

4.1 Write down a table to show that ratting weakly dominates staying silent for either prisoner in the Prisoner's Dilemma.

4.2 Find all pairs of weakly dominant strategies for both players for the given payoff matrices. Find the dominated strategy solutions of the games if they exist.

(a)

Player 2

Player 1

Strategy	X	Y
A	(4, 7)	(−2, 3)
B	(5, 1)	(0, 2)

(b)

Player 2

Player 1

Strategy	X	Y
A	(2, −5)	(6, 7)
B	(1, 5)	(4, 13)
C	(2, 9)	(3, −5)

(c)

Player 2

Player 1

Strategy	X	Y	Z
A	(6, −5)	(1, 7)	(3, 2)
B	(1, 5)	(2, 2)	(−1, 1)
C	(4, 0)	(3, 5)	(13, −1)

(d)

Player 2

Player 1

Strategy	X	Y	Z
A	(−8, 8)	(−2, 2)	(4, −4)
B	(5, −5)	(0, 0)	(3, −3)
C	(1, −1)	(−4, 4)	(7, −7)

(e)

Player 2

Player 1

Strategy	W	X	Y	Z
A	(1, −1)	(−3, 3)	(−4, 4)	(−1, 1)
B	(−2, 2)	(6, −6)	(−7, 7)	(−9, 9)
C	(3, −3)	(−5, 5)	(−6, 6)	(−12, 12)

4.3 Determine the backward-induction solutions to the sequential-move versions of the games in Exercise 4.2. Compare the solution with Player 1 moving first to the solution with Player 2 moving first for each game. When is moving first in a sequential-move game not a disadvantage?

4.4 Another famous pricing war took place in the U.S. airline industry following passage of the Airline Deregulation Act of 1978. We imagine a scenario in which two airlines are pricing seats from Philadelphia to Chicago. There are 100 customers total. Find the payoff matrix for each of the following simultaneous-move strategic games in which the airlines simultaneously choose between prices for seats and the customers buy seats as indicated. Find all pairs of weakly dominant strategies for both players for the given payoff matrices. Find the dominated-strategy solutions of the games if they exist.

(a) Airline 1: $200 $1,000
Airline 2: $200 $1,000

The customers split evenly when both airlines charge the same price. Otherwise, 80 customers fly with the cheaper airline and 20 with the more expensive airline.

(b) Airline 1: $400 $500
Airline 2: $400 $500

The customers split evenly when both airlines charge the same price. Otherwise, 60 customers fly with the cheaper airline.

(c) Airline 1: $100 $350
Airline 2: $150 $400

The customers split evenly if the prices are within $100 of each other. If one airline charges a $100 or more than the other, then 60 customers fly the cheaper airline.

(d) Airline 1: $100 $200 $300
Airline 2: $100 $200 $300

The customers split evenly when both airlines charge the same price. Otherwise, 70 customers fly with the cheaper airline.

(e) Airline 1: $100 $400 $600
Airline 2: $100 $200 $400

The customers split evenly when both airlines charge the same price. Otherwise, 80 customers fly with the cheaper airline.

4.5 Write down the payoff matrix and find the dominated-strategy solution for the following variations on the Baseball Card Auction (Example 4.4):

(a) Collector 1 values the card at $10 and can bid $5, $7, or $9. Collector 2 values the card at $12 and can bid $4, $6, or $8.
(b) Collector 1 values the card at $18 and can bid $4, $6, $7, or $10. Collector 2 values the card at $12 and can bid $2, $9 or $12.

4.6 For the two-stage sequential-move games indicated with the following game trees, find the backward induction solution of the game (Definition 4.10):

Two-Stage Game Tree

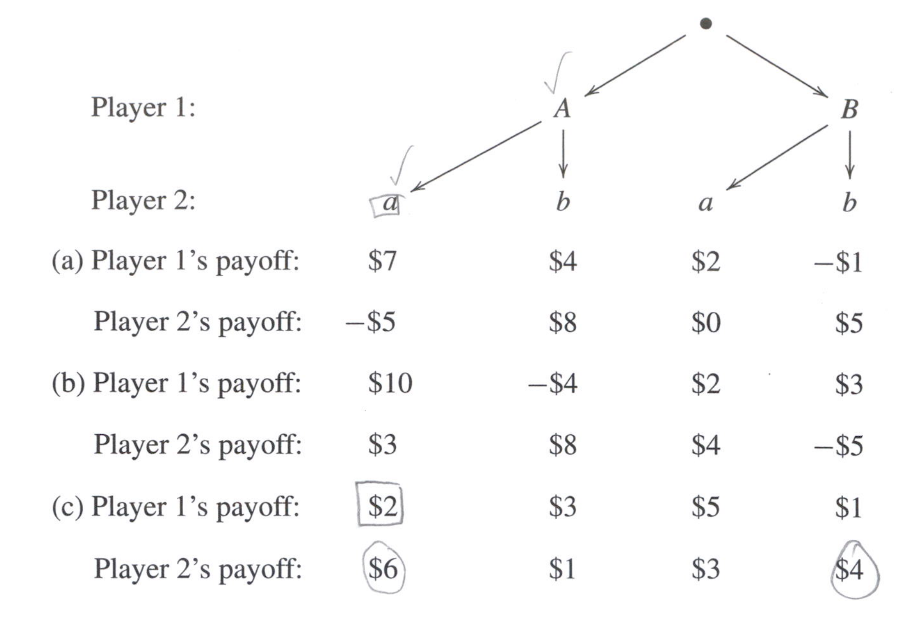

4.7 Repeat Exercise 4.6 for the following three-stage games:

Three-Stage Games

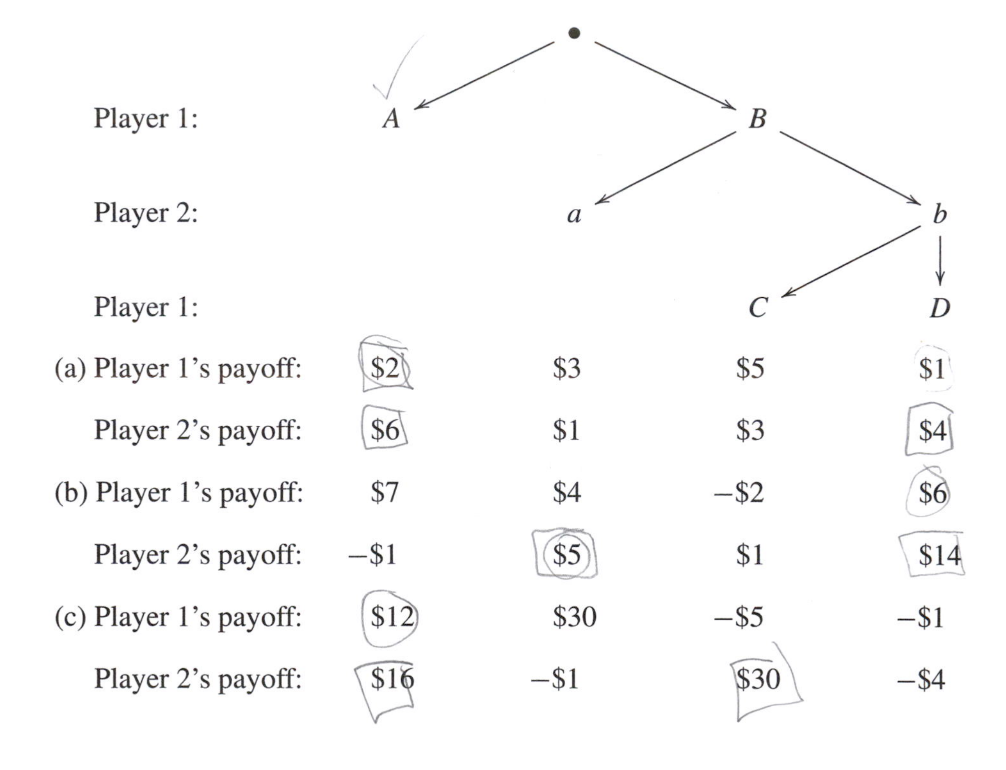

4.8 Write down the game tree for Taking Chips when $n=6$ and $k=2$. Use the Binary Labeling Rule to determine which player wins in rational play. Describe the strategy.

4.9 Write down the game tree for Taking Chips when $n=6$ and $k=3$. Use the Binary Labeling Rule to determine which player wins in rational play. Describe the strategy.

4.10 Consider the game Taking Chips when $n \geq 3$ and $k=2$. Determine which player wins in rational play based on n.

4.11 Consider the game Taking Chips when $n \geq 4$ and $k=3$. Determine which player wins in rational play based on n.

4.12 The game Three-Pile Nim is played starting with three piles. Prove that any node of the form (m,m,n) in Three-Pile Nim has binary label 1. *Hint*: Player 1 can reduce this to a game of Two-Pile Nim with binary label 0 with one move.

4.13 Use Exercise 4.12 and abridged game trees to progressively compute the binary label for the following games of Three-Pile Nim:

(a) $\text{Nim}(1,2,3)$
(b) $\text{Nim}(1,2,4)$
(c) $\text{Nim}(2,3,4)$
(d) $\text{Nim}(3,4,5)$

4.14 Investigate a real-world price war between two companies. Decide whether the game is well represented as a Prisoner's Dilemma. What was the outcome of the game? What were some of the effects for customers and other competitors in the industry?

5 Choice

The shift of focus from games to elections represents a rather dramatic change in the objects of our study. We turn from probabilities, payoffs, and winning strategies to questions of the fairness, equity, and correctness of electoral outcomes. We have seen, however, that an election is not so different from a game at the most basic level. Recall in Definition 2.7, we defined an election as a game producing an outcome called the *social choice*. In this chapter, we study the mechanism for producing this outcome, the *voting method*.

We begin with a discussion to motivate the idea of voter preferences. The summary table of voter preferences is called a *preference table*. We introduce this important device along with a number of voting methods that will be studied throughout the text. We conclude with a discussion of *social preferences* and *social welfare*.

Preference Rankings

In practice, most elections do not require voters to rank all the candidates. Usually a "vote" consists of simply an indication of first choice. As we have mentioned, the use of voter preferences is assumed to facilitate our mathematical analysis of elections. That said, we observe that there are elections that do explicitly involve preference rankings of the candidates. We give two "controversial" examples, both occurring in the year 1990.

The world of sports provides a number of examples of elections in which voters submit preference rankings for the candidates. The Heisman Trophy in college football and the Most Valuable Player (MVP) awards and Hall of Fame elections for many professional sports are notable examples. Our example is from the National Basketball Association (NBA).

Example 5.1 A Controversial MVP Award

The 1989–1990 NBA season featured several future hall-of-fame players in their prime. Among the many strong performances, three players had standout seasons: Michael Jordan, Magic Johnson, and Charles Barkley. The MVP award was decided by the votes of 92 NBA experts (i.e., writers, broadcasters, etc.).

Each voter listed five players in order from first choice to fifth choice. Here is a sample ballot:

One Possible MVP Ballot

1st choice	Charles Barkley
2nd choice	Magic Johnson
3rd choice	Michael Jordan
4th choice	Karl Malone
5th choice	Hakeem Olajuwon

The election was run by assigning points to players according to the following rule: 10 points for each first-place vote, 7 points for each second-place vote, 5 points for each third-place vote, 3 points for each fourth-place vote, and 1 point for each fifth-place vote. Thus 920 was the maximum point total for a player. The first-place votes along with the total points for the leading candidates were as follows:

Total First-Place Votes

Player	Total
Charles Barkley	38
Magic Johnson	27
Michael Jordan	21

Total Points

Player	Total
Magic Johnson	636
Charles Barkley	614
Michael Jordan	571

The MVP award for the 1989–1990 NBA season was awarded to Magic Johnson.

Our second example comes from the world of politics.

Example 5.2 A Controversial Presidential Election

The 1990 Irish presidential election featured three candidates, Brian Lenihan, Mary Robinson, and Austin Currie, representing the Independent, Fianna Fáil, and Fine Gael parties, respectively. The favorite was Lenihan, a popular and well-known figure in Irish politics. Each voter received a ballot listing the three candidates. A voter marked his or her first-choice candidate and, if desired, indicated a second-choice candidate as well. When the votes were collected, the first-place totals were as follows:

Candidate	First-place votes	Percentage
Lenihan	694,484	44%
Robinson	612,265	39%
Currie	267,902	17%

Because no candidate received a majority of the first-place votes, the candidate with the least first-place votes, Austin Currie, was eliminated from contention. The ballots were then recounted with all first-place votes for Currie transferred to the voter's second-choice candidate (if one was given). The second-round totals were as follows:

Candidate	First-place votes	Percentage
Lenihan	731,273	48%
Robinson	817,830	52%

With a majority of the votes in the second round, Mary Robinson was declared the president of Ireland.

In both cases, the candidate with the most first-place votes (Barkley and Lenihan, respectively) was not the winner of the election. Such an outcome may not be controversial, of course. One needs to examine the performances of the players in the first case and the demographics of Ireland in the second to judge the correctness of the outcome. Our main interest in these examples is the illustration they provide of preference lists in voting. A Controversial MVP Election (Example 5.1) uses a variation of the *Borda Count,* the voting method introduced by the Chevalier de Borda. The voting method for the Irish presidential election (Example 5.2) is a version of the the *Hare Method*, named for Thomas Hare. This voting system is currently used by several political bodies for electing both individual candidates and committees (see Exercise 5.13).

Voting Methods

As the preceding examples suggest, voting methods come in great variety. In fact, the selection of a voting method for a given political or social body is itself a significant social choice. We illustrate this idea with the following simple example that will serve to introduce several standard voting methods.

Example 5.3 Family Dinner Decision

A family of seven must decide on a restaurant destination. The candidates are Thai, Greek, and Mexican restaurants and a diner. Both Mom and Dad want Thai food, while the three sisters want Greek food. Big brother wants Mexican, and baby brother wants the diner. Denoting the candidates for dinner as T, G,

M, and *D*, the family members' true preferences for the dining options are summarized in the following table called a *preference table*:

Preference Table for Family Dinner Decision

Family member	Mom	Dad	Big bro.	Sister 1	Sister 2	Sister 3	Baby bro.
1st choice	*T*	*T*	*M*	*G*	*G*	*G*	*D*
2nd choice	*M*	*M*	*T*	*M*	*M*	*M*	*T*
3rd choice	*D*	*D*	*D*	*D*	*D*	*D*	*M*
4th choice	*G*	*G*	*G*	*T*	*T*	*T*	*G*

Definition 5.4 Given a group of voters and a set of candidates, the **preference table** is the listing of each voter's preferences for the candidates ranked in order from first choice to last choice.

Once the preference table is assembled, all that is needed is a voting method to make a choice. For our example, we imagine that the family members have differing opinions about voting methods, as well.

The Plurality Method

The three sisters present a united front, a voting bloc of three, in the Family Dinner Decision (Example 5.3). They propose arguably the most natural voting method.

Definition 5.5 The **Plurality Method** is the voting method in which the voters' first choices are counted for each candidate. The **Plurality social choice** is the candidate (or candidates) with the most first-place votes.

The **Plurality Method** is commonly used and easy to implement. Full rankings are not necessary, only first-place votes. For the Family Dinner Decision, the first-place vote counts are as follows:

Plurality Method for Dinner Decision

Candidate	First-place votes
G	3
T	2
M	1
D	1

The Greek restaurant is the Plurality social choice.

The Borda Count

Big brother plays the role of the Chevalier de Borda in our scenario. He proposes a natural way to take into account all voter preferences, not just the first choices. Each voter assigns 3 points, called *Borda points*, to his or her first-choice candidate, 2 points to his or her second-choice candidate, 1 point to his or her third-choice candidate, and 0 points to his or her least favorite candidate. Here is the table of points:

Borda Points for Family Dinner Decision

Candidate	Mom	Dad	Big bro.	Sister 1	Sister 2	Sister 3	Baby bro.	Points
T	3	3	2	0	0	0	2	10
G	0	0	0	3	3	3	0	9
M	2	2	3	2	2	2	1	14
D	1	1	1	1	1	1	3	9

This idea is the basis for the following voting method.

Definition 5.6 The **Borda Count** is the voting method defined as follows: suppose that there are n candidates. Each voter awards his or her first-choice candidate $n-1$ points, the second-choice candidate $n-2$ points, and so on, with 0 points awarded for last-place votes. We will refer to these points as **Borda points**. The **Borda Count social choice** is the candidate(s) with the most total Borda points as awarded by the voters.

As we have seen in Example 5.1, a version of the Borda Count is used to elect the Most Valuable Player in the NBA, where partial preference lists and a different decreasing sequence of Borda points is used, namely, 10, 7, 5, 3, and 1. For consistency, we will always use the values $n-1$ down to 0 with n voters for the Borda Count in the text. See, however, Exercise 5.8 for more on other choices of points.

For Example 5.3, the Mexican restaurant is the candidate with the most Borda Points and so is the Borda Count social choice.

The Hare Method

It is now the parents' turn to make a proposal. They suggest a voting method reminiscent of the Irish presidential election (Example 5.2).

Definition 5.7 The **Hare Method** is the voting method in which the social choice is determined by a step-by-step process of eliminating candidates until a candidate has a majority of first-place votes. If no candidate has a majority of first-place votes, then the candidate having the least first-place votes is eliminated. The preference table is then adjusted to remove this candidate. If there is a tie among candidates with the least first-place votes, we break the tie by eliminating the candidate whose name occurs last alphabetically. The process is repeated, with one candidate eliminated on each round, until some candidate has a majority of first-place votes of first-place votes. This candidate is the **Hare social choice**.

The key step in the Hare Method serves to introduce a common operation with preference tables. If candidate X is eliminated, each voter adjusts his or her personal rankings simply by removing X and using the resulting ranking of the remaining candidates. This is the idea of *transferring votes* we saw in Example 5.2. Precisely, a voter is said to **transfer votes** if, when one candidate is eliminated from contention, his or her preference ranking is automatically adjusted to reflect his or her preferences for the remaining candidates. Here is a picture of the simple operation:

A Voter Transferring Votes after Candidate X Is Eliminated

Original ranking		New ranking
A		A
B		B
C	Eliminate X	C
X		D
D		

We run the Hare Method on the preference table for Example 5.3. At the first stage, M and D each have only one first-place vote. We break the tie using alphabetical order and eliminate M. Next, D is eliminated, leaving the following preference table:

Third-Stage Preference Table for the Hare Method

Family member	Mom	Dad	Big bro.	Sister 1	Sister 2	Sister 3	Baby bro.
1st choice	T	T	T	G	G	G	T
2nd choice	G	G	G	T	T	T	G

The Thai restaurant is the Hare social choice, a victory for Mom and Dad.

We remark that it is also possible to define the Hare Method, wherein all candidates tied with the least first-place votes at a stage are eliminated at once. This alternate version of the Hare Method gives the same result in the preceding example but can produce a different social choice in general.

A Dictatorship

Finally, we come to baby brother. He is not happy with any of the proposed voting methods. Unfortunately, he does not know of a viable alternative that will deliver D as the social choice. Taking advantage of his position as the youngest, he simply insists that the diner should be the social choice. Baby brother is proposing the following voting method:

Definition 5.8 A **Dictatorship** is the voting method in which one fixed voter chooses the social choice according to his or her first choice regardless of the preferences of the other voters. The voter making the social choice is called the **dictator**.

A Dictatorship represents an extreme version of a voting method. One voter makes the decision. The other voters register their preferences as an exercise in futility. Notice that we must be clear about who the dictator is when identifying a dictatorship! With baby brother as dictator, the diner is the social choice for the Family Dinner Decision.

The Family Dinner Decision reveals an interesting phenomenon. As summarized in the following diagram, the problem of choosing a voting method has simply replaced the original problem of choosing a restaurant:

Family Dinner Decision

Family member		Voting method		Social choice
Sisters	⟶	Plurality Method	⟶	Greek
Big brother	⟶	Borda Count	⟶	Mexican
Mom and Dad	⟶	Hare Method	⟶	Thai
Baby brother	⟶	Dictatorship	⟶	Diner

We remark that each of our voting methods other than a Dictatorship is capable of producing ties. In practice, the problem of breaking ties is a difficult one to resolve. As we did in the definition of the Hare Method (Definition 5.7), we use a simple device – alphabetical order – to break ties in this text. We make this official by introducing the following convention:

Convention 5.9 *When the outcome of an election is a tie, we determine the social choice from the finalists using alphabetical order of candidate names. For example, if, say, A, B, and C are tied, we declare A the winner.* ■

Elections often involve quite a few more voters than the Family Dinner Decision. In this case, it is natural to group voters by voting blocs (Definition 2.11) and keep track of the number of voters in each bloc. Recall that a voting bloc is a collection of voters with exactly the same preference order for the candidates. Here is an example:

Example 5.10 Camp Activities Election

A camp counselor offers her 68 campers the opportunity to choose their activity for the day. The candidates are Archery (A), Basketball (B), Capture the Flag (C) and Dodgeball (D). Each camper ranks the activities according to his or her individual preferences, giving the following preference table:

Preference Table for Camp Activities Election

Number of voters	22	20	15	10	1
1st choice	A	D	C	B	D
2nd choice	B	B	D	C	C
3rd choice	D	C	B	D	B
4th choice	C	A	A	A	A

We run this election using the Plurality, Borda Count, and Hare Methods. We also introduce yet another voting method called the *Coombs Method.*

In the Camp Activities Election preference table, the first-place votes are as follows:

First-Place Votes for Camp Activities Election

Candidate	First-place votes
A	22
B	10
C	15
D	21

Archery is the Plurality Method social choice.

For the Borda Count, we have the following table:

Candidate	22	20	15	10	1	Total
A	3(22)	0(20)	0(15)	0(10)	0(1)	66
B	2(22)	2(20)	1(15)	3(10)	1(1)	130
C	0(22)	1(20)	3(15)	2(10)	2(1)	87
D	1(22)	3(20)	2(15)	1(10)	3(1)	125

Using the Borda Count, the social choice is Basketball.

We next apply the Hare Method. Candidate B has the least first-place votes and so is eliminated from contention. We obtain a new preference table:

Preference Table with B Eliminated

Number of voters	22	20	15	10	1
1st choice	A	D	C	C	D
2nd choice	D	C	D	D	C
3rd choice	C	A	A	A	A

The next candidate eliminated is D, yielding

Preference Table with B and D Eliminated

Number of voters	22	20	15	10	1
1st choice	A	C	C	C	C
2nd choice	C	A	A	A	A

The Hare social choice is candidate Capture the Flag.

The Coombs Method

We introduce one further voting method representing a counterpart to the Hare Method. The Hare Method narrows the search for a majority candidate by eliminating candidates that have the least support. The Coombs Method takes the opposite approach, eliminating candidates who are unacceptable to the most voters.

efinition 5.11

The **Coombs Method** is the voting method defined by a step-by-step process of eliminating candidates. At each stage, if no candidate has the majority of first-place votes, the candidate with the most last-place votes is eliminated. As with the Hare Method, ties at each stage for the most last-place votes are broken by alphabetical order. The preference table is then adjusted to remove this candidate. The process is repeated until a surviving candidate receives a majority of first-place votes. That candidate is the **Coombs social choice.**

We apply the Coombs Method to Example 5.10. Since no candidate has a majority of first-place votes, we look at the last-place votes and see that candidate A has the most of these. Eliminating A yields the following preference table:

Preference Table with A Eliminated

Number of voters	22	20	15	10	1
1st choice	B	D	C	B	D
2nd choice	D	B	D	C	C
3rd choice	C	C	B	D	B

Again, no candidate has a majority of first-place votes, so we eliminate C with 42 last-place votes, yielding the following two candidate election:

Preference Table with A and C Eliminated

Number of voters	22	20	15	10	1
1st choice	B	D	D	B	D
2nd choice	D	B	B	D	B

Candidate D now has a majority of first-place votes, so the Coombs social choice in the Camp Activities Election is Dodgeball.

Social Preferences

The Plurality, Borda Count, Hare, and Coombs Methods all may be described as "democratic" voting methods. One way to distinguish these methods from a Dictatorship is seen by applying the voting methods to an election with just two candidates.

Definition 5.12 The voting method for two candidates in which the candidate with the most votes wins is called the **Simple Majority Method**.

The Plurality, Borda Count, Hare, and Coombs Methods each reduce to the Simple Majority Method for elections with only two candidates. The Dictatorship, of course, does not. The Simple Majority Method, in turn, leads to the following important notion:

Definition 5.13 Suppose that we are given a preference table for a group of voters each ranking a set of candidates including candidate X and candidate Y. If X beats Y in a head-to-head election using the Simple Majority Method, we say that X is **socially preferred** to Y. If X and Y tie in the Simple Majority Method, we say that **society has no preference** between X and Y.

We emphasize that all voters in a head-to-head election vote even if their favorite candidate is not involved. In the Camp Activities Election, for example, we can determine the social preference for A versus B by eliminating candidates C and D.

We obtain the following:

Preference Table for Camp Activities Election with *C* and *D* Eliminated

Number of voters	22	20	15	10	1
1st choice	*A*	*B*	*B*	*B*	*B*
2nd choice	*B*	*A*	*A*	*A*	*A*

We conclude that *B* is socially preferred to *A*. We can determine the other social preferences using the same method. Here are the results:

Head-to-Head Results for Camp Activities Election

A	*B*	Winner
22	46	*B*

A	*C*	Winner
22	46	*C*

A	*D*	Winner
22	46	*D*

B	*C*	Winner
52	16	*B*

B	*D*	Winner
32	36	*D*

C	*D*	Winner
25	43	*D*

We next introduce a device called a *directed graph* to keep track of social preferences.

efinition 5.14

A **directed graph** is a system of nodes and branches in which the branches have arrows or **directions**. Notice that we have written game trees as directed graphs in which arrows always point in the direction of the flow of the game from the root node to the end nodes.

Here we use the arrows on branches to indicate social preference. We express the fact that *X* is socially preferred to *Y* by including an arrow from *X* to *Y*:

X is socially preferred to *Y*.

$$X \longrightarrow Y$$

When we assemble all the social preferences between candidates in an election, we obtain a mathematical object called a *complete directed graph*. Formally:

efinition 5.15

A **complete directed graph** is a collection of nodes and arrows such that each pair of nodes is connected by a unique arrow.

Given a preference table, we consider the complete directed graph with nodes corresponding to the candidates and the directions of the branches determined by social preference, as earlier. We call the result the **social preference graph** for the preference table. Here is the graph for the Camp Activities Election:

Social Preference Graph for the Camp Activities Election

(5.1)

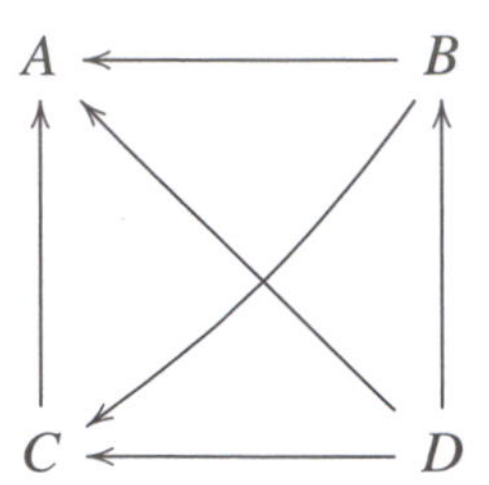

We note that, strictly speaking, the social preference graph may include branches without arrows. This will occur when there are ties in the Simple Majority runoffs. In some cases, it will be useful to keep track of ties at this level.

Pairwise Comparisons

We define a voting method based on the Simple Majority Method. The key idea is an **agenda**, which is an ordering of the candidates established before the election. For convenience, we will use reverse alphabetical order of the candidates to establish an agenda. Thus, if our candidates are A, B, C and D the agenda is

Agenda: $D \quad C \quad B \quad A$

In this case, the first candidate on the agenda is D and the last is A.

Definition 5.16 The **Pairwise Comparisons with Agenda** voting method is run using a given agenda as follows: the first candidate on the agenda runs against the second candidate. The winner of this head-to-head election faces the third candidate on the agenda. Continue in this manner until one candidate survives to run against the last candidate on the agenda. The winner of the Simple Majority election between this surviving candidate and the last candidate on the agenda is then the **Pairwise Comparisons with Agenda** social choice.

We determine the Pairwise Comparisons with Agenda social choice for the Camp Activities Election. Here we begin by running D versus C. As seen earlier, D wins this Simple Majority Election and so survives to face B. We then run B versus D. Since D beats B head to head, D survives to the final round. We run A versus D. Since D beats A head to head, D is the Pairwise Comparisons social choice. We explore some other possibilities for establishing an agenda next.

Social Welfare

Decision making for a group or society often goes beyond electing candidates. Whether it be a local government balancing limited financial resources between schools and police and fire departments or a newly elected Congress setting an agenda for the term, ranking competing alternatives is an important activity for any governing body. We have the following underlying question:

How should a society rank competing alternatives?

As in the case of social choice, we will see a number of possible methods for producing a ranking of candidates from preference tables. We introduce some here, but first we give the official definition.

Definition 5.17 A ranking of a set of candidates or alternatives from first to last by a group of voters is called a **social welfare**. A **social welfare method** is a mechanism that uses the rankings of the candidates by individual voters in a society to produce a social welfare.

Notice that a social welfare method determines, in particular, a social choice method. The candidate ranked first by society is the social choice. Conversely, all our social choice methods extend naturally to social welfare methods. We explain this with an example.

Example 5.18 Student Activities Election

The student government of a small college must decide how to prioritize its entertainment budget. The events to be planned are as follows: an art appreciation trip (*A*), a bar-BQ (*B*), a concert (*C*), and a dance (*D*). The students have all been polled. Their individual preferences are given in the following table:

Student Activities Election Preference Table

Number of voters	46	120	150	250	84
1st choice	*A*	*D*	*C*	*B*	*A*
2nd choice	*B*	*C*	*A*	*D*	*C*
3rd choice	*C*	*B*	*D*	*C*	*D*
4th choice	*D*	*A*	*B*	*A*	*B*

We use our various social choice methods, extended to social welfare methods, to rank these candidates.

Plurality Social Welfare Method

Here the candidates are ranked according to the number of first-place votes. In the Student Activities Election, the result is as follows:

First-Place Votes for Student Activities Election

First-place votes	Candidate
130	*A*
250	*B*
150	*C*
120	*D*

$\Longrightarrow$ Plurality social welfare: *B* *C* *A* *D*

Borda Count

The Borda Count naturally extends to a social welfare method: the candidates are ranked according to their Borda points. The calculation is as follows:

Borda Count for Student Activities Election

Candidate	46	120	150	250	84	Total
A	3(46)	0(120)	2(150)	0(250)	3(84)	690
B	2(46)	1(120)	0(150)	3(250)	0(84)	962
C	1(46)	2(120)	3(150)	1(250)	2(84)	1154
D	0(46)	3(120)	1(150)	2(250)	1(84)	1094

We conclude

Borda Count social welfare: *C* *D* *B* *A*

Hare Social Welfare Method

We follow the procedure of the Hare Method. The first candidate eliminated is given the lowest rank, the second eliminated is given the next lowest rank, and so on until we reach the Hare social choice, who is ranked first. For the Student Activities Election, the Hare social welfare is obtained as follows: candidate *D* is eliminated first and so ranked last. We then eliminate *D* and obtain the following preference table:

Preference Table for Student Activities Election with *D* Eliminated

Number of voters	46	120	150	250	84
1st choice	*A*	*C*	*C*	*B*	*A*
2nd choice	*B*	*B*	*A*	*C*	*C*
3rd choice	*C*	*A*	*B*	*A*	*B*

We next eliminate candidate A, leaving B and C. We quickly check that B beats C head to head and so obtain

Hare social welfare: $C \quad B \quad A \quad D$

Coombs Social Welfare Method

We follow the procedure of the Coombs Method. The first candidate eliminated is given the lowest rank, the second eliminated is given the next lowest rank, and so on. For the Student Activities Election the procedure goes as follows: we first eliminate A, the candidate with the most last-place votes. We obtain a new preference table:

Preference Table for Student Activities Election with *A* Eliminated

Number of voters	**46**	**120**	**150**	**250**	**84**
1st choice	B	D	C	B	C
2nd choice	C	C	D	D	D
3rd choice	D	B	B	C	B

We next eliminate B, leaving C and D. We see that D beats C head to head by a count of 370 to 280, so

Coombs social welfare: $D \quad C \quad B \quad A$

The Dictatorship also extends to a social welfare method, although it hardly seems to deserve the name. The **Dictatorship social welfare** is just a ranking of the candidates submitted by the dictator.

We conclude by remarking that the problem of ties arises for social welfare methods as it does for social choice methods. We use our alphabetical convention (Convention 5.9) to break all ties in the social welfare.

Exercises

5.1 For each of the following preference tables, determine the social welfare ranking using the following voting methods:

(a) Plurality Method
(b) Borda Count Method
(c) Hare Method
(d) Coombs Method

(1)

Number of voters	100	60	70	50	80
1st choice	A	B	C	A	B
2nd choice	B	A	B	C	C
3rd choice	C	C	A	B	A

(2)

Number of voters	33	40	16	22	14
1st choice	A	B	C	D	A
2nd choice	B	A	B	A	B
3rd choice	C	D	D	B	C
4th choice	D	C	A	C	D

(3)

Number of voters	12	10	17	11	5	13
1st choice	A	E	B	C	D	D
2nd choice	B	A	C	D	E	E
3rd choice	D	C	A	A	C	A
4th choice	C	D	E	E	B	B
5th choice	E	B	D	B	A	C

(4)

Number of voters	18	17	12	23	20
1st choice	A	E	A	D	C
2nd choice	B	B	C	B	B
3rd choice	D	C	D	C	D
4th choice	C	A	E	A	E
5th choice	E	D	B	E	A

(5)

Number of voters	10	20	14	18	16	11	13
1st choice	A	B	E	C	D	E	F
2nd choice	B	E	D	D	A	F	A
3rd choice	D	F	F	A	B	C	C
4th choice	C	D	A	B	C	A	B
5th choice	E	C	B	F	F	B	E
6th choice	F	A	C	E	E	D	D

5.2 Find the social preference graphs for preference tables (1)–(5) in Exercise 5.1.

5.3 Find the Pairwise Comparisons with Agenda social choice for each of preference tables (1)–(5) in Exercise 5.1 using the following agendas:

(a) Take the agenda to be the reverse alphabetical order ranking.
(b) Notice that choosing an agenda for a Pairwise Comparisons election is the same as choosing a social welfare. Use the Plurality Social Welfare Method to establish an agenda for each of preference tables (1)–(5) and determine the Pairwise Comparisons with Agenda social choice in each case. Be sure to reverse the order so that the winner of the Plurality Method is last on the agenda.
(c) Repeat (b) using the Borda Count to establish the agenda.

5.4 Construct a preference table with 3 candidates A, B, and C and 20 voters satisfying the following requirements. You may use a different preference table for each problem.

(a) The Plurality social choice is not the same candidate as the Borda Count social choice.
(b) The Borda Count social choice is not the same candidate as the Hare social choice.
(c) The Plurality social choice is not the same candidate as the Coombs social choice.
(d) The Coombs social choice is not the same candidate as the Hare social choice.

5.5 Give an example to show that a candidate can be ranked first in the Borda Count social welfare and last in the Plurality social welfare.

5.6 Give an example to show that a candidate can be ranked first in the Plurality social welfare and last in the Borda Count social welfare.

5.7 Give an example to show that a candidate can be ranked first in the Plurality social welfare and last in the Coombs social welfare. Explain why such an example is impossible when the Coombs Method is replaced by the Hare Method.

5.8 In a Controversial MVP Election (Example 5.1), the Borda points are assigned according to the system 10, 7, 5, 3, 1, with 0 points awarded for any ranking below fifth. We call the Borda point assignments a **Borda weight system**. For the NBA MVP election, the Borda weight system is 10, 7, 5, 3, 1, 0, 0, 0, ...

(a) Run the Borda Count on the following preference table first with the weight system 10, 7, 5, 3, 1 and then with the weight system 10, 5, 3, 1, 0 to show that the outcome of a Borda Count election depends on the Borda weight system.

Number of voters	**39**	**19**	**18**	**10**	**6**
1st choice	A	B	C	D	B
2nd choice	B	E	D	C	C
3rd choice	C	C	E	A	A
4th choice	D	D	A	B	D
5th choice	E	A	B	E	E

(b) Find a preference table with three candidates so that the Borda Count with the usual weight system 2, 1, 0 and the Borda count with the weight system 6, 3, 0 produce different social welfare rankings.

(c) Find a preference table with four candidates so that the Borda Count with weighted system 3, 2, 1, 0 and the Borda count with weight system 7, 3, 2, 1 produce different social welfare rankings.

5.9 (Copeland Method) The **Copeland Method** is defined by a round-robin tournament. Each pair of candidates runs head to head using the Simple Majority method. In each such pairing, the winning candidate receives 1 point and the losing candidate 0 points. If two candidates tie in a head-to-head match, each candidate receives 1/2 point. The candidate with the most overall points is the **Copeland social choice.** Determine the Copeland social choice for the preference tables in Exercise 5.1.

5.10 Show that in an election with three candidates total, if one candidate is the outright winner of the Copeland Method, then that candidate wins the Pairwise Comparisons Method with any agenda. Is the same true with four candidates?

5.11 Show that a candidate can win a Copeland Method election outright and yet lose in a Plurality Method election.

5.12 (The Smith Set) Given a preference table with candidates A, B, C, ..., say that candidate X **dominates** a subset of candidates, say, $\{A, B, C\}$ if X beats A, B, and C head to head. More generally, say that a set of candidates $\{X, Y\}$ dominates a set $\{A, B, C\}$ of candidates if X and Y both beat each of A, B, and C in head-to-head elections.

(a) Show that $\{X, Y\}$ dominates $\{A, B, C\}$ for the following preference table:

Number of voters	20	25	27	30	31	37
1st choice	A	D	C	D	Y	D
2nd choice	B	C	A	Y	X	Y
3rd choice	C	A	B	X	D	X
4th choice	D	B	D	B	B	A
5th choice	X	X	X	C	C	B
6th choice	Y	Y	Y	A	A	C

(b) Given a preference table, the **Smith set** is the smallest subset $\mathcal{S}$ of the candidates such that the set $\mathcal{S}$ dominates the set of candidates not in $\mathcal{S}$. Find the Smith set for the preference table in part (a).

(c) Find the Smith set for the preference tables in Exercise 5.1.

5.13 A version of the Hare Method called the **Single Transferable Vote Method** is used by Ireland, Australia, and other countries to fill multiple seats from a set of candidates. Here is a simplified version of the method: each voter lists his or her preferences for all candidates. The **Droop quota** is given by the formula

$$\text{Quota} = \frac{\text{total votes cast}}{\text{seats left to fill} + 1} + 1$$

All candidates whose first-place votes total to at least the quota are elected. These candidates and the open seats are eliminated and the election is repeated until all the seats are filled. If, at some stage, no candidate meets the Droop quota, then the candidate with the least first-place votes is eliminated. When a candidates is eliminated, his or her votes are transferred to the remaining candidates in the usual way. Run the Single Transferable Vote Method in the following cases:

(a) Fill two seats from the five candidates in Exercise 5.1(3).
(b) Fill three seats from the five candidates in Exercise 5.1(4).
(c) Fill four seats from the six candidates in Exercise 5.1(5).

6 Strategy and Choice

In this chapter, we delve deeper into the idea of an election as a game of strategy (Definition 2.9). The implications of this definition are somewhat unsettling. Is a popular election a game to be won? In a democracy, we expect social choice to be a faithful accounting of the true preferences of the people rather than an outcome arrived at by the strategic cleverness of certain of the voters. Unfortunately, the ideal of a strategy-proof, democratic election is an impossible dream. We state the *Gibbard-Satterthwaite Theorem* (Theorem 6.4), which asserts that with three or more candidates, every nondictatorial voting method is vulnerable to strategic manipulation. We verify the theorem for each of the democratic voting methods introduced in Chapter 5. We also show how strategic considerations in social choice can elect a third-party candidate in a popular election. The latter result is an example of the *Paradox of the Chair*. Finally, we give an example of the *No-Show Paradox*, a scenario in which voters in an election better serve their own interests by staying home as opposed to participating in the election.

The Condorcet Paradox

The developments in this chapter will all flow out of arguably the central example in social choice theory. We introduce this important example by means of the following simple puzzle:

Example 6.1 Choosing the Final

Twenty-five students in a game theory seminar are given the option of choosing the format for their final assignment. The options are to write a term paper (T), take an oral exam (O), or take a written exam (W). The professor proposes that the students vote on each of the options in head-to-head Simple Majority elections and thereby see which option is preferred to the other two. First, the professor pits term paper versus oral exam. The winner is term paper, by a vote of 18 to 7. Next, the students vote on writing a term paper versus taking a written exam. Here the written exam wins by a vote of 15 to 10.

The professor is now prepared to assign a written exam as the chosen option. One student speaks up and suggests that the class should run the remaining head-to-head election: oral exam versus written exam. The outcome of this election is a surprise to almost everyone. Taking an oral exam beats taking a written exam by a count of 17 to 8. How can this happen?

To answer this question, we unveil the preference table for the game theory students:

Preference Table for Choosing the Final

Number of voters	10	8	7
1st choice	T	W	O
2nd choice	O	T	W
3rd choice	W	O	T

We check directly that when all participants vote sincerely, T beats O, O beats W, and W beats T in head-to-head elections.

The Choosing the Final election is an instance of the *Condorcet Paradox.* We define this term precisely now. Notice that each individual voter in the preceding election has perfectly reasonable preferences. For example, the first bloc of voters prefers T, then O, and lastly W. We say that the true preferences of each voter for the candidates are *transitive*. Precisely, given a list of candidates, a **transitive** preference order is a ranking of the candidates from first (most preferred) to last (least preferred). Notice that in a transitive preference order, if candidate A is preferred to B and B is preferred to C, then necessarily A is preferred to C.

Recall from Chapter 5 that given any pair of candidates X and Y, we defined the social preference of X versus Y to be the candidate who wins the head-to-head election by the Simple Majority Method (Definition 5.13). This definition led, in turn, to the social preference graph. For example, in the camp Camp Activities Election (Example 5.10), society has transitive preferences for the candidates A, B, C, and D. Candidate D is socially preferred to A, B, and C. Candidate C is socially preferred to B and A, and candidate B is socially preferred to A. The social ranking of these candidates is D first, B second, C third, and A last.

For the Choosing the Final election (Example 6.1), society prefers T to O, O to W, and W to T. Here is the graph:

Social Preference Graph for Choosing the Final

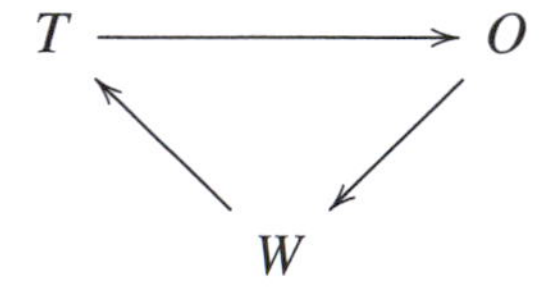

Society does not have transitive preferences for the candidates even though each voter does. We capsulize this phenomenon as follows:

Definition 6.2 A **Condorcet Paradox** occurs when a group of voters submits transitive preferences for a slate of candidates but running the Simple Majority Method for one candidate versus another does not result in a transitive social preference order of the candidates.

We note that a Condorcet Paradox can be detected by looking at the social preference graph. If there is a path in the graph beginning at one node and ending at the same node, then there is a Condorcet Paradox. A path that begins and ends at the same node in a directed graph is called a **cycle**. Notice that there are no cycles in the social preference graph for the Camp Activities Election (5.1). The Condorcet Paradox is the principal obstruction to achieving fairness in multiple-candidate elections. When the voters submit a preference table with a Condorcet Paradox, the voting method must untangle a basic social conflict to make the social choice.

The Gibbard-Satterthwaite Theorem

We use the Condorcet Paradox to explore the role of strategy in elections. Recall that voters are said to vote sincerely when they submit their true preferences for the candidates as their ranking. By contrast, we say that voters vote **strategically** if they submit a ranking different from their true preference. The notions of sincere versus strategic voting extend naturally to voting blocs.

Focusing now on a particular voting method, we consider the effect of strategic voting on the outcome. We certainly expect that when the voters change their votes, the outcome changes. However, we might hope that the voters have no incentive to vote other than sincerely. We formalize this ideal as follows:

Definition 6.3 A voting method is said to be **manipulable** if there is an election in which some voter or voting bloc can improve its payoff by voting strategically.

In two-candidate elections, the Simple Majority Method is not manipulable (see Exercise 6.2). For multiple candidate elections, we have the following remarkable theorem:

Theorem 6.4 The Gibbard-Satterthwaite Theorem *For elections with three or more candidates, the only voting method that is not manipulable is a Dictatorship.* ■

The Gibbard-Satterthwaite Theorem is easy to believe and was long conjectured to be true. The proof, however, is quite advanced, taking as a starting place the last theorem we will prove in this book, Arrow's Impossibility Theorem. We can easily verify that all the nondictatorial voting methods introduced in Chapter 5 satisfy the conclusion of Theorem 6.4. In fact, the preference table for the Choosing the Final election can be used to show manipulability in each case.

We will compare the outcomes of sincere voting versus strategic voting in elections. This comparison requires that we assume that each voter has true preferences. We name the voting blocs P, Q, and R and write

(6.1) **True Payoff Matrix for Choosing a Final**

Voting bloc	***P***	***Q***	***R***
Number of voters	**10**	**8**	**7**
1st choice	*T*	*W*	*O*
2nd choice	*O*	*T*	*W*
3rd choice	*W*	*O*	*T*

We begin with the easiest case.

Theorem 6.5 *The Plurality Method is manipulable.*

Proof The Plurality social choice is T. This is the least desirable outcome for the bloc R. If R submits the ranking

Strategic Vote for *R*
W
O
T

instead and the other blocs stay with their true preferences, the election becomes

New Preference Table when *R* Votes Strategically

Voting bloc	***P***	***Q***	***R***
Number of voters	**10**	**8**	**7**
1st choice	*T*	*W*	*W*
2nd choice	*O*	*T*	*O*
3rd choice	*W*	*O*	*T*

The social choice is now W, which is a better outcome for voting bloc R than the outcome T. ■

The other cases are not much harder.

Theorem 6.6 *The Borda Count is manipulable.*

Proof The Borda points in the original election (6.1) are

$$T:28 \qquad O:24 \qquad W:23$$

Again, this is the least desirable outcome for the bloc R. If bloc R submits the ranking

Strategic Vote for R
W
O
T

instead and the other blocs stay with their true preferences, the Borda Count becomes

$$T : 28 \qquad O : 17 \qquad W : 30$$

The social choice is now W, which is, again, a better outcome for voting bloc R. ■

Theorem 6.7 *The Hare Method is manipulable.*

Proof To apply the Hare Method to the true preference table (6.1), we first eliminate O, the candidate with the least first-place votes. Then we observe that W beats T head to head. Thus W is the true Hare social choice. Now suppose that bloc P changes its ranking to

Strategic Vote for P
O
T
W

so that the new preference table is

Voting bloc	P	Q	R
Number of voters	**10**	**8**	**7**
1st choice	O	W	O
2nd choice	T	T	W
3rd choice	W	O	T

In this case, O is the Hare social choice, which is a better outcome for voting bloc P. ■

Theorem 6.8 *The Coombs Method is manipulable.*

Proof We leave this proof for you in Exercise 6.3. ■

Electing a Third-Party Candidate

We next apply the Condorcet Paradox to address the political science question

How can a third-party candidate win an election?

Here we assume that our third-party candidate has the least built-in support. The surprising mathematical answer we give to this question is contained in the following example:

Example 6.9 Third-Party Candidate Election

We imagine a congressional district somewhere in the United States in which 45 percent of the voters are Republican, 35 percent are Democratics and the other 20 percent are Independents. We assume that each group has a representative candidate named R, D, and I, respectively, and that each group votes as a bloc. The voting method is the Plurality Method. We assume that the preference lists for the three blocs are summarized by the following table. For simplicity, we assume that the district has exactly 100 voters so that we may treat the percentages as voter counts.

True Preference Table for Third-Party Candidate Election

Voting bloc	**Republican**	**Democrat**	**Independent**
Number of voters	**40**	**35**	**25**
1st choice	R	D	I
2nd choice	D	I	R
3rd choice	I	R	D

If each bloc votes for its preferred candidate, then R wins. This is the expected outcome. Suppose, however, that the voting blocs treat the election as a game. Recall from Definition 2.9 that an election may be viewed as a simultaneous-move game of strategy. Our solution concept for such games is the notion of a dominated-strategy solution (Definition 4.6). We prove the following:

Theorem 6.10 *The Independent candidate I wins the Third-Party Candidate Election if the voters play the election as a game and follow the dominated-strategy solution.*

Proof We look for weakly dominated strategies as in Definition 4.6. Note that the strategies for our players (or voting blocs) are just votes for one of the three candidates. Let us denote these strategies by R, D, and I. We show that the strategy of voting for R is a weakly dominant strategy for the Republican bloc. Let us check that R weakly dominates D for the Republican bloc:

(6.2) **R Weakly Dominates D for the Republican Bloc**

Republican	Democrat	Independent	Winner
R	*R*	*R*	*R*
D	*R*	*R*	*R*
R	*R*	*D*	*R* ✓
D	*R*	*D*	*D*
R	*R*	*I*	*R*✓
D	*R*	*I*	*D*
R	*D*	*R*	*R* ✓
D	*D*	*R*	*D*
R	*I*	*R*	*R* ✓
D	*I*	*R*	*D*
R	*D*	*D*	*D*
D	*D*	*D*	*D*
R	*D*	*I*	*R* ✓
D	*D*	*I*	*D*
R	*I*	*D*	*R* ✓
D	*I*	*D*	*D*
R	*I*	*I*	*I*
D	*I*	*I*	*I*

As usual, we have checked the outcome that the Republican bloc prefers. In each case, where the outcomes are different, the Republicans prefer the outcome when they vote for R versus when they vote for D. The proof that R weakly dominates I for the Republican bloc is similar and left to Exercise 6.6.

We may now assume that the Republican bloc follows its weakly dominant strategy and votes sincerely for R. We compare I and D for the Independent voting bloc. We first have the following:

I Weakly Dominates R for the Independent Bloc when the Republicans Vote for R

Independent	Republican	Democratic	Winner
I	*R*	*R*	*R*
R	*R*	*R*	*R*
I	*R*	*D*	*R*
R	*R*	*D*	*R*
I	*R*	*I*	*I*✓
R	*R*	*I*	*R*

Next, we observe the following:

I* Weakly Dominates *D* for the Independent Bloc when Republicans Vote for *R

Independent	Republican	Democratic	Winner
I	R	R	R
D	R	R	R
I	R	D	R ✓
D	R	D	D
I	R	I	I ✓
D	R	I	R

Here the case of interest is when the Democratic bloc votes for D. In this case, the winner is R when the Independents vote for I and D when the Independents vote for D. Looking back at the true preference table for the voting blocs, we see that the Independent bloc prefers R to D.

We conclude, then, that using the logic of weakly dominated strategies, the Republican bloc will vote for R and the Independent bloc will vote for I. How should the Democrats vote? If the Democrats vote for D or R, the outcome will be R, which is their least preferred outcome. If they vote for I, on the other hand, the outcome will be I. Thus the Democrats should vote for I. Summarizing, we have shown that the dominated-strategy solution occurs when the Republicans vote for R, whereas the Democrats and Independents both vote for I. The outcome is that the Independent candidate I wins! ■

The Republican bloc has the most votes and yet fails to elect its preferred candidate R. In fact, when the voting blocs follow the logic of the dominated-strategy solution, the Republican bloc's least-preferred option I is elected! This remarkable conclusion is known as the *Paradox of the Chair*. The standard formulation is given in Exercise 6.9.

The No-Show Paradox

We conclude with one further example of a strange phenomenon in elections. We have seen that voters may have an incentive to vote strategically in an election. We now ask whether there is ever an incentive for voters to withhold their vote in an election. Could staying home and skipping an election ever be in a voter's best interest? The idea is formalized in the following definition:

Definition 6.11 A voting method exhibits a **No-Show Paradox** if there is a preference table and voting bloc P such that the outcome of the election when P does not vote in the election is preferred by P to the outcome when P does vote.

We give an example of the No-Show Paradox for the Hare Method.

Theorem 6.12 *The Hare Method exhibits the No-Show Paradox.*

Proof Consider the following preference table with four candidates (A, B, C, D) and 27 voters:

True Preference Table for a Hare Method Election

Voting bloc			P	
Number of voters	**10**	**5**	**4**	**8**
1st choice	A	B	B	C
2nd choice	B	C	C	D
3rd choice	C	A	D	A
4th choice	D	D	A	B

Running the Hare Method, we first eliminate D and then C. In the head-to-head A versus B election, we see that A wins 18 to 9. Thus A is the true Hare social choice. Now suppose that voting bloc P with four voters does not participate in the election. Then the preference table becomes

Preference Table with Voting Bloc P Removed

Number of voters	10	5	8
1st choice	A	B	C
2nd choice	B	C	D
3rd choice	C	A	A
4th choice	D	D	B

In this case, we eliminate candidate D and then candidate B. Now, in the head-to-head A versus C election, we see that C wins 13 to 10. Thus C is the new Hare social choice. Finally, observe that voting bloc P prefers candidate C to candidate A according to the true preference table. We have the needed example of a No-Show Paradox. ■

The No-Show Paradox is an example of a *fairness criterion* for a voting method. By its design, the Hare Method is vulnerable to this peculiar and arguably undesirable phenomenon. The Coombs Method also can be shown to exhibit this paradox (see Exercise 6.12). Voting methods based on direct counts such as the Plurality Method and the Borda Count are not vulnerable to this paradox (see Exercise 6.11). We will study many more fairness criteria beginning in Chapter 12.

Exercises

6.1 Explain why a Dictatorship is a nonmanipulable voting method.

6.2 Explain why the Simple Majority Method is a nonmanipulable voting method for two-candidate elections.

6.3 Use the Choosing the Final preference table (6.1) to prove that the Coombs Method is manipulable.

6.4 Consider the following preference table:

Number of voters	**5**	**7**	**8**	**3**	**2**
1st choice	A	B	C	D	D
2nd choice	B	A	D	A	A
3rd choice	C	D	B	B	C
4th choice	D	C	A	C	B

Use this preference table to show that

(a) The Plurality Method is manipulable.
(b) The Borda Count Method is manipulable.
(c) The Hare Method is manipulable.
(d) The Coombs Method is manipulable.

6.5 Write down the social preference graph for the following preference tables. Decide whether or not these preference tables contain a Condorcet Paradox (see Definition 6.2).

(a)

Number of voters	**8**	**9**	**7**	**5**
1st choice	A	C	B	D
2nd choice	B	D	C	A
3rd choice	C	B	A	B
4th choice	D	A	D	C

(b)

Number of voters	**4**	**6**	**8**	**3**	**2**
1st choice	A	B	C	D	A
2nd choice	B	D	D	A	D
3rd choice	C	C	B	B	C
4th choice	D	A	A	C	B

(c)

Number of voters	**12**	**13**	**17**	**10**
1st choice	A	E	B	D
2nd choice	B	C	D	E
3rd choice	C	B	E	B
4th choice	D	A	A	C
5th choice	E	D	C	A

6.6 Make a table like (6.2) to prove that R weakly dominates voting for I for the Republican bloc in the Third-Party Candidate Election. This completes the proof of the Theorem 6.10.

6.7 Decide whether the Independent bloc has a weakly dominant strategy in the Third-Party Candidate Election *without* the assumption that the Republican bloc will vote for R.

6.8 Consider the following preference table with three voting blocs P, Q, and R and three candidates A, B, and C:

Voting bloc	P	Q	R
Number of voters	**14**	**6**	**9**
1st choice	A	B	C
2nd choice	B	C	B
3rd choice	C	A	A

(a) Use the Plurality Method. Decide whether placing A first (voting sincerely) weakly dominates placing B first for voting bloc P.

(b) Use the Hare Method. Suppose that voting bloc P votes sincerely. Consider voting bloc Q. Decide whether, for Q, voting sincerely weakly dominates voting insincerely in the form Q: $C \quad B \quad A$. Notice that voting bloc R has six strategy choices corresponding to the six possible orderings of the candidates.

(c) Repeat (b) but use the Coombs Method.

6.9 (Paradox of the Chair) Three committee members P, Q, and R vote on three alternatives A, B, and C. Voter P is the chair of the committee, so P's vote breaks all three-way ties. Otherwise, the majority wins. Show that there is an election in which the dominated-strategy solution results in P's last choice winning the election.

6.10 Use the Choosing the Final election (Example 6.1) to prove the following:

(a) The Pairwise Comparisons with Agenda Method is manipulable.

(b) For each candidate T, O, and W, there is an agenda so that this candidate is the Pairwise Comparisons with Agenda social choice.

6.11 Explain why there can never be a No-Show Paradox in an election using the following voting methods:

(a) The Plurality Method
(b) Borda Count
(c) Dictatorship

6.12 Give an example of the No-Show Paradox (Definition 6.11) for a Coombs Method election. *Hint*: This can be done with three candidates and four voting blocs.

7 Choice and Chance

We have seen the role that strategy can play in making a social choice. What role is there for chance? In a democracy, we do not expect chance to be involved in the mechanism of an election. Yet, despite the seeming disconnect between chance and social choice, modern electoral politics has become a showcase for probability theory. Each major election is extensively analyzed as experts compete to correctly predict voter preferences, turnout, and ultimately, the final outcome. The U.S. Electoral College is perhaps the most famous example of a *weighted voting system*. We introduce weighted voting systems in this chapter and explore the connection between these voting systems and probability theory. We define the key notion of the *power index*, which provides a measure of the power of voting blocs in a weighted voting system. We also consider a direct connection between choice and chance. We prove a special case of Condorcet's *Jury Theorem*.

We begin with a motivating example inspired by the 2012 U.S. presidential election between Barack Obama and Mitt Romney.

Example 7.1 Swing-States Election

In the months leading up to election day for the 2012 U.S. presidential election, pundits and pollsters narrowed their focus to the possible outcomes of several **swing states**, states whose elections were too close to call and whose electoral votes would decide the outcome. Romney's chances were slim. He seemed to have 206 safe electoral votes compared with 253 safe votes for Obama. The swing states were Florida (FL), Ohio (OH), Virginia (VA), Colorado (CO), Iowa (IA), and New Hampshire (NH). Romney needed 64 more electoral votes to reach the winning total of 270 electoral votes. The outcome depended on these six swing states. The electoral votes for the swing states are shown in the following table. We have also included a probability for each state to vote for Romney. These probabilities are just conjectures included to allow for a calculation; a careful study of polling data is generally needed to predict election outcomes.

Swing state	**Electoral votes**	**Probability for Romney**
FL	29	0.47
OH	18	0.48
VA	13	0.55
CO	9	0.53
IA	6	0.50
NH	4	0.52

Given these data, we ask

What is the probability that Romney wins the election?

To win, Romney needed to secure a combination of states whose electoral votes would add to at least 64. We will call such a combination of states a *winning coalition.* The electoral votes assigned to each state will be called the *weight* of the state, and the sum of 64 needed by Romney to win will be called the *quota*. To compute the probability, we may assume that the states recorded their votes sequentially. We can then use a probability tree in which each stage corresponds to the given state going for Romney (R) or for Obama (O).

Since the full probability tree will be quite large, it will be useful to include only the branches that lead to winning coalitions for Romney. For example, observe that if Romney lost either Florida or Ohio, then he could not reach the needed 270 total votes. We thus include only the branches corresponding to Romney winning these two swing states. Further abridgments are included as appropriate. The abridged probability tree is as follows:

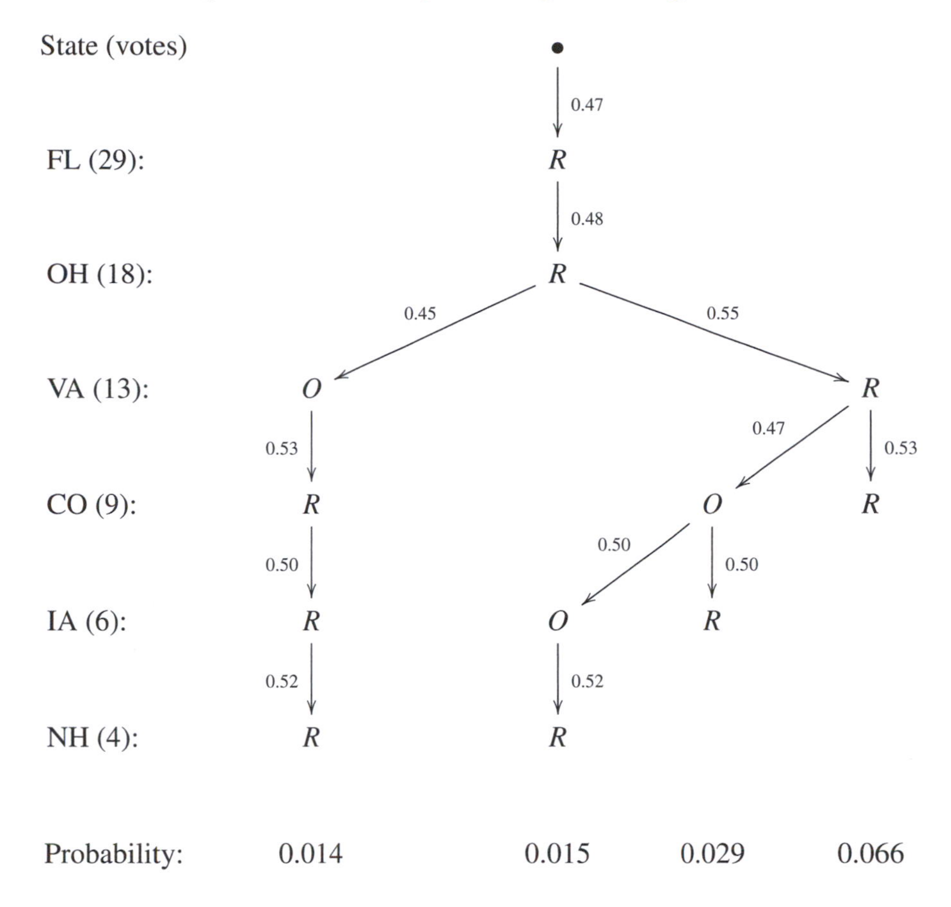

As usual, we multiply the branch probabilities to determine the probability of each type of winning coalition in accordance with the Law of the Probability Tree (Theorem 3.6). To compute the probability that Romney wins the election, we add the probabilities in the last row:

$$\text{Probability Romney wins} = 0.014 + 0.015 + 0.029 + 0.066 = 0.12$$

Romney has a 12 percent chance to win with these assumptions.

Example 7.1 motivates a new type of voting system in which the individual voters are assigned different roles in determining the outcome. Examples of this type of voting system include "proportional" systems such as the U.S. Electoral College and various legislative systems in which different branches have different powers. A simple example of a voting system in which voters have unequal power was given in Example 6.1. There we saw that the weakest voting bloc, the Independent voters, ended up having a strategic advantage because of the respective preferences of the voting blocs. While this type of paradox will be of great interest to us in subsequent chapters, we here eliminate the complications arising from multiple candidates by considering only voting systems with two possible alternatives. For convenience, we refer to these possibilities as "yes" and "no."

Definition 7.2 A **yes-no voting system** is a voting system in which there are only two candidates or alternatives in the election.

The simplest mechanism for a yes-no voting system is the Simple Majority Method. Just count the votes for "yes" and those for "no." The majority wins. Of course, in this case, all the voters have the same amount of power. To introduce differences in power, we assign individual voters weights like the electoral votes for each state in Example 7.1. The passage of a measure is then determined by comparing the sum of the weights of those voters voting "yes" with the quota. We summarize this as follows:

Definition 7.3 A **weighted voting system** is a yes-no voting system in which each voter is assigned a **weight** that is a nonnegative integer. A measure passes if the total weight of "yes" votes equals or exceeds a fixed **quota**. Here the quota is a positive integer q. We usually require that q is at least as large as half the total weight of all voters.

Notationally, a weighted voting system involving voters $P_1, P_2, \ldots, P_n$ consists of a table

Voter	Weight
P_1	w_1
P_2	w_2
$\vdots$	$\vdots$
P_n	w_n

and a number

$$q = \text{the quota}$$

Our assumption about the quota is that

$$q > \frac{w_1 + w_2 + \cdots + w_n}{2}$$

This assumption ensures that a "yes" decision is a positive decision, whereas "no" is the default or negative outcome. A measure passes if the total weight of voters voting "yes" is greater than or equal to q.

In practice, we will not need to keep track of the names of the voters, so will simply order them in decreasing order of their weight. We may then abbreviate our notation to the following:

$$[q\colon w_1, w_2, \ldots, w_n]$$

For example, the weighted voting system in Example 7.1 is given, in this notation, as

$$[64: 29, 18, 13, 9, 6, 4]$$

Power in Elections

The Swing-States Election (Example 7.1) provides a glimpse of the complexity involved in predicting the outcome of an Electoral College election. Switching our focus from the candidates to an individual state, say, Virginia, we are faced with a different sort of question. Namely, we can ask

> What is the probability that Virginia's vote is decisive in the Swing-States Election?

We are now led to consider the electoral votes of the states exclusively. We quickly see that the states of Florida and Ohio have more say in the final decision than Virginia. In fact, these states can effectively eliminate Romney by voting for Obama. A voter in a weighted voting system whose "no" vote implies that the measure does not pass is said to have **veto power**. We can also see that Virginia has more say in the final decision than New Hampshire, although this requires a closer look. We conclude that Ohio and Florida have more *power* than Virginia, which, in turn, has more power than New Hampshire in the Swing-States Election. Using an approach introduced by John Banzhaf in a famous legal case (see Exercise 7.3), we can make this notion of power precise. The key ideas are contained in the following definitions:

Definition 7.4 A **winning coalition** in a yes-no voting system is a collection of voters whose "yes" votes are sufficient to pass a measure even if all other voters vote "no."

Definition 7.5 Suppose that there is a winning coalition in a yes-no voting system. A voter in this coalition is said to be a **critical voter** if the collection is no longer a winning coalition without this voter.

In other words, a voter P is a critical voter in a collection $\mathcal{C}$ of voters if (1) $\mathcal{C}$ is a winning coalition and (2) the collection $\mathcal{D}$ obtained by removing P from $\mathcal{C}$ is not a winning coalition. Notice that a voter P with veto power is a critical voter in every winning coalition.

We list the winning coalitions and critical voters for the weighted voting system

$$[64: 29, 18, 13, 9, 6, 4]$$

corresponding to the Swing-States Election as follows:

Critical Voters for Swing-States Election

Winning coalition	Critical voters
FL, OH, VA, CO	FL, OH, VA, CO
FL, OH, VA, IA	FL, OH, VA, IA
FL, OH, VA, NH	FL, OH, VA, NH
FL, OH, VA, CO, IA	FL, OH, VA
FL, OH, VA, CO, NH	FL, OH, VA
FL, OH, VA, IA, NH	FL, OH, VA
FL, OH, CO, IA, NH	FL, OH, CO, IA, NH
FL, OH, VA, CO, IA, NH	FL, OH

We now count the number of appearance of each state in the critical voter lists:

FL: 8 OH: 8 VA: 6 CO: 2 IA: 2 NH: 2

We let C denote the sum of these numbers, that is, $C = 8+8+6+2+2+2 = 28$. We then define the *power index* for this voting system to be the list

Power Index for Swing-States Election

$$\left[\frac{8}{28}, \frac{8}{28}, \frac{6}{28}, \frac{2}{28}, \frac{2}{28}, \frac{2}{28}\right]$$

According to this analysis, Florida has four times as much power as New Hampshire in this election, whereas Virginia has three times as much power as New Hampshire.

Definition 7.6 The **power index** for a yes-no voting system is the list of fractions, one for each voter in the system, obtained as follows: let C denote the total number of all critical voters in all winning coalitions. Given a voter P_i, let C_i denote the number of times P_i is a critical voter. Then the fraction of power for P_i is C_i/C. Thus the power index for the yes-no voting system is the list

$$\left[\frac{C_1}{C}, \frac{C_2}{C}, \ldots, \frac{C_n}{C}\right]$$

While the power index nicely quantifies the differences in power between the swing states, it does not directly address our original question. We asked for the probability that Virginia's vote matters in the election. Philip Straffin proved that if the denominator C is replaced with 2^{n-1}, the resulting fractions are in fact the probabilities that the individual states play a critical role in the outcome. It is customary to use the denominator C, and we will continue to do so here.

We remark that the power index just defined is known as the **Banzhaf power index** Banzhaf's method is not the only way to define a power index. Another famous method, developed by Lloyd Shapley and Martin Shubik, is introduced in Exercise 7.5. We give an example with three parts to illustrate calculation of the power index.

Example 7.7 Computing the Power Index

When the voting system has a manageably small number of voters, we can compute the power index of the system by "brute force" as we did earlier for the Swing-States Election. That is, we simply write down all the winning coalitions and list, for each one, the critical voters. When doing this, it is easiest to start with the smallest winning coalitions and work our way up to the largest one (i.e., the coalition consisting of all the voters). We give three examples with four, five, and six voters, respectively. As we will see, the amount of work grows steadily with the number of voters in the system.

(1) $$[100\colon 60, 50, 35, 15]$$

Here we have four voters, which we denote P_1, P_2, P_3, and P_4 with weights $60, 50, 35$, and 15, respectively. We obtain the following table of winning coalitions and critical voters:

Critical Voters (1)

Winning coalition	Critical voters
P_1, P_2	P_1, P_2
P_1, P_2, P_3	P_1, P_2
P_1, P_2, P_4	P_1, P_2
P_1, P_3, P_4	P_1, P_3, P_4
P_2, P_3, P_4	P_2, P_3, P_4
P_1, P_2, P_3, P_4	None

We conclude that the power index for this system is as follows:

$$\text{Power index for (1)} = \left[\frac{4}{12}, \frac{4}{12}, \frac{2}{12}, \frac{2}{12}\right]$$

(2) $$[10\colon 7, 5, 3, 2, 1]$$

Here we have five voters, which we denote P_1, P_2, P_3, P_4, and P_5 with weights $7, 5, 3, 2$, and 1, respectively. We then obtain the following table of winning coalitions and critical voters:

Critical Voters (2)

Winning coalition	Critical voters
P_1, P_2	P_1, P_2
P_1, P_3	P_1, P_3
P_1, P_2, P_3	P_1
P_1, P_2, P_4	P_1, P_2
P_1, P_2, P_5	P_1, P_2
P_1, P_3, P_4	P_1, P_3
P_1, P_3, P_5	P_1, P_3
P_1, P_4, P_5	P_1, P_4, P_5
P_2, P_3, P_4	P_2, P_3, P_4
P_1, P_2, P_3, P_4	None
P_1, P_2, P_3, P_5	P_1
P_1, P_2, P_4, P_5	P_1
P_1, P_3, P_4, P_5	P_1
P_2, P_3, P_4, P_5	P_2, P_3, P_4
P_1, P_2, P_3, P_4, P_5	None

Thus the power index for this voting system is as follows:

$$\text{Power index for (2)} = \left[\frac{11}{25}, \frac{5}{25}, \frac{5}{25}, \frac{2}{25}, \frac{1}{25}\right]$$

(3) $[20\colon\ 10, 10, 5, 5, 2, 3]$

This time we have six voters, denoted P_1, P_2, P_3, P_4, P_5, and P_6 with weights $10, 10, 5, 5, 2$, and 3, respectively. We compute the power index, as earlier, by making the following (rather large) table:

Critical Voters (3)

Winning coalition	Critical voters
P_1, P_2	P_1, P_2
P_1, P_2, P_3	P_1, P_2
P_1, P_2, P_4	P_1, P_2
P_1, P_2, P_5	P_1, P_2
P_1, P_2, P_6	P_1, P_2
P_1, P_3, P_4	P_1, P_3, P_4
P_2, P_3, P_4	P_2, P_3, P_4
P_1, P_2, P_3, P_4	None
P_1, P_2, P_3, P_5	P_1, P_2
P_1, P_2, P_3, P_6	P_1, P_2
P_1, P_2, P_4, P_5	P_1, P_2
P_1, P_2, P_4, P_6	P_1, P_2
P_1, P_2, P_5, P_6	P_1, P_2
P_1, P_3, P_4, P_5	P_1, P_3, P_4
P_1, P_3, P_4, P_6	P_1, P_3, P_4
P_2, P_3, P_4, P_5	P_2, P_3, P_4
P_2, P_3, P_4, P_6	P_2, P_3, P_4
P_1, P_3, P_5, P_6	P_1, P_3, P_5, P_6
P_1, P_4, P_5, P_6	P_1, P_4, P_5, P_6
P_2, P_3, P_5, P_6	P_2, P_3, P_5, P_6
P_2, P_4, P_5, P_6	P_2, P_4, P_5, P_6
P_1, P_2, P_3, P_4, P_5	None
P_1, P_2, P_3, P_4, P_6	None
P_1, P_2, P_3, P_5, P_6	None
P_1, P_2, P_4, P_5, P_6	None
P_1, P_3, P_4, P_5, P_6	P_1
P_2, P_3, P_4, P_5, P_6	P_2
$P_1, P_2, P_3, P_4, P_5, P_6$	None

We conclude

$$\text{Power index for (3)} = \left[\frac{16}{64}, \frac{16}{64}, \frac{8}{64}, \frac{8}{64}, \frac{4}{64}, \frac{4}{64}\right]$$

This example indicates the computational difficulty inherent in computing power indices in a weighted system with many voters with different weights. Imagine, for example, the work involved in computing the power index for the U.S. Electoral College! The result in this case requires extensive computer calculations. When the number of voters is (relatively) large but there are only several different weights, we can make the computation using counting techniques. We return to this topic in Chapter 13.

We conclude with two examples of probability calculations arising from elections.

Example 7.8 Passing the Bill

Senator Rich has almost achieved his election promise of passing a major bank subsidy bill. One last step remains: the bill must meet the approval of a five-member executive committee. Senator Rich happily has a seat on this committee. Unfortunately for him, he has no idea how the other four senators will vote. What is the probability that the bill passes?

We may assume, of course, that Senator Rich (R) will vote "yes." Let us denote the other for senators by S_1, S_2, S_3, and S_4. Since each of these four can vote either "yes" or "no," there are $2 \times 2 \times 2 \times 2 = 16$ possible ways the votes can come out. We must count how many of these 16 ways result in passage of the bill. Here is a list of the winning coalitions that include R:

Winning Coalitions with *R* for Passing the Bill
S_1, S_2, S_3, S_4
S_1, S_2, S_3
S_1, S_2, S_4
S_1, S_3, S_4
S_2, S_3, S_4
S_1, S_2
S_1, S_3
S_1, S_4
S_2, S_3
S_2, S_4
S_3, S_4

Because there are 11 winning coalitions out of 16 possibilities, we conclude the following:

$$\text{Probability the bill passes} = \frac{11}{16}$$

Senator Rich's support of the bill ensures that the probability of the bill passing is better than 50-50. This result certainly conforms to our intuition.

Our second development confirms a related intuition. The result we prove is a special case of the famous Jury Theorem first proved by the Marquise de Condorcet in the eighteenth century.

Example 7.9 Hung Jury

The guilt or innocence of a defendant is determined by a jury of 12 of his peers. The jurors have come down 6 for guilty and 6 for innocent. The judge offers the defendant two options to break the tie. The judge will add one new juror or she will add three new jurors to the original 12. The Defendant believes that each juror has a 60 percent chance of voting "innocent." How many jurors should he request the judge to add, one or three?

We make some simplifying assumptions to allow us to compute probabilities. We assume that the original 12 jurors remain deadlocked at 6 to 6, whereas the new juror or jurors will vote "innocent" with probability $p = 0.6$ irrespective of what the other jurors do. We can then easily see that the defendant will be set free with probability $p = 0.6$ when he requests one extra juror.

The probability that the defendant will be set free with three extra jurors is computed using the partial probability tree corresponding to the cases where at least two of the three jurors vote "innocent":

Innocent Verdict with Three New Jurors

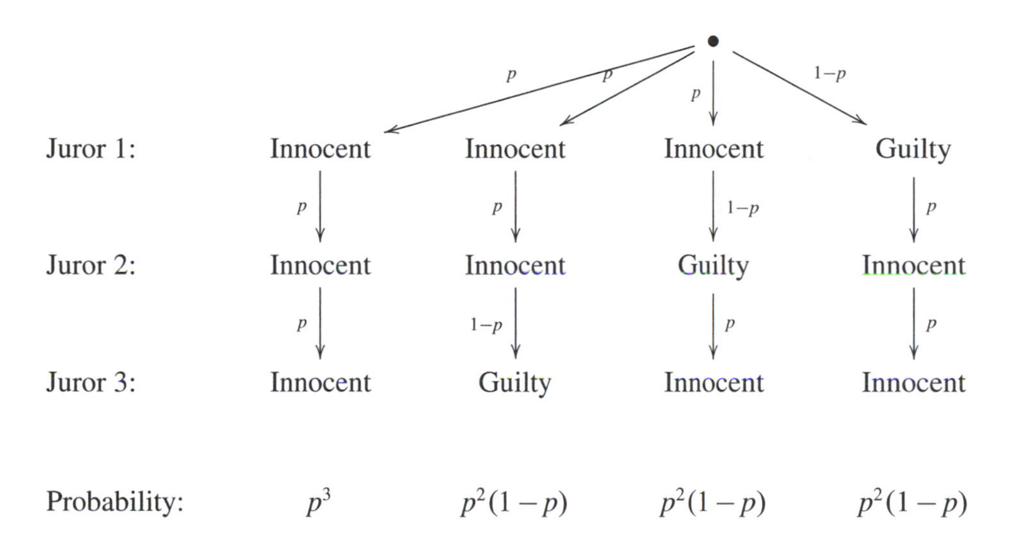

We conclude the following:

$$P(\text{free with 3 extra jurors}) = p^3 + 3p^2(1-p)$$

Plugging in $p = 0.6$ gives P(free with 3 extra jurors) $= 0.648$. The defendant will have an almost 5 percent better chance with three jurors.

In fact, there is always an advantage to the extra jurors, provided that the probability p that a given juror votes "innocent" satisfies $p > 1/2$. We can prove this by comparing the probabilities P(free with 1 extra juror) $= p$ and P(free with 3 extra jurors) $= p^3 + 3p^2(1-p)$. It is helpful to take the difference, which we denote $D(p)$. We then perform some basic algebra:

$$\begin{aligned} D(p) &= P(\text{free with 3 extra jurors}) - P(\text{free with 1 extra juror}) \\ &= p^3 + 3p^2(1-p) - p \\ &= p^3 + 3p^2 - 3p^3 - p \\ &= -2p^3 - 3p^2 - p \\ &= p(1-p)(2p-1) \end{aligned}$$

Since $0 \leq p \leq 1$, the terms p and $1 - p$ are nonnegative. Thus $D(p)$ is positive precisely when $2p - 1 > 0$ or when $p > 1/2$.

Our calculation is a special case of the following general result proved by Condorcet in 1789:

Theorem 7.10 Jury Theorem *Suppose that we are given a jury pool for which the probability that each juror will vote correctly is a fixed probability* $p > 1/2$. *Suppose that the jurors vote without consulting. Then the probability that the jury will return a correct verdict approaches* 1 *as the number of jurors increases.* ■

The Jury Theorem represents arguably the first major result in the modern theory of social choice. Condorcet offered his theorem as mathematical evidence for the wisdom of obeying the majority will in elections. We will revisit this theorem in Chapter 21.

Exercises

7.1 Consider the following variations on the Swing-States Election (Example 7.1). In these cases, we imagine that a candidate Y is running for president. The swing states, electoral votes, and probabilities of voting for candidate Y are given along with the number of electoral votes needed by Y to win. Compute the probability that Y wins the election.

(a)

State	Electoral votes	Probability of Y winning
PA	20	0.60
WA	12	0.45
WI	10	0.40
VA	9	0.70

26 electoral votes needed by Y to win

(b)

State	Electoral votes	Probability of Y winning
AZ	11	0.40
MN	10	0.65
CO	9	0.55
NM	5	0.60

18 electoral votes needed by Y to win

(c)

State	Electoral votes	Probability of Y winning
IL	20	0.50
NC	15	0.45
IN	11	0.40
MA	11	0.60
AL	9	.55

40 electoral votes needed by Y to win

(d)

State	Electoral votes	Probability of Y winning
FL	29	0.60
MI	16	0.52
WA	12	0.44
MD	10	0.42
LA	8	0.45
OK	7	0.58

52 electoral votes needed by Y to win

7.2 Compute the power index for the following weighted systems:

(a) $[12: 7, 5, 3, 2]$
(b) $[20: 18, 10, 5, 2]$
(c) $[100: 50, 40, 30, 20, 10]$
(d) $[20: 14, 7, 6, 5, 3]$
(e) $[40: 20, 10, 10, 10, 5, 5]$
(f) $[75: 25, 25, 25, 15, 10, 10]$

7.3 John Banzhaf used the power index to argue that a voting method used by the Nassau County Board of Supervisors in the 1960s was unfair. The voting system was defined as follows: the voters and weights are

Township	Weight
Hempstead 1	9
Hempstead 2	9
North Hempstead	7
Oyster Bay	3
Glen Cove	1
Long Beach	1

The quota was $q = 16$. Compute the power index for this system. Explain why this voting system is unfair.

7.4 The power index (Definition 7.6) applies to any yes-no voting system. Write down the winning coalitions and identify the critical voters for the following committees to compute the power index:

(a) A student government committee consists of two members and three officers. To pass, a measure must have the support of at least one member and all three officers.
(b) A congressional committee consists of three congressmen and two senators. To pass, a measure must have the support of at least three members of the committee and at least one senator.
(c) A university committee consists of four faculty members, one dean, and the provost. To pass, a measure must either have the support of at least two faculty members and the dean, or it must have the support of both the dean and the provost.
(d) A corporate board consists of five vice presidents and the president. The vice presidents use the Simple Majority Method to pass measures up to the president. The president has veto power.

7.5 The Shapley-Shubik power index provides an alternative way to measure the power of voters in a yes-no voting system. The index is defined as follows: consider all possible orderings of the voters, called the **ordered coalitions**. For each ordered coalition, we define the **pivotal voter** as follows: first consider the leftmost voter in this ordered coalition. If his or her "yes" vote is sufficient to pass the measure, then he or she is the pivotal voter for this ordered coalition. Otherwise, consider the two leftmost voters in the given ordered coalition. If these two voters are together a winning coalition, then the second voter is the pivotal voter. Continue in this way, moving from left to right in the given ordered coalition, until enough voters are included to create a winning coalition. The last voter needed is the pivotal voter. The Shapley-Shubik power index for each voter is then the number of times that the voter is pivotal divided by the total number of ordered coalitions. Here is an example with three voters $[10:8,6,2]$:

Power Index Example

Ordered coalition	Pivotal player
$P_1\,P_2\,P_3$	P_2
$P_1\,P_3\,P_2$	P_3
$P_2\,P_1\,P_3$	P_1
$P_2\,P_3\,P_1$	P_1
$P_3\,P_1\,P_2$	P_1
$P_3\,P_2\,P_1$	P_1

Thus the Shapley-Shubik power index is

$$\left[\frac{4}{6},\frac{1}{6},\frac{1}{6}\right]$$

Compute the Shapley-Shubik power index for the following weighted systems:

(a) $[11:\ 8,6,5]$
(b) $[12:\ 10,5,2]$
(c) $[20:\ 10,8,6,5]$ (Note that there are $4\cdot 3\cdot 2\cdot 1 = 24$ ordered coalitions to consider.)
(d) $[12:\ 10,7,2,1]$

7.6 Suppose in Passing a Bill (Example 7.8) that the other committee members vote "yes" with probability $p{=}3/5$. Compute the probability that the bill passes assuming, as before, that Senator Rich votes "yes."

7.7 Suppose in Passing a Bill (Example 7.8) that the other committee members vote "yes" with probability p. Find p so that the probability q that the bill passes assuming that Senator Rich votes "yes" is $q = 1/2$.

7.8 Suppose that the committee in Passing a Bill (Example 7.8) consists of seven members including Senator Rich. Compute the probability that the bill passes assuming that Senator Rich votes "yes." Assume, as in the original example, that the other committee members vote "yes" with probability $p = 1/2$.

7.9 Compute the probability of a defendant going free if five jurors are added to the Hung Jury (Example 7.9).

7.10 Prove that if the added jurors all vote "innocent" with probability $p > 1/2$, then

$$P(\text{Defendant goes free with 5 extra jurors})$$
$$> P(\text{Defendant goes free with 3 extra jurors})$$

8 Chance and Strategy

Perhaps the most entertaining games to play are those involving both chance and strategy. This class includes popular board games such as Monopoly, Backgammon, and Scrabble, as well as a whole range of card games from Blackjack to Bridge. The roles of chance and strategy vary from game to game. For all these games, winning requires some combination of good fortune and strategic skill.

In this chapter, we analyze a simplified version of the game Poker (Example 8.1). Our example features the strategies of bluffing and calling that arise in the general game. We use this example to introduce two themes that we will return to in Part II of this text. First, we introduce the idea of treating an expected value as the payoff of a game. Second, we introduce the idea of *mixed strategies*. We conclude by considering a somewhat more complicated version of the original game (Example 8.8). This game includes the strategy of raising a bet in addition to the strategies of bluffing and calling.

Example 8.1 Betting on Diamonds

This is a two-player game with many possible variations. Each player antes \$2. Player 1 chooses a random card from a deck and looks at it but does not show the card to Player 2. Player 1 may now either bet or fold. If Player 1 elects to bet, he or she bets \$8. Player 2 may now either call or fold. If Player 2 calls and the card is a diamond, then Player 1 wins the pot; if the card is not a diamond, then Player 2 wins the pot. If either player folds, he or she loses the ante of \$2 to the other player.

A first observation is that Player 1 should bet whenever he or she gets a diamond. There is no incentive to fold with the winning card. We remove the possibility and write the game tree for Betting on Diamonds as follows:

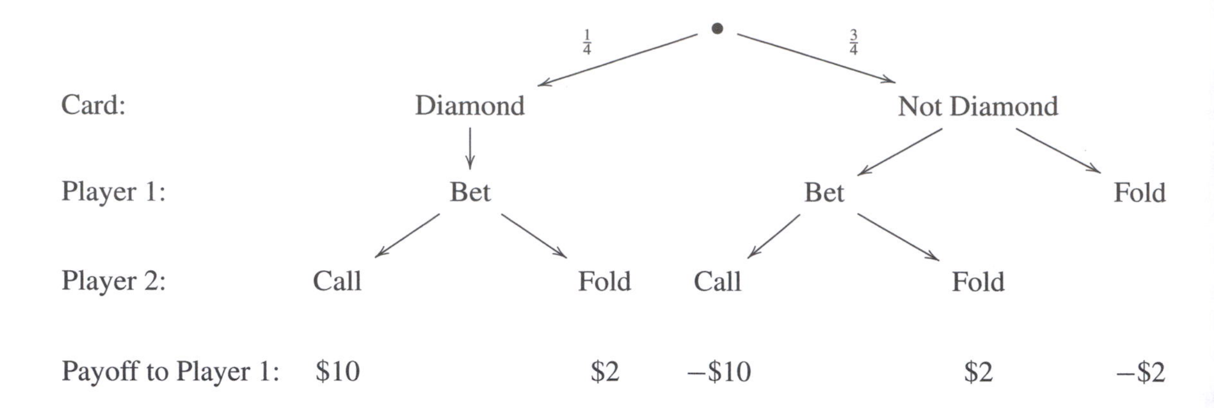

Betting on Diamonds is a game without perfect information. In fact, it is precisely Player 2's lack of information that creates the intrigue in this game. Player 1 has an incentive to bet without a diamond because his or her bet may persuade Player 2 to fold. This is the strategic device of bluffing familiar from Poker. By definition, a **bluff** is a bet by a player with a losing hand in a game without perfect information aimed at inducing the other player(s) to fold.

We turn to the natural question

Which player does Betting on Diamonds favor?

Obviously, the odds of the draw favor Player 2. However, Player 1 has the advantage of information and, arguably, the better strategic position. How do we weigh these two competing advantages? Games of chance and strategy present a formidable challenge in our pursuit of a notion of "fairness" for games. The severe restrictions we have placed on betting in the game Betting on Diamonds will allow us to make some headway on this problem for this game.

Let us recall that our model for a working definition of fairness is in the realm of games of chance. In this setting, we may compute the expected value E of the game and thereby quantify the expected payoff to Player 1. Assigning a "value" to a game such as Betting on Diamonds is clearly a somewhat different problem from the case of a game of pure chance. The idea of an expected value is still a useful one, however. In fact, expected values highlight the strategic aspect of the game.

As usual, when we consider expected values, we imagine playing the game Betting on Diamonds for multiple rounds. We have the following definition:

Definition 8.2 The game obtained by playing multiple rounds of a simple game involving strategy is called a **repeated game.**

The repeated game is often a much richer strategic encounter than a single round. Certainly this is the case with Betting on Diamonds, where knowledge of past behavior can influence strategy choices in the current round. We first consider the simplest possible behavior by the players in a repeated game.

Pure Strategies

Consider a repeated game of Betting on Diamonds with many rounds. We consider the situation in which the players play completely predictably. That is, we assume each player chooses the same strategy over and over again in each round.

Definition 8.3 When a player follows the exact same strategy in each round of a repeated game, we call this a **pure strategy.**

First, take the repeated game when Player 1 chooses the pure strategy "Always bluff" and Player 2 chooses the pure strategy "Always call." The game tree reduces to a probability tree. As usual, we keep track of the payoff to Player 1.

Player 1 Always Bluffs and Player 2 Always Calls in Betting on Diamonds

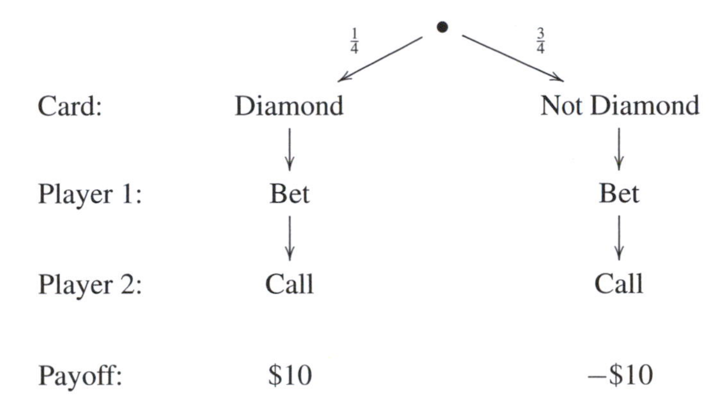

The expected value of this game is

$$E = \$10 \cdot \frac{1}{4} - \$10 \cdot \frac{3}{4} = -\$5$$

Next, notice that if Player 2 chooses the pure strategy "Always call," then Player 1 would be wise to follow the strategy "Never bluff." Here is the probability tree when the players follow these pure strategies:

Player 1 Never Bluffs and Player 2 Always Calls

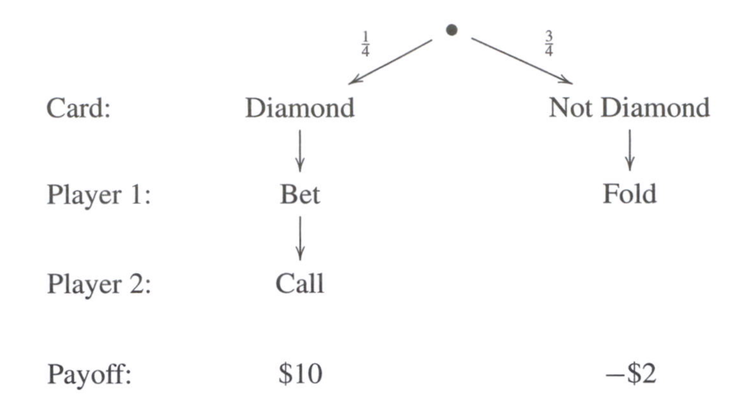

Player 1's expected value improves to

$$E = \$10 \cdot \frac{1}{4} - \$2 \cdot \frac{3}{4} = \$1$$

in this scenario.

We have two more combinations to consider. When Player 1 follows the pure strategy "Never bluff" and Player 2 follows the pure strategy "Always call," the repeated game favors Player 1:

$$E = \$8\frac{1}{4} - \$2\frac{3}{4} = \$2$$

Finally, when Player 1 follows the pure strategy "Never bluff" and Player 2 follows the pure strategy "Never call," then the expected value of the game is

$$E = \$2 \cdot \frac{1}{4} - \$2 \cdot \frac{3}{4} = -\$1$$

When the players in a repeated game of Betting on Diamonds are restricted to pure strategy choices, the game becomes a two-player, two-strategy zero-sum game. Player 1 chooses between the pure strategies "Never bluff" and "Always bluff," whereas Player 2 chooses between the pure strategies "Never call" and "Always call." Here is the payoff matrix:

Payoff Matrix for Betting on Diamonds with Pure Strategies

		Player 2	
	Strategy	Always call	Never call
Player 1	Always bluff	(−\$5, \$5)	(\$2, −\$2)
	Never bluff	(\$1, −\$1)	(−\$1, \$1)

We refer to this game as the *pure-strategy game*.

Definition 8.4 The **pure-strategy game** associated with a repeated game is the game in which the players choose pure strategies.

Observe that neither player has a weakly dominant strategy in the pure-strategy game Betting on Diamonds. Even in this simplified form, Betting on Diamonds represents a true conflict. While we have not yet addressed the question of the advantaged player in Betting on Diamonds, we have introduced an important technique: the idea of treating expected value as the payoffs of a repeated game is a key move in the study of strategic games.

Mixed Strategies

We allow the players more flexibility in how they play the repeated game of Betting for Diamonds. Suppose that Players 1 and 2 split time between their two strategic options. For Player 1, the choices are "Bluff" and "Not bluff." Player 1 may plan to bluff, say, one-third of the time that he or she draws a non-diamond and fold the other two-thirds of the time. Of course, he or she must draw a non-Diamond for the question of bluffing to arise. We will say that Player 1 will choose the probability $p = 1/3$ for "Bluff." The probability that Player 1 folds a losing hand is then $1 - p = 2/3$.

Independently, Player 2 decides how often to call a bet by Player 1 and how often to fold when Player 1 bets. Again, if Player 1 folds, then this decision does not need to be made. We introduce the probability q to measure Player 2's choice between "Call" and "Fold." With this view, Player 1 chooses a probability

$$p = P(\text{"Bluff"}) = \text{probability Player 1 bluffs with non-Diamond}$$

and Player 2 separately chooses a probability

$$q = P(\text{"Call"}) = \text{probability Player 2 calls a bet}$$

These choices are called *mixed strategies.*

Definition 8.5 A player in a repeated game who splits time between his or her strategy options according to a fixed probability or probabilities is said to choose a **mixed strategy**.

In practice, the players do more than decide how often to bluff or call, respectively. They decide when to bluff and when to call. We will assume that this decision has been automated for both players. That is, the players choose how often to bluff and call, respectively, but the choice of when is determined by a random process.

Example 8.6

Suppose that Player 1 chooses to bluff with probability $p=1/3$ and Player 2 chooses to call with probability $q=1/2$. What is the expected value of the game?

Notice that these choices of mixed strategies become branch probabilities in our original game tree. We can thus further label our game tree to obtain a genuine probability tree as follows:

Probability Tree for Betting on Diamonds with Mixed Strategies $p=1/3$ and $q=1/2$

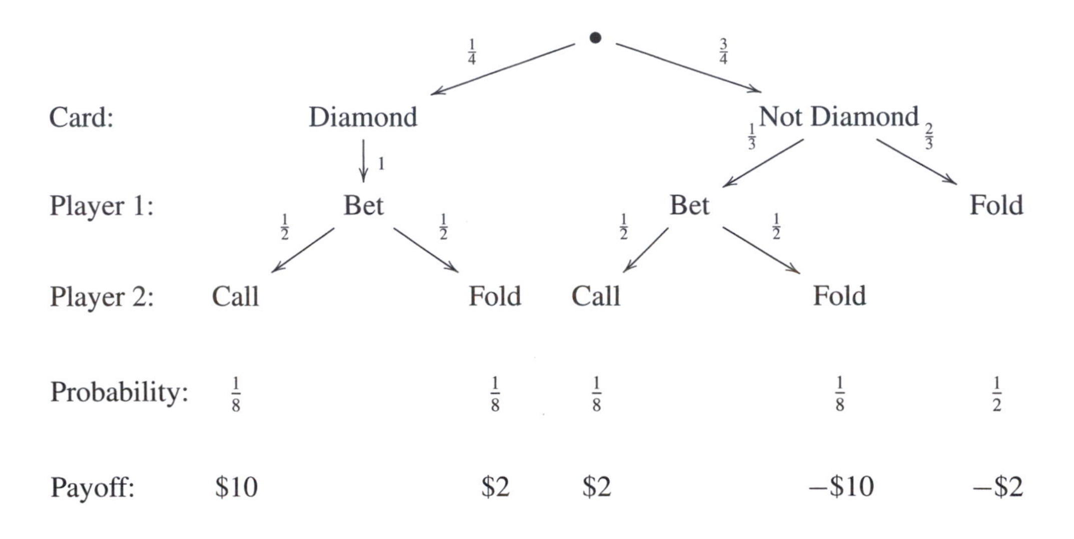

The expected value is then

$$E=\$10\cdot\frac{1}{8}+\$2\cdot\frac{1}{8}+\$2\cdot\frac{1}{8}-\$10\cdot\frac{1}{8}-\$2\cdot\frac{1}{2}=-\$0.50$$

We can continue to compute the expected value for different values of p and q to get a sense of how the players should split their time between their different strategy choices. In fact, it is not difficult to find a formula for the expected value for any mixed-strategy choices of a probability p by Player 1 and probability q by Player 2. First, here is the probability tree:

Probability Tree for Betting on Diamonds with Mixed Strategies *p* and *q*

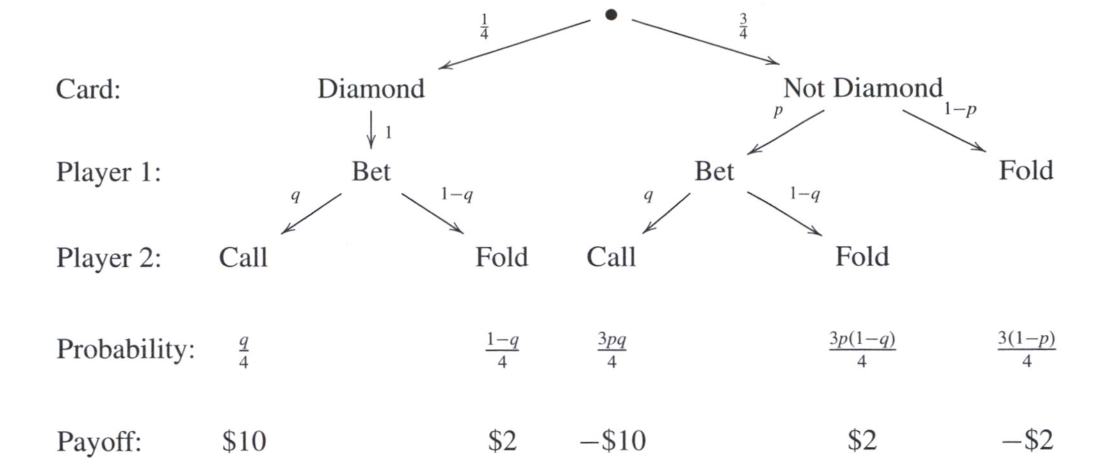

Let us write

$$E(p,q) = \text{expected value when Player 1 chooses } p \text{ and Player 2 chooses } q$$

We compute

$$\begin{aligned} E(p,q) &= \$10 \cdot \frac{q}{4} + \$2 \cdot \frac{1-q}{4} - \$10 \cdot \frac{3pq}{4} - \$2 \cdot \frac{3(1-p)}{4} + \$2 \cdot \frac{3p(1-q)}{4} \\ &= \$2q + \$3p - \$9pq - \$1 \end{aligned}$$

We have created a new zero-sum game called the *mixed-strategy game*:

(8.1) **Payoff Matrix for Betting on Diamonds with Mixed Strategies *p* and *q***

Player 2

Player 1

Strategy	q
p	$(\$2q + \$3p - \$9pq - \$1, -\$2q - \$3p + \$9pq + \$1)$

The game is a simultaneous-move strategic game in which Player 1 chooses p and Player 2 chooses q, both probabilities. The payoff to Player 1 is then computed as $E(p,q)$. The game here is zero sum, so Player 2 receives $-E(p,q)$ as payoff.

Definition 8.7 The **mixed-strategy game** associated with a simultaneous-move game is the game in which the players choose mixed strategies and the payoffs are the expected values for these mixed-strategy choices.

Notice that if we restrict the players to the choices $p = 0$ or 1 and $q = 0$ or 1, we obtain the original (pure) strategy game. For example, $p = 0$ and $q = 1$ correspond to Player 1 playing "Never bluff" and Player 2 playing "Always call."

It may seem that we have only succeeded in further complicating the game Betting on Diamonds. Actually, we will be able to use this analysis to assign a *value* to the game Betting on Diamonds in Chapter 9 (see also Exercises 8.5 and 8.6).

The game Betting on Diamonds is maybe the simplest possible game involving the strategy of bluffing. Despite this simplicity, our analysis has involved some fairly complicated mathematics. It is perhaps now clear why a theory of Poker represents such a formidable challenge. We consider one further game here involving another strategic feature of Poker.

Example 8.8 Bluffing, Calling, and Raising

The game begins the same as Betting on Diamonds. Both players ante \$2. Player 1 draws a card that he or she keeps secret and either bets \$8 or folds. If Player 1 bets, Player 2 now has three options. He or she can (1) raise the bet by \$8, which entails putting \$16 dollars in the pot, \$8 to match Player 1's bet and then \$8 more; (2) he or she can call Player 1's bet by putting \$8 in the pot; or (3) he or she can fold. If Player 2 raises Player 1's bet, Player 1 must match the extra bet and put \$8 more into the pot. Player 1 does not have the option to fold when Player 1 raises. As before, when neither player folds, the game is determined by Player 1's card: if the card drawn is a diamond, Player 1 wins the pot, and if not, Player 2 wins the pot.

Here is the game tree:

Game Tree for Bluffing, Calling, and Raising

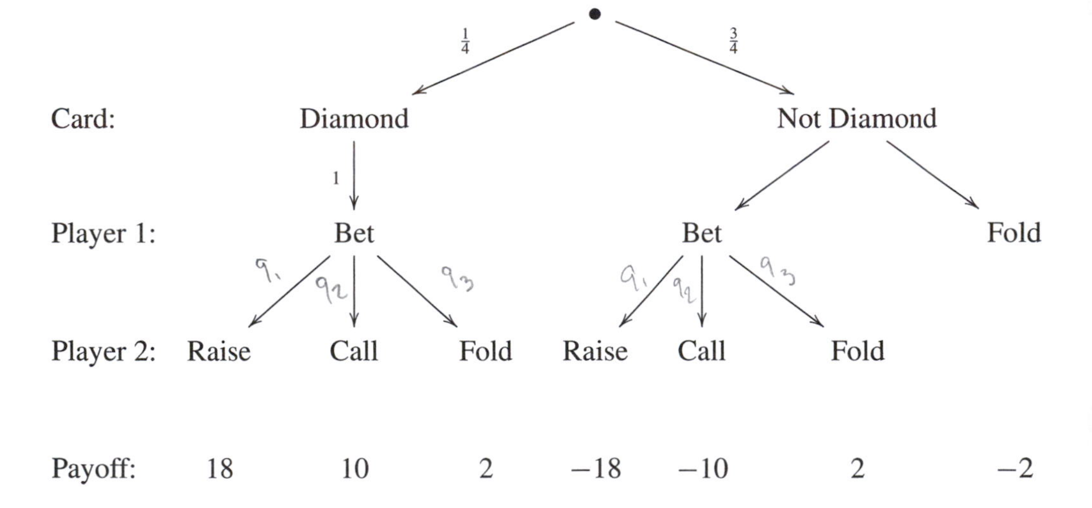

We consider the repeated game with mixed strategies for Bluffing, Calling, and Raising. Again, Player 1 chooses between bluffing and not bluffing. We set

$$p = \text{probability Player 1 bluffs}$$

Player 2 can split his or her time among three strategies. We introduce three variables:

$$q_1 = \text{probability Player 2 raises}$$
$$q_2 = \text{probability Player 2 calls}$$
$$q_3 = \text{probability Player 2 folds}$$

Notice, however, that we must have

$$q_1 + q_2 + q_3 = 1$$

so, actually, Player 2 is free to choose two of these probabilities, say, q_1 and q_2, as long as $q_1 + q_2 \leq 1$. Then q_3 is automatically given by $q_3 = 1 - q_2 - q_3$. It will be convenient notationally to continue to write all three probabilities for Player 2. The resulting probability tree when the players choose mixed strategies is as follows:

Mixed Strategies in Bluffing, Calling, and Raising

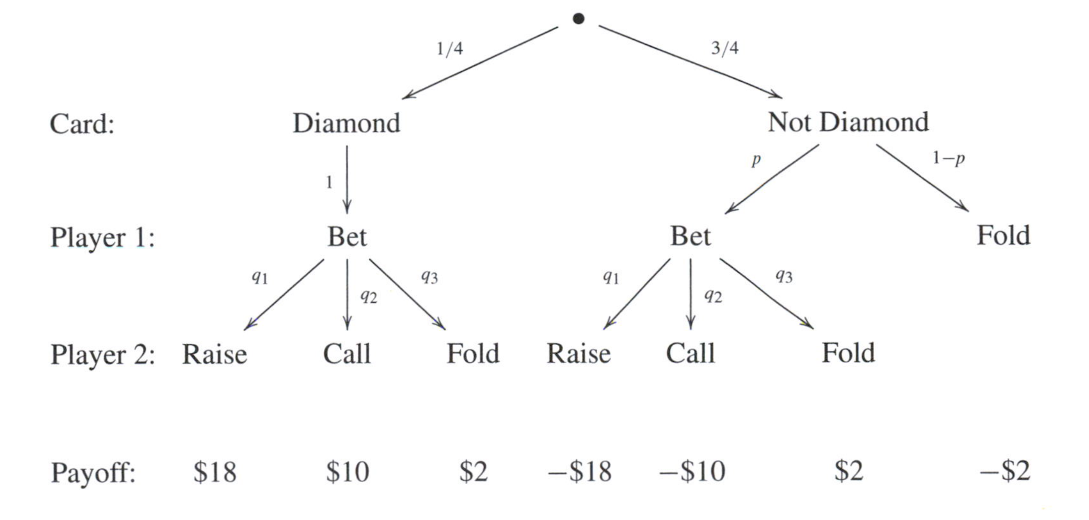

Here is a sample calculation:

Example 8.9

Suppose that Player 1 bluffs with probability $p = 1/3$ and Player 2 raises with probability $q_1 = 1/4$ and calls with probability $q_2 = 1/2$. Then the probability that Player 2 folds is $q_3 = 1/4$. We can compute E directly from the tree

$$\begin{aligned} E = {} & \$18\left(\frac{1}{4}\cdot\frac{1}{4}\right) + \$10\left(\frac{1}{4}\cdot\frac{1}{2}\right) + \$2\left(\frac{1}{4}\cdot\frac{1}{4}\right) \\ & - \$18\left(\frac{3}{4}\cdot\frac{1}{3}\cdot\frac{1}{4}\right) - \$10\left(\frac{3}{4}\cdot\frac{1}{3}\cdot\frac{1}{2}\right) + \$2\left(\frac{3}{4}\cdot\frac{1}{3}\cdot\frac{1}{4}\right) - \$2\left(\frac{3}{4}\cdot\frac{2}{3}\right) \\ = {} & -\$0.75 \end{aligned}$$

Playing the game with these mixed strategies, Player 1 will lose an average of 75 cents per round to Player 2.

The pure-strategy expected values are computed as before, giving

(8.2) **Payoff Matrix for Bluffing, Calling, and Raising with Pure Strategies**

Player 2

Player 1

Strategy	Always raise	Always call	Always fold
Always bluff	(−\$9, \$9)	(−\$5, \$5)	(\$2, \$2)
Never bluff	(\$3, −\$3)	(\$1, −\$1)	(−\$1, \$1)

Notice that neither player has a weakly dominant strategy in this pure-strategy game, as was the case with Betting on Diamonds. We will return to this zero-sum game in Chapter 15.

Finally, we observe that the mixed-strategy game here has payoff given as an expected value $E(p, q_1, q_2)$. We will leave the derivation of this formula for you in Exercise 8.8.

Exercises

8.1 Consider the game Betting on Heads with the same rules as Betting on Diamonds except that now Player 1 flips a coin but does not show the result to Player 2.

Player 1 wins if the flip is "Heads." Here is the game tree:

Game Tree for Betting on Heads

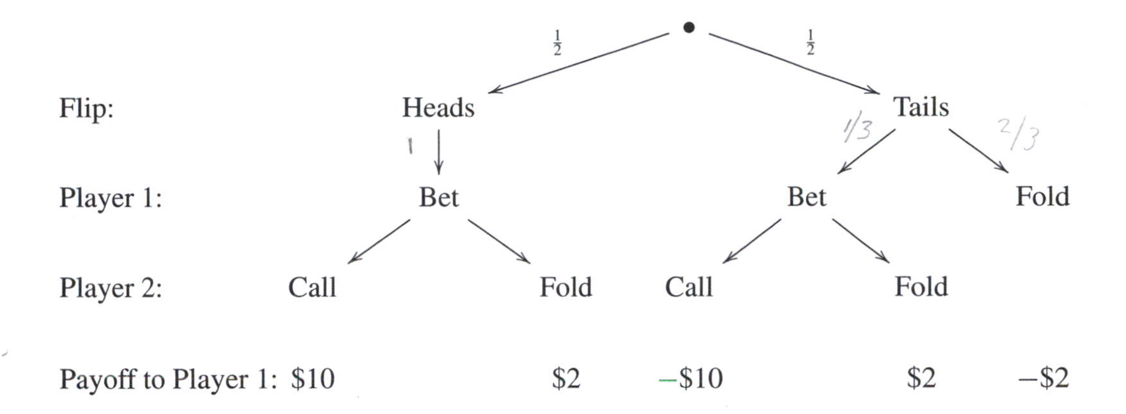

Fill in the expected values for the pure-strategy zero-sum game:

Payoff Matrix for Betting on Heads with Pure Strategies

Player 2

Player 1

Strategy	Always call	Never call
Always bluff	(_, _)	(_, _)
Never bluff	(_, _)	(_, _)

Does either player in this zero-sum game have a weakly dominant strategy? Which player (if either) does this game favor?

8.2 Define the game Betting on Doubles using the same rules as Betting on Diamonds where Player 1 secretly rolls a pair of dice. Player 1 wins if the roll is doubles. Find the pure-strategy zero-sum game as in Exercise 8.1. Decide whether either player has a weakly dominant strategy in the pure-strategy game.

8.3 Suppose that you are given a simple act of chance with two outcomes, "Success" and "Failure," with probability r of "Success" and probability of $1 - r$ of "Failure." Use this act of chance to define a game Betting on Success. The rules are again the same as Betting on Diamonds except that here Player 1 wins if the act of chance is "Success." Here is the game tree:

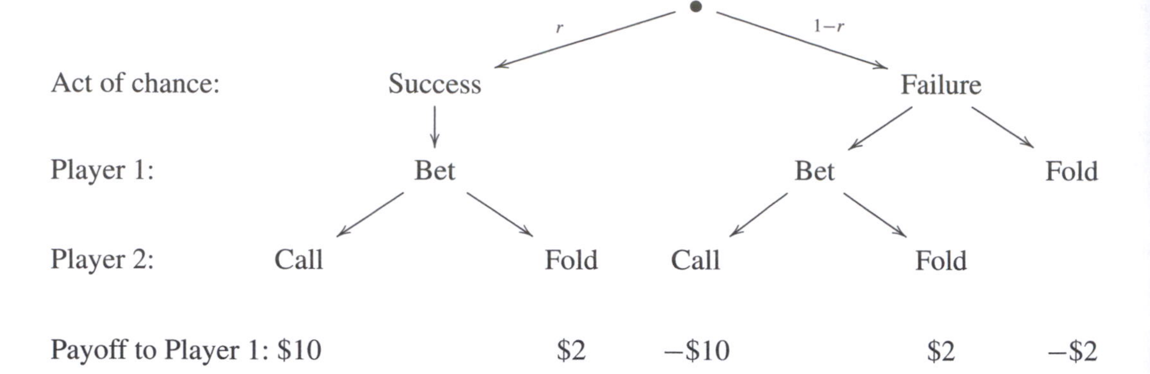

(a) Write down the pure-strategy zero-sum game payoff matrix for this game. Three of the four entries will involve the variable r.
(b) Prove that Player 2 has a weakly dominant strategy "Never call" when $r > 3/5$.
(c) Prove that Player 1 does not have weakly dominant strategy when $r < 1$.

8.4 This exercise concerns the payoff function

$$E(p,q) = \$2q + \$3p - \$9pq - \$1$$

for the mixed-strategy game of Betting on Diamonds.

(a) Evaluate $E(p,q)$ for the following pairs of probabilities:

$$\begin{aligned} p &= 3/4 \quad \text{and} \quad q = 1/5 \\ p &= 1/2 \quad \text{and} \quad q = 1/3 \\ p &= 1/8 \quad \text{and} \quad q = 7/8 \\ p &= 1/3 \quad \text{and} \quad q = 4/5 \end{aligned}$$

(b) Suppose that Player 1 chooses the value $p = 1/2$. What should Player 2 choose for q to maximize his or her expected value?
(c) If Player 2 chooses the value $q = 1/3$, what should Player 1 choose for p to maximize his or her expected value?

8.5 Find the value of p such that the expected value

$$E(p,q) = \$2q + \$3p - \$9pq - \$1$$

from the mixed-strategy game Betting on Diamonds depends only on p. That is, find p such that $E(p,q_1) = E(p,q_2)$ for all $0 \leq q_1, q_2 \leq 1$. What is the expected value of the game when Player 1 chooses this value of p? *Hint:* Set the terms involving q in $E(p,q)$ to zero and solve for p.

8.6 Find the value of q such that the expected value

$$E(p,q) = \$2q + \$3p - \$9pq - \$1$$

from the mixed-strategy game of Betting on Diamonds depends only on q . What is the expected value of the game when Player 2 chooses this value of q? Compare your answer with the expected value in Exercise 8.5. Can you explain what you found?

8.7 Consider a variation on Betting on Diamonds with the same rules except that now the fixed-bet amount it \$6 (the ante is still \$2).

(a) Write down the game tree for this game.
(b) Find the expected-value payoff matrix for the zero-sum game when both players choose from pure strategies: Player 1 between "Always bluff" and "Never bluff" and Player 2 between "Always call" and "Never call."
(c) Write down the game tree when Player 1 chooses the mixed strategy $p =$ probability Player 1 bluffs and Player 2 chooses $q =$ probability Player 2 calls. Compute the expected value in the mixed-strategy game for $p = 1/4$ and $q = 1/5$.
(d) Find a formula for $E(p,q)$.
(e) If Player 1 chooses the value $p = 2/3$, what should Player 2 choose for p to maximize his or her expected value?
(f) If Player 2 chooses the value $q = 3/4$, what should Player 1 choose for p to maximize his or her expected value?
(e) Repeat Exercises 8.5 and 8.6 for this payoff function.

8.8 This exercise concerns the mixed-strategy game associated with Bluffing, Calling, and Raising (Example 8.8).

(a) Find the expected value for the mixed-strategy game when $p = 1/3$, $q_1 = 1/8$, and $q_2 = 1/4$. Note that $q_3 = 1 - q_1 - q_2$.
(b) Find the expected value for the mixed-strategy game when $p = 1/6$, $q_1 = 1/6$, and $q_2 = 2/3$.
(c) Find the formula for the general expected value $E(p, q_1, q_2)$.

8.9 Consider a variation on Bluffing, Calling, and Raising (Example 8.8) in which Player 1 is allowed to fold after Player 2 raises. Assuming that Player 1 always folds when he or she has a non-Diamond, write down the game tree for this version of the game. Find the pure-strategy zero-sum game payoff matrix. Does either player have a weakly dominant strategy? Who does this game favor?

8.10 Consider a variation on Bluffing, Calling, and Raising (Example 8.8) in which, instead of drawing a card, Player 1 flips a coin and wins if the coin comes up "Heads." Write down the game tree for this version of the game, and find the pure-strategy zero-sum game payoff matrix. Does either player have a weakly dominant strategy? Who does this game favor?

9 Nash Equilibria

In his 1950 doctoral thesis, John Nash established the existence of stable outcomes, now called *Nash equilibria*, for a large class of strategic games. The Nash Equilibrium Theorem is the centerpiece of modern game theory with applications ranging from evolutionary biology to economics to psychology. Nash was awarded the 1994 Nobel Prize in Economics for this foundational work.

Nash's proof of the Equilibrium Theorem was a feat of technical ingenuity. The definition of a Nash equilibrium, however, is quite simple and elegant and will serve to unify several themes introduced in this first part of this book. We begin with partial-conflict games, where we have already seen an example of a Nash equilibrium in the form of the dominant-strategy solution (Definition 4.6). We observe that the notion of a Nash equilibrium offers a more general solution concept. In the context of zero-sum games, Nash equilibria correspond to *saddle points* for payoff matrices. We prove this equivalence in Theorem 9.8. We then return to the simplified Poker game Betting on Diamonds (Example 8.1). We identify a Nash equilibrium for the mixed-strategy version of this game. Our analysis here will lay the groundwork for our proof of the Minimax Theorem (Theorem 15.8) for two-strategy zero-sum games in Chapter 15. Finally, we consider two examples in the realm of elections.

Partial-Conflict Games

We begin our discussion with the Prisoner's Dilemma. Recall the scenario in Example 4.1. Two prisoners are interrogated in separate cells. Each has the choice between "Staying silent" and "Ratting." The payoff matrix for the simultaneous-move game is as follows. The negative payoffs correspond to years in prison.

Payoff Matrix for Prisoner's Dilemma

		Prisoner 2	
	Strategy	Staying silent	Rating
Prisoner 1	Staying silent	$(-1,-1)$	$(-20,0)$
	Rating	$(0,-20)$	$(-10,-10)$

Recall that we refer to the "Staying silent" strategy as a strategy of cooperation, whereas the "Ratting" strategy represents defection. The dilemma is seen by comparing the outcomes of **mutual cooperation**, when both players cooperate, and **mutual defection**, when both players defect. The outcome of mutual cooperation results in minimal prison time for both prisoners, and yet defection is a dominant strategy for each. When playing this game in repeated rounds, mutual cooperation represents a fragile alliance, whereas mutual defection is a more stable state.

The definition of a Nash equilibrium formalizes this idea of stability. When one player changes his or her strategy in a game while the other player stays fixed on a given strategy, we say the first player performs a **unilateral change in strategy**. When there are more than two players, then all but one remain fixed on a strategy, whereas a single player performs a unilateral change in strategy. This idea of unilateral action is the key to the following definition:

Definition 9.1 An outcome of a simultaneous-move strategic game is said to be a **Nash equilibrium** if no player can improve his or her payoff by a unilateral change in strategy.

The Prisoner's Dilemma gives, perhaps, the most famous example of a Nash equilibrium. From our point of view, however, the focus on mutual defection is nothing new because we have already seen that mutual defection is a dominated-strategy solution for the game. The next example reveals a difference between the concepts of Nash equilibria and dominated-strategy solutions.

Example 9.2 Chicken

Two teenagers drive their cars toward each other on a narrow road. Each driver has two options: to "Swerve" and thus avoid a collision or to "Drive straight." The consequences are obvious. If one player swerves and the other does not, the swerving driver is embarrassed (he or she is the "chicken"), and the driver who did not swerve is triumphant. If neither player swerves, then the worst-case scenario occurs for both, a head-on collision. Finally, if both players swerve, they are both a bit embarrassed, but neither has been shown up.

Since there are no obvious payoffs for this game, we will simply rank the four possible outcomes for both players. We assign the payoff 4 for the best outcome and 1 for the worst, giving

Payoff Matrix for Chicken

Teenager 2

Teenager 1

Strategy	Swerve	Drive straight
Swerve	(3,3)	(2,4)
Drive straight	(4,2)	(1,1)

Now consider the situation that occurs when Teenager 1 drives straight and Teenager 2 swerves. Does either teen regret his or her decision? Certainly, Teenager 1 does not, because he or she has received the highest payoff possible. Teenager 2 does not regret his or her decision either. If Teenager 2 had driven straight instead, the teens would have crashed. Similarly, when Teenager 1 swerves and Teenager 2 drives straight, neither player regrets his or her strategy choice given the choice of the other. The outcomes when one player swerves and the other drives straight [with payoffs (4,2) and (2,4) in the payoff matrix] are both Nash equilibria for Chicken.

Regarding the relationship between Nash equilibria and dominated strategies, we have the following theorem:

Theorem 9.3 *Suppose that we are given a two-player simultaneous-move game of pure strategy and a particular outcome. If the outcome is a dominated-strategy solution then the outcome is a Nash equilibrium.*

Proof Suppose that, say, Player 1 has a strategy that weakly dominates his or her other strategies. The dominated-strategy solution corresponds to Player 1 choosing this weakly dominant strategy and Player 2 choosing the strategy that maximizes his or her payoff given this strategy choice by Player 1. Certainly, Player 1 cannot improve by a unilateral change in strategy because Player 1 has chosen his or her weakly dominant strategy. Player 2 cannot improve by unilateral change in strategy either because he or she has already chosen the best payoff against this strategy choice by Player 1. Thus the outcome is a Nash equilibrium by definition. ■

Theorem 9.3 is an example of a *conditional statement.* Notice that the game Chicken has two Nash equilibria (when one player swerves and the other drives straight). Yet Chicken has no weakly dominated strategies. Chicken proves that the *converse* statement to Theorem 9.3 is false. We discuss conditional statements and their converses further in Chapter 10.

Here is an effective procedure for finding a Nash equilibrium in a general payoff matrix. In each column, mark the entry (or entries) with the maximum payoff to the row player. We call this value the **column maximum**. Since the row player controls whether we move up or down the column, for an entry to be a

Nash equilibrium, the row entry must be a column maximum. Next, in each row, mark the entry with the maximum payoff to the column player, called the **row maximum**. The Nash equilbria then correspond to any outcomes with both entries marked. Here is an example:

Locating a Nash Equilibrium

☐ = column maximum ▬ = row maximum

Player 2

Player 1

Strategy	*W*	*X*	*Y*	*Z*
A	(3, 2)	(1, 14)	(−5, 6)	(9, −6)
B	(7, 4)	(2, 1)	(3, 5)	(6, 3)
C	(6, 3)	(3, 7)	(8, 2)	(−4, 4)

This game has one Nash equilibrium at the outcome (C, X).

Zero-Sum Games

We now turn to the class of zero-sum games. We motivate both a notational convention for these games and the characterization of Nash equilibria with the following example:

Example 9.4 Paying a Debt

Carla owes $7 to Rose. Rather than pay her, Carla proposes an alternative. She produces a 3 × 3 array of numbers and suggests that they play a game. Rose will choose a row and Carla will choose a column simultaneously. Carla will pay Rose the amount in that row and column. Here is the array:

1	4	18
9	5	15
12	3	2

Should Rose play the game proposed by Carla to settle the debt?

Before addressing this question, we observe that Carla has introduced an important notational convention for zero-sum games. Notice that the payoff matrix for Paying a Debt is given as

Payoff Matrix for Paying a Debt

Carla

Rose

Strategy	C_1	C_2	C_3
R_1	$(1,-1)$	$(4,-4)$	$(18,-18)$
R_2	$(9,-9)$	$(5,-5)$	$(15,-15)$
R_3	$(12,-12)$	$(3,-3)$	$(2,-2)$

However, the inclusion of the column player's (Carla's) payoffs is redundant. They are always the negatives of the row player's (Rose's) payoffs. So writing Rose's payoffs as Carla did in Paying a Debt is sufficient. We hereby introduce this as a formal convention.

Convention 9.5 When writing down the payoff matrix for a two-player zero-sum game, we will only write the payoffs for the row player. The payoffs for the column player are then the negatives of these payoffs. ■

Following this convention, we now write

Payoff Matrix for Paying a Debt

Carla

Rose

Strategy	C_1	C_2	C_3
R_1	1	4	18
R_2	9	5	15
R_3	12	3	2

Notice that in this case the entries are all positive numbers. Negative numbers in a zero-game payoff matrix correspond to the row player paying money to the column player.

Returning to the game Paying a Debt, we can imagine that Rose is skeptical. Carla proposes to sweeten the deal. She tells Rose that she will reveal her choice first. Rose can then choose her strategy afterwards. Carla's proposal seems generous. Generally, there is an advantage to playing a simultaneous-move game sequentially if you are allowed to go last.

Carla's proposal is to play the simultaneous-move game sequentially with Rose as the second player. As in Chapter 4, we may consider the Backward Induction Solution (Definition 4.10) for Carla as follows:

Backward Induction Solution of Paying the Debt for Carla

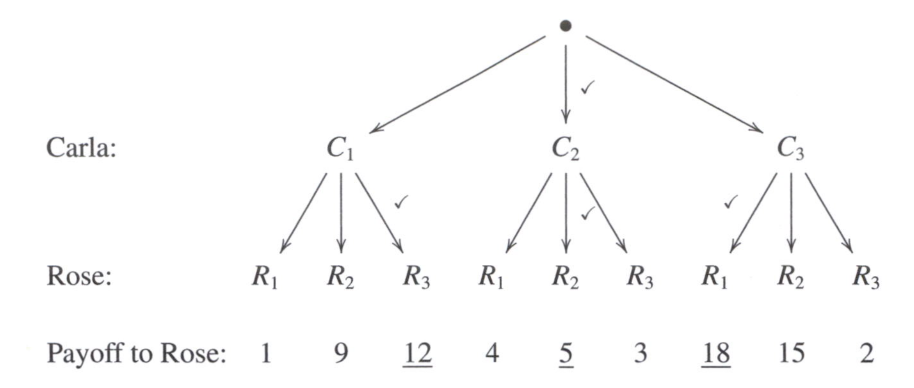

Now let us observe that Carla has located the maximum value (underlined) in each column. She then chooses the column C_2, whose maximum value (5) is the minimum of the column maxima (12, 5, 18). We call this the *minimax strategy* for the column player because it minimizes the maximum amount she will have to pay with each strategy.

We have already seen the idea of a maximin strategy (Definition 4.12). Our convention of listing the payoffs only for the row player leads to the need for a new name: minimax. For Rose, the maximin strategy is found as follows: she examines her minimum payment for each row:

$$R_1 \text{ minimum} = 1, \quad R_2 \text{ minimum} = 5, \quad \text{and} \quad R_3 \text{ minimum} = 2$$

Rose chooses her maximin strategy when she selects the row R_2 with the largest minimum (5) and thereby maximizes her minimum payoff. The minimax and maximin can be found easily for any payoff matrix using a table as follows:

Maximin and Minimax for Paying a Debt

		Carla			
	Strategy	C_1	C_2	C_3	**Row minimum**
	R_1	1	4	18	1
Rose	R_2	9	5	15	5[a]
	R_3	12	3	2	2
	Column maximum	12	5[b]	18	

[a] Maximin.
[b] Minimax.

We notice that

$$\text{maximin} = \text{minimax} = 5$$

This is called a *saddle point*. Summarizing, we have the following definition:

Definition 9.6 Assume that you are given a zero-sum game with a payoff matrix presented as in Convention 9.5. The **maximin strategy** for the row player is the choice of the row whose minimum value is at least as large as the other row minima. This value is called the **maximin**. The **minimax strategy** for the column player is the column whose maximum value is at least as small as the other column maxima. The **minimax** is the value of this column maximum. If maximin = minimax, the common value is called a **saddle point**.

We have the following general fact:

Theorem 9.7 *For any payoff matrix for a zero-sum game,*

$$\textit{minimax} \geq \textit{maximin}$$

Proof Let R be the row for the maximin strategy and C the column for the minimax strategy. Then

$$\text{maximin} = \text{minimum value in row } R$$

$$\text{minimax} = \text{maximum value in column } C$$

Let a denote the payoff entry in row R, column C. Now a is an entry in row R. Thus a is no smaller than the minimum entry in row R and so maximin $\leq a$. However, a is also an entry in column C, so a is no larger than the maximum in column C and so $a \leq$ minimax. Combining these inequalities gives

$$\text{maximin} \leq a \leq \text{minimax}. \qquad \blacksquare$$

Returning to the game Paying a Debt, observe that if Carla announces that she will choose C_2, then the best Rose can do is choose R_2. Even if Rose goes first and chooses R_2, Carla would be best choosing C_2. A unilateral change by Carla would only increase her payment to Rose. The saddle point at 5 is a Nash equilibrium. In fact, we have the following general result:

Theorem 9.8 *An outcome of a zero-sum game is a Nash equilibrium if and only if that outcome is a saddle point of the payoff matrix.*

Proof We first suppose that a given zero-sum game has a saddle point

$$m = \text{minimax} = \text{maximin}$$

at row R and column C. Now the minimax is chosen from among the column maxima. Thus $m =$ minimax is, in particular, the maximum for column C. If the column player stays fixed at C, then the row player can do no better than the maximum value m of column C, which occurs at row R. The row player does not improve by a unilateral change of strategy. Now use the fact that $m =$ maximin to deduce that m is, in particular, the minimum value in row R. It follows, then, that the column player cannot improve by a unilateral change of strategy when the row player is fixed at row R. We conclude that the saddle point corresponds to a Nash equilibrium.

Now suppose that a given zero-sum game has a Nash equilibrium. Again, let R denote the row, C the column, and m the value. We prove that

$$m = \text{minimax} = \text{maximin}$$

Since the row player cannot improve by a unilateral change in strategy from row R, we must have that m is the maximum value in column C. Now the minimax is the minimum of all the column maxima, and m is one such column maxima. It follows that $m \geq$ minimax.

Similarly, the column player cannot improve by switching from column C when the row player stays fixed at R. Thus m is the row R minimum. It follows that m is no larger than the maximum of all row maximin: $m \leq$ maximin. We have now shown that

$$\text{minimax} \leq m \leq \text{maximin} \tag{9.1}$$

By Theorem 9.7, we always have minimax $\geq$ maximin. We conclude, then, that the inequalities in (9.1) are equalities and that $m =$ maximin $=$ minimax is a saddle point. ■

Theorem 9.8 is an example of a *biconditional statement*. It establishes an equivalence between two seemingly distinct mathematical notions. We will return to these statements in Chapter 10. With Theorem 9.8, we can quickly produce many examples of zero-sum games that have a Nash equilibrium but no dominated strategies. The payoff matrix for Paying a Debt is one example. Nash equilibria in zero-sum games are essentially unique (see Exercise 9.6) unlike the case for partial-conflict games. Finally, we remark that saddle points in payoff matrices are special occurrences. If we write down an arbitrary matrix of numbers, we are unlikely to see a saddle point.

Mixed-Strategy Games

We now return to the simplified Poker game Betting on Diamonds (Example 8.1). Recall the rules: Both players ante \$2. Player 1 then draws a card and does not show it to Player 2. Player 1 either bets \$8 or folds. If Player 1 bets, Player 2 can call or fold. If either player folds, the other gets the ante. Otherwise, Player 1 wins the pot if the card drawn is a diamond, and Player 2 wins the pot if the card is not a diamond.

Rather than playing one round of the game, we imagined playing multiple rounds. Player 1 chooses a probability for playing $p = P(\text{"Bluff"})$, and Player 2 chooses $q = P(\text{"Call"})$. We then have

Game Tree for the Mixed-Strategy Game for Betting on Diamonds

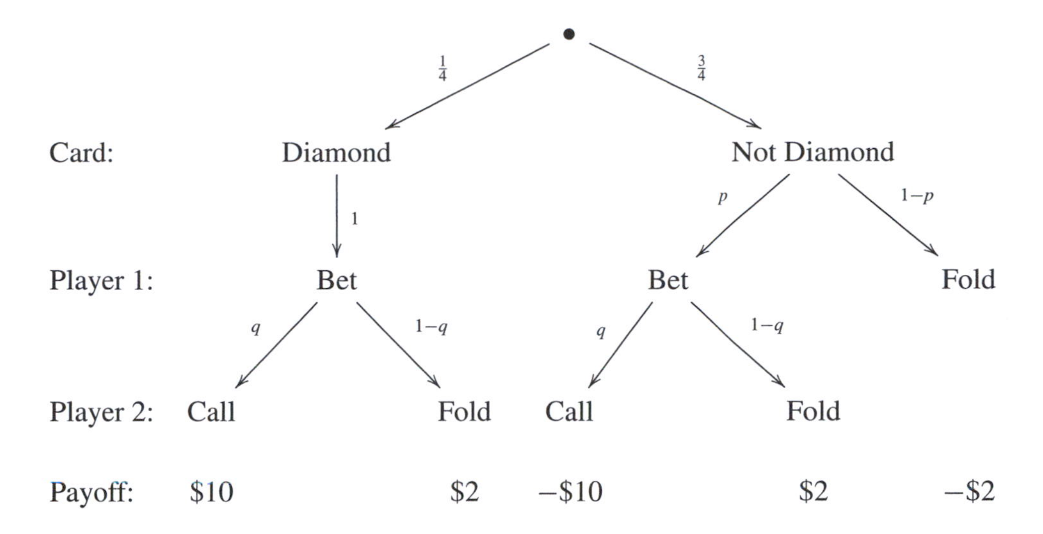

We computed the expected value to be the payoff function

$$E(p,q) = 2q + 3p - 9pq - 1$$

The mixed-strategy version of Betting on Diamonds is then the simultaneous-move game with the following payoff matrix:

Payoff Matrix for Mixed-Strategy Version of Betting on Diamonds

Player 2

Player 1

Strategy	q
p	$2q+3p-9pq-1$

Some remarks about this game are in order. First, it is a stretch to call the preceding table a "matrix." The players are not choosing from a finite set of strategies, so there are no rows or columns. Player 1 can choose any probability p with $0 \leq p \leq 1$, and Player 2 can choose any probability q with $0 \leq q \leq 1$. The payoff for the game is obtained by plugging these chosen values for p and q into the formula $E(p,q)$. While the game is not finite, it is a zero-sum game, and we are following Convention 9.5. The value of $E(p,q)$ is the payoff to Player 1. The payoff to Player 2 in this game is $-E(p,q)$.

The mixed-strategy game appears to take a rather simple game and produce something much more complicated. The advantage of this process is that the mixed-strategy game has a Nash equilibrium. In this setting, a Nash equilibrium is a pair of probabilities p^* and q^* for Players 1 and 2, respectively, such that neither player can improve his or her expected payoff by unilaterally switching to another probability. We find these probabilities as follows:

Nash Equilibrium for the Mixed-Strategy Game for Betting on Diamonds

Step 1: Find p^* so that $E(p^*,q)$ does not depend on q. We accomplish this by first locating the terms in $E(p,q) = 2q + 3p - 9pq - 1$ that involve q. These are $2q$ and $-9pq$. We then set the sum of these terms to zero and solve for p^*:

$$2q - 9pq = 0 \quad \text{implies} \quad p^* = \frac{2q}{9q} = \frac{2}{9}$$

Dividing by q here technically requires that we ensure that $q \neq 0$, but this will not be a problem.

Step 2: Find q^* so that $E(p,q^*)$ does not depend on p. Here we observe that the terms in $E(p,q)$ that involve p are $3p$ and $-9pq$. Again, we set the sum of these terms to zero and solve, this time, for q^*:

$$3p - 9pq = 0 \quad \text{implies} \quad q^* = \frac{3p}{9p} = \frac{1}{3}$$

Steps 1 and 2 have produced probabilities $p^* = 2/9$ and $q^* = 1/3$ and a corresponding payoff

$$E\left(\frac{2}{9}, \frac{1}{3}\right) = -\frac{1}{3}$$

The expected value at these probabilities has Player 1 losing 33 cents per round to Player 2. We prove that these probabilities correspond to a Nash equilibrium of the randomized game.

Theorem 9.9 *The pair* $(p^*,q^*) = (2/9, 1/3)$ *with expected value* $E = -\$.33$ *is a Nash equilibrium for the mixed-strategy game Betting for Diamonds.*

Proof First, suppose that Player 1 fixes on $p^* = 2/9$, whereas Player 2 considers a unilateral change in strategy to a probability q. We compute the payoff as

$$E\left(\frac{2}{9}, q\right) = 2q + 3\left(\frac{2}{9}\right) - 9q\left(\frac{2}{9}\right) - 1 = -\frac{1}{3}$$

This shows that when Player 1 chooses his or her strategy to be $p^* = 2/9$, the expected value payoff is -33 cents regardless of what value Player 2 chooses for q. Player 2 cannot unilaterally improve his or her payoff – in fact, Player 2 cannot even change the payoff!

Similarly, notice that

$$E\left(p, \frac{1}{3}\right) = 2\left(\frac{1}{3}\right) + 3p - 9p\left(\frac{1}{3}\right) - 1 = -\frac{1}{3}$$

Player 1 cannot alter the payoff by a unilateral change of strategy when Player 2 chooses his or her strategy to be $q^* = 1/3$. Since neither player can improve his or her payoff by a unilateral change of strategy, this pair of strategies yields a Nash equilibrium by definition. ■

The expected value $E(2/9, 1/3) = -\$0.33$ is called the *value* of the game Betting on Diamonds. We return to the problem of assigning values to zero-sum games in Chapter 15.

Electoral Politics

The concept of a Nash equilibrium applies directly to our definition of an election as a game of strategy played by voters. For example, consider the Plurality Method election in Example 6.9:

Preference Table for the Third-Party Candidate Election

Voting bloc	Republican	Democrat	Independent
Number of voters	**40**	**35**	**25**
1st choice	R	D	I
2nd choice	D	I	R
3rd choice	I	R	D

Observe that sincere voting is not a Nash equilibrium: if the Republican and Independent blocs vote sincerely, then the Democratic bloc is better off voting for I than for D. It is also easy to see that each unanimous decision is a Nash equilibrium because no voting bloc can unilaterally change the outcome. In order to focus on more realistic outcomes, we consider only votes in which a bloc votes for either its first or second choice. We then obtain eight possible voting outcomes, as in the following table:

Nash Equilibria for the Third-Party Candidate Election

Republican	Democrat	Independent	Winner	Nash equilibrium?
R	D	I	R	No
D	D	I	D	No
R	I	I	I	Yes
D	I	I	I	No
R	D	R	R	Yes
D	D	R	D	No
R	I	R	R	No
D	I	R	D	No

We notice that the first Nash equilibrium corresponds to the outcome found in Theorem 6.10, whereas the second corresponds to the Independent bloc compromising with the Republican bloc in the game Battle of the Sexes (Example 2.10).

We conclude with one final example of a Nash equilibrium in the setting of elections. We consider a model for the game played by the candidates. Our example is a simplified version of a model for electoral politics constructed by Harold Hotelling and Arthur Downs.

Example 9.10 Political Positioning

Two candidates stake out positions on a political spectrum. For simplicity, we view the possible political positions as ranging from 0 = extreme liberal position to 1 = extreme conservative position. We further assume that the voters are evenly distributed in this range from 0 to 1 and that a voter will vote for a candidate whose political position is closest to his or hers. If a voter's position is the same distance from both candidates, that voter flips a coin to make his or her choice. The game is a simultaneous-move strategic game between the two candidates. Each candidate chooses his or her position on the political spectrum with the goal of gaining more voters than his or her opponent and so winning the election.

Here is a picture of the game with two possible candidate positions X and Y shown:

We define the payoffs for each candidate by the formula

$$\text{Payoff to candidate} = \text{fraction of voters won} + \frac{\text{fraction of voters split}}{2}$$

We then observe that the only Nash equilibrium for this game occurs when both candidates take the position in the exact center of the political spectrum, where they each have a 50–50 chance of winning the election. Suppose that both candidates begin in this center position, and X stays fixed at the center while Y moves to a position left of center (a unilateral change in strategy). Then X wins all voters to the right of center plus half the voters between X and Y. Thus X now wins the election outright, and Y has not improved by his or her unilateral change.

It is easy to see that no other outcome represents a Nash equilibrium: a candidate always has incentive to move to the center to increase his or her fraction of the voters. While this example may not represent much of a surprise, the analysis of "positioning games" along the lines of Example 9.10 are of considerable interest in economics and politics.

One interesting consequence of this model is that when we start with three candidates, there is no Nash equilibrium. We leave the details for you in Exercise 9.12.

The preceding examples hint at the diverse contexts in which the notion of a Nash Equilibrium arises. The definition captures a far-reaching and often significant phenomenon of stability in conflict and competition.

Exercises

9.1 Find the Nash equilibrium for the game Battle of the Sexes (Example 2.6).

9.2 Find all Nash equilibria (if any) for the following partial-conflict games:

(a)

Player 2 (columns); **Player 1** (rows)

Strategy	*W*	*X*	*Y*	*Z*
A	(−7, 2)	(1, 5)	(−5, 4)	(3, 3)
B	(3, 6)	(4, 1)	(3, 9)	(13, −4)
C	(−6, 1)	(1, 3)	(−3, 2)	(4, 2)

(b)

Player 2 (columns); **Player 1** (rows)

Strategy	*W*	*X*	*Y*	*Z*
A	(−3, 12)	(2, 0)	(−5, 4)	(3, 3)
B	(−2, 6)	(−4, −11)	(2, 8)	(3, −4)
C	(6, 1)	(4, 2)	(−13, 2)	(7, 1)
D	(8, 2)	(3, 5)	(6, 1)	(4, 4)

(c)

Player 2 (columns); **Player 1** (rows)

Strategy	*W*	*X*	*Y*	*Z*
A	(2, 1)	(3, 0)	(7, −1)	(8, 0)
B	(4, −1)	(−4, 1)	(0, 2)	(6, 5)
C	(1, 3)	(8, −2)	(4, 4)	(−1, −1)
D	(5, −2)	(2, −5)	(4, 0)	(−5, −3)

9.3 Consider the following partial-conflict auction game: two players bid simultaneously for a dollar. The high bidder gets the dollar but must pay his or her bid. The low bidder must pay his or her bid and gets nothing (thus has a zero or negative payoff). Player 1 can only bid the amounts \$0.50, \$0.95, \$1.10, and \$1.20, and Player 2 can only bid the amounts \$0.40, \$0.80, \$1.00, and \$1.50. Write down the payoff matrix for this game. Identify all weakly dominant pairs of strategies for both players, and find all Nash equilibria.

9.4 Define a simultaneous-move game of strategy as follows: each player names an integer between 1 and 10 inclusive. If the players name the same number, they get paid nothing. If the players name different numbers, then each player is paid the dollar amount he or she named. Find the Nash equilibria for this game.

9.5 Find the maximin and minimax strategies and payoffs for the following zero-sum game payoff matrices. Decide in each case if there is a saddle point.

(a) **Column player**

Row player

Strategy	T_1	T_2	T_3
S_1	5	−1	6
S_2	4	15	22
S_3	−8	−3	6

(b) **Column player**

Row player

Strategy	T_1	T_2	T_3
S_1	−2	−1	−6
S_2	5	2	1
S_3	−6	1	−3

(c) **Column player**

Row player

Strategy	T_1	T_2	T_3	T_4
S_1	2	−7	4	6
S_2	15	6	10	14
S_3	6	−12	−3	2

(d) **Column player**

Row player

Strategy	T_1	T_2	T_3	T_4	T_5
S_1	−3	−17	3	6	12
S_2	5	−6	−10	4	−7
S_3	−4	11	−3	2	−1
S_4	8	−12	−30	12	21

9.6 Suppose that a zero-sum game has a payoff matrix with two saddle points, m and m'. Prove that $m = m'$.

9.7 Consider the mixed-strategy Games with expected payoff $E(p,q)$ given by the following formulas. As usual, Players 1 and 2 choose probabilities p and q, and $E(p,q)$ is the payoff to Player 1. Find the Nash equilibrium (p^*,q^*) to the game for the following equations for $E(p,q)$. Compute the expected value, $E(p^*,q^*)$, of the game at the Nash equilibrium.

(a) $E(p,q) = 3p + 2q - 4pq - 7$
(b) $E(p,q) = 4p + q - 8pq + 3$
(c) $E(p,q) = \frac{3}{2}p + \frac{1}{4}q - 3pq + 1$

9.8 Consider the mixed-strategy game in Problem 9.7 for the function

$$E(p,q) = p + q + 2pq + 4$$

Note that since we are only allowed to consider $0 \leq p^* \leq 1$ and $0 \leq q^* \leq 1$, our algebraic method for finding the Nash equilibrium (p^*,q^*) fails us here. Does this game have a Nash equilibrium? Prove you answer.

9.9 Consider the variation on Betting on Diamonds in which Player 1 wins if the card is a jack. The ante and bet are still \$2 and \$8, respectively. Find the Nash equilibrium of the mixed-strategy game and the expected value at this outcome.

9.10 Determine the Nash equilibria for the voting blocs in the following elections. Assume that no voting bloc votes for its least preferred choice. For (c), assume that no bloc votes for its two least preferred choices. Use the Plurality Method.

(a)

Voting bloc	**Republican**	**Democrat**	**Independent**
Number of voters	**40**	**35**	**25**
1st choice	R	D	I
2nd choice	D	R	R
3rd choice	I	I	D

(b)

Voting bloc	**Republican 1**	**Republican 2**	**Democrat**	**Independent**
Number of voters	**28**	**12**	**35**	**25**
1st choice	R	R	D	I
2nd choice	D	I	R	D
3rd choice	I	D	I	R

(c)

Voting bloc	**Republican**	**Democrat**	**Independent**	**Green**
Number of voters	**28**	**32**	**15**	**25**
1st choice	R	D	I	G
2nd choice	D	G	R	I
3rd choice	I	R	D	D
4th choice	G	I	G	R

9.11 Decide whether sincere voting is a Nash equilibrium for the elections in Exercise 9.10 assuming that the elections are run using the following voting methods:

(a) Borda Count
(b) Hare Method
(c) Coombs Method

9.12 Suppose that there are three candidates X, Y, and Z in the Political Positioning Game (Example 9.10). Define the payoff to a candidate by

$$\begin{aligned} \text{Payoff to candidate} = {} & \text{fraction of voters closest to candidate} \\ & + \frac{\text{fraction of voters equidistant to this candidate and one other}}{2} \\ & + \frac{\text{fraction of voters equally close to all three candidates}}{3}. \end{aligned}$$

Show that this positioning game has no Nash equilibrium.

Web Resources

The Web has an enormous number of free sites with games, blogs, and articles relevant to our study. We mention some sites featuring future topics.

Apps

Games of chance including Poker, Blackjack, and *Craps* can be played at the site wizardofodds.com.

We will introduce the *Iterated Prisoner's Dilemma* in Chapter 16. You can play the game at the website s3.boskent.com/prisoners-dilemma/fixed.html. This app implements state-of-the-art strategies for the game discovered by William Press and Freeman Dyson. There is a link to the article as well.

Nim can be played at www.kongregate.com/games/ricardorix/nim.

We will introduce a number of Nim-like games in Part II. You can try some of these now. The site www.cut-the-knot.org has, in particular, versions of *Northcott's Game* (Example 20.3) and *Turning Turtles* (Exercise 20.5).

We discuss the combinatorial games *Chomp* and *Hex* in Chapter 20. These games are available online as well. Chomp can be played www.math.ucla.edu/~tom/Games/chomp.html. Hex is available at www.mazeworks.com/hex7/.

The site egwald.ca/operationsresearch/cooperative.php has a number of partial-conflict and zero-sum games.

The site www.270towin.com has interactive electoral maps.

Content

The *Stanford Encyclopedia of Philosophy* at plato.stanford.edu has extensive, accessible articles on game theory and social choice theory with references.

The organization FairVote.org is devoted to promoting fair elections. It advocates for the use of *Instant Runoff Voting* (the Hare Method).

The site rangevoting.org advocates for *Range Voting*. The site contains helpful information about other voting methods as well.

Part II

Basic Theory

10 Proofs and Counterexamples

In this chapter, we digress from our study of games and elections to focus more directly on the logic of mathematical proof. Our goal is to identify some proof methods that we have seen in the first part of the text and to introduce some that we will use in Parts II and III. We begin with the idea of a *mathematical statement* and the basic problem of constructing proofs and finding counterexamples. We then introduce proof methods for two types of mathematical statements: *conditional statements* and *universal statements.* We conclude with an introduction to the method of *proof by induction.* We apply this method to prove a version of Zermelo's Theorem (Theorem 4.17) for takeaway games.

Mathematical Statements

Like a coin flip, which offers two possible outcomes – a basic branching in a game of chance – the building blocks for a mathematical theory are statements with two possible values: true or false.

Definition 10.1 A **mathematical statement** is a statement that, by its wording, only admits two possible interpretations. The statement is true or the statement is false.

Mathematical statements arise frequently in ordinary language. For example,

"Today is Tuesday" or "It is raining"

Both are statements that can only be interpreted as true or false. Of course, there may be gray areas in such phrasings. Here are some examples of mathematical statements related to our topics of study:

Statement 1: "The Borda Count social choice always has a plurality of the first place votes."
Statement 2: "A zero-sum game with a saddle point has a Nash equilibrium."
Statement 3: "There is a guaranteed winning strategy for Player 1 in Chess."

There is no gray area for these statements. Statements 1 is a false statement. We can quickly check, for example, that the Democratic candidate D is the Borda Count social choice for the Third-Party Candidate Election (Example 6.9), whereas the Republican candidate R has the most first-place votes. The Third-Party

Candidate Election is called a *counterexample* to Statement 1. Generally, we have the following definition:

Definition 10.2 An example that proves a mathematical statement false is called a **counterexample**.

Statement 2, however, is true. We gave a proof of this statement in Theorem 9.8. We assumed that a given zero-sum game had a saddle point and then unraveled the meaning of saddle point to prove that the saddle point is, in fact, a Nash equilibrium.

Statements 1 and 2 illustrate the basic dichotomy between proofs and counterexamples. Given a mathematical statement, we can try to produce a proof and so establish that the statement is true, or we can search for a counterexample to establish that statement is false. In other words, the goal is to determine the correct branching.

Proof or Counterexample?

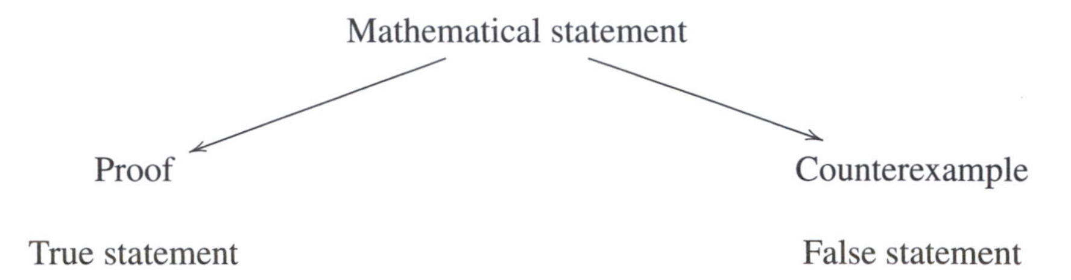

The correct branching for Statement 3 is an open question. It is not known whether Player 1 can guarantee victory in Chess. Nonetheless, by its phrasing, Statement 3 must be either true or false, even if we do not know which it is. Indeed, this is the only requirement of a mathematical statement. A mathematical statement whose truth value is not known but is believed to be true is called a **conjecture**.

Conditional Statements

Many theorems in mathematics are presented as statements of the form

"If P, then Q."

These are called **conditional statements.** Some examples in ordinary language are

"If it is Tuesday, then the train arrives at 9:00 AM" and

"If it is raining, then the game is canceled"

The phrase P is called the **premise** and Q the **conclusion** of the conditional statement. The statement "If P, then Q" is a true statement if Q is true whenever P is true. Here are two examples of true conditional statements we have seen in this text:

Statement 4: "If a simultaneous-move game has a dominated strategy, then it has a Nash equilibrium."
Statement 5: "If a zero-sum game has a Nash equilibrium, then it has a saddle point."

To prove a conditional statement "If P, then Q" we assume that the premise P is true and use this assumption to prove that the conclusion Q is true as a consequence. We give a simple example with integers. We say that an integer n is **even** if there is an integer k with $n = 2k$. We say that an integer n is **odd** if there is an integer k with $n = 2k + 1$.

Theorem 10.3 *If an integer n is even, then n^2 is even.*

Proof We assume that we are given an integer n that satisfies the premise; that is, we assume that n is even. Then, by definition, $n = 2k$ for some integer k. Observe that

$$n^2 = (2k)^2 = 2\left(2k^2\right)$$

Thus n^2 is even by definition. ■

Having proved a conditional statement "If P, then Q," it is natural to consider the *converse* statement. The **converse** to a conditional statement "If P, then Q" is the statement "If Q, then P". For example, we have:

Statement: "If n is even, then n^2 is even."
Converse: "If n^2 is even, then n is even."

Notice that our proof of Theorem 10.3 does not shed light on whether the converse statement is true or false. In general, a statement and its converse can have different truth values. For instance, in ordinary language, consider the statements

Statement: "If it is snowing, then it is cold."
Converse: "If it is cold, then it is snowing."

While "If it is snowing, then it is cold" is arguably true, its converse is certainly false.

Given a conditional statement "If P, then Q," the **contrapositive** is the statement "If not Q, then not P". For example,

Statement: "If it is snowing, then it is cold."
Contrapositive: "If it is not cold, then it is not snowing."

Notice that both statements are true. However, if we start with a false conditional statement such as, "If it is raining, then it is snowing," the contrapositive is also false.

Statement: "If it is raining, then it is snowing."
Contrapositive: "If it is not snowing, then it is not raining."

This agreement is simply a matter of logic and does not depend on the phrases involved. We say that a conditional statement "If P, then Q" and its contrapositive "If Q, then P" are **logically equivalent**, meaning that they have the same truth values regardless of the phrases involved.

The contrapositive offers an alternate approach to proving a conditional statement. We use this approach to prove the converse to Theorem 10.3.

Theorem 10.4 *If the square n^2 of an integer n is even, then n is even.*

Proof We prove the following:
Contrapositive: "If an integer n is not even, then n^2 is not even."
Assume that we are given an integer n that is not even. Then n is odd. By definition, there is some integer k such that $n = 2k + 1$. Now we observe that

$$n^2 = (2k+1)^2 = 4k^2 + 4k + 1 = 2(2k^2 + 2) + 1$$

is odd by definition. ■

Theorems 10.3 and 10.4 combine to establish the truth of an important type of mathematical statement. A **biconditional** is a mathematical statement of the form

"P if and only if Q"

The biconditional "P if and only if Q" is true if P and Q have exactly the same truth value or, alternately, if both "If P, then Q" and "If Q, then P" are true statements. We have established the following:

Corollary 10.5 *An integer n is even if and only if n^2 is even.* ■

A biconditional establishes a direct correspondence or **equivalence** between two distinct concepts. For example, the converse for Statement 4 earlier is also true. The corresponding biconditional statement was proved in Chapter 9:

Theorem 9.8 *An outcome of a zero-sum game is a Nash equilibrium if and only if it is a saddle point.* ■

It is customary to use the symbol $\Longleftrightarrow$ to indicate an equivalence. For zero-sum games, we may write

$$\text{Nash equilibrium} \Longleftrightarrow \text{saddle point}$$

Let us now return to Statement 5:

Statement 5: "If a simultaneous-move game has a dominated strategy, then it has a Nash equilibrium."
Converse to Statement 5: "If a simultaneous-move game has a Nash equilibrium, then it has a dominated strategy."

We have observed that the game Chicken is a counterexample to the Converse to Statement 5. Generally, a counterexample to the conditional "If P, then Q" is an example with P true and Q false. We express the one-way conditional writing

$$\text{Dominated-strategy solution} \Longrightarrow \text{Nash equilibrium}$$

The symbol $\Longrightarrow$ stands in for the word "implies." We may say that a dominated-strategy solution is a "stronger" condition than a Nash equilibrium because a dominated-strategy solution implies a Nash equilibrium. In this sense, conditional statements allow for a comparison of mathematical concepts.

Universal and Existential Statements

The Holy Grail of a mathematical theory is the general fact. As case in point, we have the Gibbard-Satterthwaite Theorem:

Theorem 6.4 Gibbard-Satterthwaite *All nondictatorial voting methods with three or more candidates are manipulable.* ■

The result is significant because it reveals a common feature satisfied by all relevant examples. The Gibbard-Satterthwaite Theorem is an example of a *universal statement* about elections. We focus on these statements now.

Definition 10.6 A **universal statement** is a mathematical statement of the form "All objects x of a particular type have a specific property."

Here are two examples in ordinary language:

"All U.S. presidents were 50 years or older at their death."
"All stars are further than 1 million miles from Earth."

For a more mathematical example, we formulate the following version of Zermelo's Theorem:

Statement 6: "In every takeaway game, one player has a strategy that guarantees victory."

A direct proof of a universal statement takes the following form: an object x of the specified type is named. Nothing else about x is assumed. The object x is then shown to have the desired property. Here is an example with integers:

Theorem 10.7 *For all integers n, the integer $n^2 + n$ is even.*

Proof Let n be any integer. We assume nothing about n, and so n is either even or odd. If n is odd, then $n + 1$ is even, of course. Observe that

$$n^2 + n = n(n + 1)$$

Either n or $n+1$ is even (we do not know which). Multiplying an even integer by any integer gives an even. Thus n^2+n is even. ■

A counterexample to a universal statement is called an *existential statement.* Consider the following (false) universal statement:

"For all positive integers n, the integer $2n^2+5$ is prime."

The integer $n=5$ is a counterexample to this statement. There are many others, in fact. The integer $n=5$ proves the truth of the following existential statement:

"There exists a positive integer n such that the integer $2n^2+5$ is not prime."

Definition 10.8 An **existential statement** is a statement of the form "There exists an object x of a particular type such that x has a specific property."

The proof of an existential statement is the construction of one object x of the given type having the desired property. While universal statements assert a general fact, existential statements assert an isolated instance. They are opposites, in this sense. The precise relationship is provided by **negation**, which means adding the word "not" to a mathematical statement. The negation of a universal statement is an existential statement, and vice versa. For instance,

Not "All cars have a radio" = "There is some car without a radio"

Not "There is a green house" = "All houses are not green"

Here are some mathematical examples:

Not "All triangles have at least two unequal sides"

= "There is a triangle with three equal sides"

Not "All simultaneous-move games have a Nash equilibrium"

= "There is a simultaneous-move game without any Nash equilibria"

Universal and existential statements present a compelling version of the problem of proofs and counterexamples. Given a universal statement, we are faced with two possibilities: the statement is true, and there is a proof that holds for all objects described. The other possibility is that the statement is false, and there is a counterexample. We will explore the dichotomy for universal statements when we study questions of fairness for voting methods in Chapter 12.

Mathematical Induction

Experimentation with number patterns can quickly yield some interesting conjectures in the form of universal statements about integers $n=1,2,3,\ldots$. Here is an example: add the first n consecutive odd numbers and consider the sum.

Sums of Consecutive Odd Integers Are Perfect Squares

$$\begin{gathered} 1 = 1 \\ 1+3 = 4 \\ 1+3+5 = 9 \\ 1+3+5+7 = 16 \\ 1+3+5+7+9 = 25 \\ \vdots \end{gathered}$$

We may now believe that

$$1+3+5+7+\cdots+1{,}995+1{,}997+1{,}999 = 1{,}000^2$$

and, similarly, that

$$1+3+5+7+\cdots+9{,}995+9{,}997+9{,}999 = 5{,}000^2$$

although these are more tedious to check. We have arrived at a conjecture:

"For all positive integers n, the sum of the first n odd numbers equals n^2"

The *Principle of Induction* provides a method for proving this type of statement. We explain the method using this example: consider the two sides of the conjectured equality for a fixed integer n.

Left-hand side for $n = 1+3+5+\cdots+(2n-1)$ and Right-hand side for $n = n^2$

Let us add the next odd number $2n+1$ to both terms:

$$\begin{aligned} (\text{Left-hand side for } n)+(2n+1) &= 1+3+5+\cdots+(2n-1)+(2n+1) \\ &= \text{Left-hand side for } n+1 \\ (\text{Right-hand side for } n)+(2n+1) &= n^2+(2n+1) \\ &= (n+1)^2 \\ &= \text{Right-hand side for } n+1 \end{aligned}$$

What have we shown? Suppose that we know that for some particular value of n,

$$\text{Left-hand side for } n = \text{Right-hand side for } n$$

Using the preceding, we could then conclude that

$$\text{Left-hand side for } n+1 = \text{Right-hand side for } n+1$$

Let us write $S(n)$ for the expression

$$S(n): \text{“}1+3+5+\cdots+(2n-3)+(2n-1)=n^2.\text{”}$$

Although we have not directly proved $S(n)$ for all n, we have proved that the following conditional statement is true:

“If $S(n)$ is true, then $S(n+1)$ is true”

The truth of this conditional sets up a chain of deductions. We know that $S(1)$ is true. Since $S(1)$ is true, $S(2)$ is true. Since $S(2)$ is true, $S(3)$ is true, and so on. Given any positive integer n, following this chain, we will eventually reach the conclusion that $S(n)$ is true. In this way, we convert the proof of a conditional statement into a proof of a universal statement.

We summarize this important proof method as follows:

(10.1) **Principle of Induction**

Let $S(n)$ be any mathematical statement about the integer n. To prove the universal statement

For all $n = 1, 2, 3, \ldots$, the statement $S(n)$ is true

it is sufficient to follow these steps:

1. **Base Case:** Prove that the statement $S(1)$ is true.
2. **Induction Hypothesis:** Assume that the statement $S(n)$ is true for some positive integer n.
3. **Induction Step:** Prove that $S(n+1)$ is true as a consequence.

The Principle of Induction may be used to prove many beautiful formulas with numbers. Our interest in induction is in the realm of games and their game trees. We leave some further exploration of numeric induction to the exercises and turn to an adaptation of the method for trees.

Induction with Trees: Takeaway Games

Recall that a takeaway game is finite two-player sequential-move game of pure strategy in which the last player to move wins the game. Examples include Taking Chips (Example 4.14) and Nim (Definition 4.18). Statement 3 earlier is the following version of Zermelo's Theorem:

Theorem 10.9 *For every takeaway game, one player has a strategy that is guaranteed to win.*

We prove Theorem 10.9 now. Our proof is based on the Binary Labeling Rule introduced in Chapter 4 and will serve to introduce the *Principle of Induction for Finite Rooted Trees.* We begin by observing an "equivalence" between the class of takeaway games and a class of games we call *Tree Games*.

First, recall that a takeaway game has a game tree that is a finite, rooted tree indicating all the stages (nodes) and moves (branches) of the game. The last move of the game is a move to an end node of the tree. The player making this move is the winner.

Now suppose, conversely, that we are given a rooted tree. For example,

A Rooted Tree

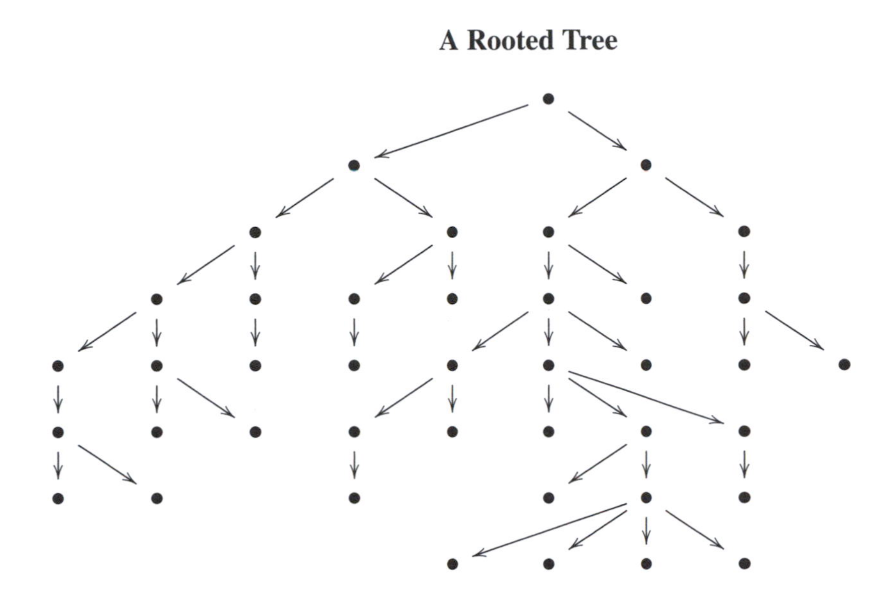

We define the Tree Game as follows:

Definition 10.10 Given a finite, rooted tree, the corresponding **Tree Game** is the takeaway game played as follows: Player 1 starts at the root node. On his or her turn, each player must move to a node in the tree directly adjacent to the current node. If a player has no move (i.e., the current node is an end node), that player loses.

If we write down the game tree for a takeaway game such as Taking Chips or Nim, we see that the players are simply playing the corresponding Tree Game. This correspondence between takeaway games and Tree Games is one key to our proof of Theorem 10.9.

Next, we recall the Binary Labeling Rule from Definition 4.16. Given a rooted tree, start by labeling the end nodes 0. Then, working backward up the tree toward the root node, nodes that branch to at least one node labeled 0 are labeled 1, whereas nodes only branching to a 1 are labeled 0. The end result for A Rooted Tree is as follows:

A Rooted Tree with Binary Labels

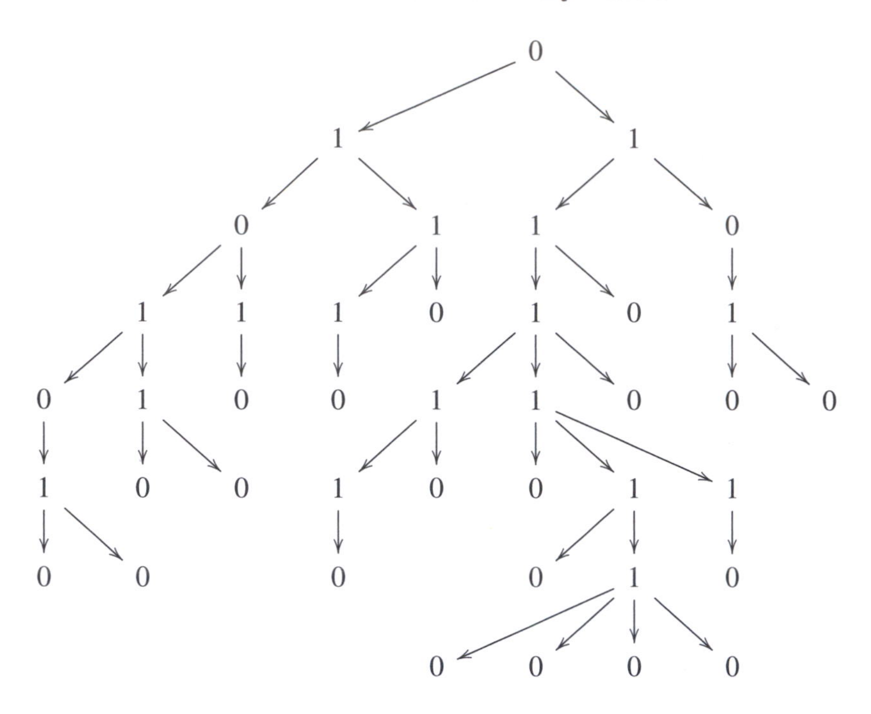

The Binary Labels reveal the solution of the game. Since the root node is labeled 0, Player 2 wins the Tree Game in rational play. Recall that the strategy to win is "Move to a 0 Node." Player 1 has no choice but to place Player 2 at a node labeled 1 for his or her first turn. Player 2 then moves to a node labeled 0 on his or her turn. Player 1 thus always starts his or her turn at a node labeled 0, and the pattern continues.

We seek to prove that we can apply this same logic to any Tree Game and so, by our equivalence, to any takeaway game. We introduce the notion of the *depth* of a rooted tree to create a method of induction. First, define a **path** in a tree to be a sequence of edges starting at one node and ending at some later node in the tree.

Definition 10.11 The **depth** of a finite, rooted tree is the number of edges in the longest possible path from root node to an end node.

Here are some examples:

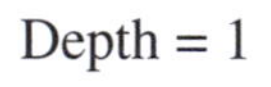

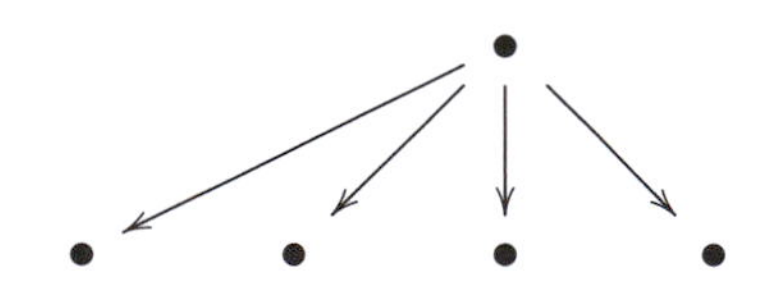

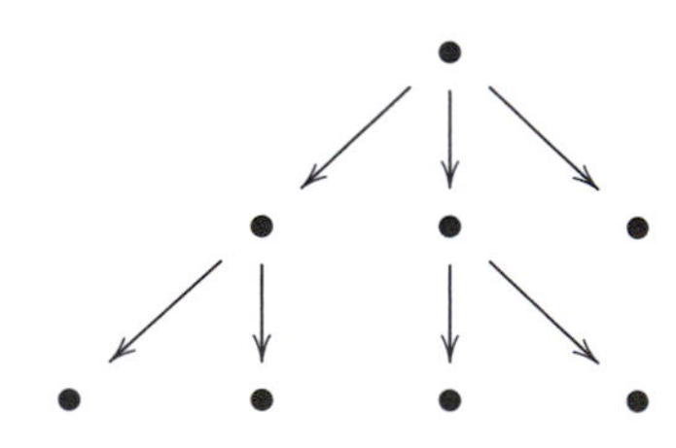

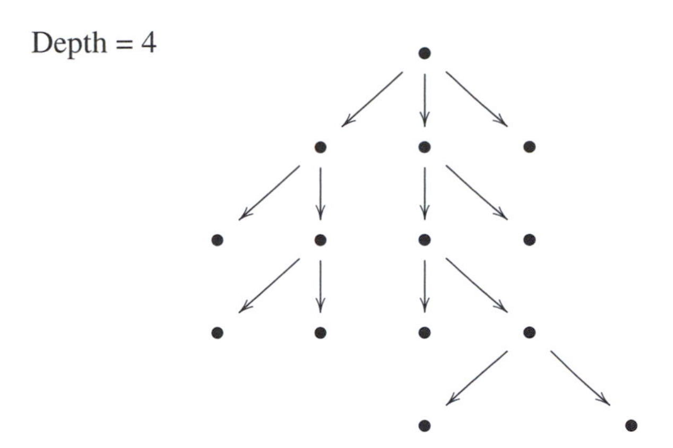

Of course, there are many possible rooted trees of each given depth. Nevertheless, the notion serves to classify all rooted trees according to whether they are of depth $= 1, 2, 3, \ldots$. This classification leads, in turn, to the following adaptation of the Principle of Induction.

(10.2) **Principle of Induction for Rooted Trees**

To prove the universal statement

"For all finite rooted trees, the statement S is true"

it suffices to prove the following:

1. **Base Case:** Prove the statement S is true for all trees of depth = 1.
2. **Induction Hypothesis:** Assume that there is an integer n such the statement S is true for all trees of depth $\leq n$.
3. **Induction Step:** Prove that the statement S is true for all trees of depth $= n+1$.

We remark that the Induction Hypothesis here is a bit different from before. Here we assume the statement S is true for all trees of depth $\leq n$ as opposed to for just depth $= n$. This is referred to as a **strong induction hypothesis**.

Applying the Principle of Induction for Rooted Trees requires a strategy for producing a *subtree* from a given tree. Generally, a **subtree** of a given tree $\mathcal{T}$ is a tree $\mathcal{T}'$ built from $\mathcal{T}$ using some of the edges and nodes but perhaps omitting some. The following definition provides one direct method for producing subtrees:

inition 10.12 Given any rooted tree and a node A, the A-**rooted subtree** is the subtree of the original obtained by taking A to be the root node and including all edges and nodes reachable from A.

When the tree is a game tree, the A-rooted subtree corresponds to the subgame starting at outcome A. Here is an illustration:

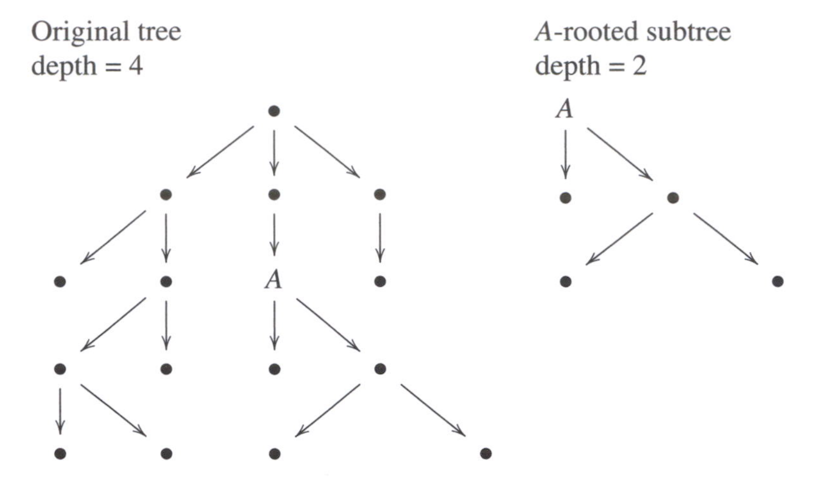

The A-rooted subtree always has strictly smaller depth than the original tree (unless A is the root, in which case they are the same tree). Recall from Chapter 4 that the Binary Labels for subtrees, treated as rooted trees in their own right, are the same labels as for the original tree. We are now prepared to give our proof of Theorem 10.9.

Proof of Theorem 10.9 We may assume that the takeaway game is the Tree Game corresponding to a rooted tree $\mathcal{T}$. We prove this result by induction on the depth of $\mathcal{T}$.

Base Case: Observe that when $\mathcal{T}$ has depth = 1, the root node is labeled 1 automatically. Indeed, Player 1 wins the resulting Tree Game with any move because all nodes except the root are end nodes.

Induction Hypothesis: We assume that the following statement of the theorem is true for every rooted tree $\mathcal{T}'$ of depth $\leq n$. If the root node of $\mathcal{T}'$ is labeled 1, then Player 1 has a guaranteed winning strategy obtained by moving to a 0 on each turn. If the root node of $\mathcal{T}'$ is labeled 0, then Player 2 has this guaranteed winning strategy beginning on his or her turn.

Induction Step: Now suppose that we are given a rooted tree $\mathcal{T}$ with depth $= n+1$. Apply the Binary Labeling Rule. We have two cases: if the root node for $\mathcal{T}$ is labeled with a 1, then, by the definition of the labeling, there is a node, say, A, adjacent to the root node of $\mathcal{T}$ such that A is labeled with a 0. Now consider the A-rooted subtree of $\mathcal{T}$. Call this subtree $\mathcal{T}'$. The root node A of $\mathcal{T}'$ is labeled 0. Further, $\mathcal{T}'$ has depth $\leq n$. Our induction hypothesis tells us that the player whose turn starts at node A loses the Tree Game for $\mathcal{T}'$. Thus Player 1 should move to A, and then he or she is guaranteed victory.

If the root node of $\mathcal{T}$ is labeled 0, then every node adjacent to the root is labeled with a 1. This means that whatever move Player 1 makes, he or she places Player 2 in the position of going first in a Tree Game with tree $\mathcal{T}'$ of depth $\leq n$ such that the root node of $\mathcal{T}'$ is labeled 1. Player 2 can guarantee victory in each such game by our induction hypothesis. Thus Player 2 can guarantee victory regardless of Player 1's first move, completing the proof. ■

The Principle of Induction for Trees allows us to extend our idea for solving takeaway games with small trees to a proof that applies to any takeaway game. We conclude by mentioning that the Principle of Induction often allows for a similar extension, converting a fact we have obtained for $n=2$ objects to a general rule. Here is an example: suppose that we have to make two choices, say, first picking a shirt to wear and then selecting a tie. The number of different outfits we have is then the number of shirts we own multiplied by the number of ties. This simple fact directly extends to the following theorem:

Theorem 10.13 Principle of Counting *Suppose that a sequence of n choices is to be completed, in which there are C_1 possibilities for the first choice, C_2 possibilities for the second choice, C_3 for the third, and so on, with C_n possibilities for the nth choice. Then there are*

$$C(n) = C_1 \cdot C_2 \cdots C_{n-1} \cdot C_n$$

ways to complete the sequence of choices.

Proof Base Case: Making one choice with C_1 possibilities gives $C(1) = C_1$ ways to make the choice.

Induction Hypothesis: Fix a positive integer n, and assume that in any sequence of n choices for which there are C_1 possibilities for the first choice, C_2 for the second, and so on, the total number of ways to make the sequence of n choices is

$$C(n) = C_1 \cdot C_2 \cdots C_{n-1} \cdot C_n$$

Induction Step: Suppose that we now confront a sequence of $n+1$ choices. We will think of it as two choices: the first choice consists of making the initial n choices – we know that there are $C(n) = C_1 \cdot C_2 \cdots C_{n-1} \cdot C_n$ ways to do this. The second choice is the $n+1$st. There are C_{n+1} possibilities for this choice. Then, by the case of two choices,

$$C(n+1) = C(n) \cdot C_{n+1} = C_1 \cdot C_2 \cdots C_{n-1} \cdot C_n \cdot C_{n+1}$$

as needed. ■

Exercises

10.1 For the following conditional statements, state the converse and the contrapositive:

(a) "If it rains, then we will stay home."
(b) "If it rains or it is cold, then we will stay home."

(c) "If I go to the dance, then Jane and Sarah will go. "
(d) "If I don't go to the dance, then neither Jane nor Sarah will go."

10.2 For the following conditional statements, decide whether the statement, the converse, both, or neither are true.

(a) "If an integer n is even, then $n+3$ is odd."
(b) "If an integer $n < 3$, then $n^2 < 9$.
(c) "If an integer n is negative, then $n < n^2$. "
(d) "If one player in a two-player simultaneous-move strategic game has a dominated strategy, then the game has a dominated-strategy solution."

10.3 Find the negation of the following universal and existential statements.

(a) "All men are wise."
(b) "Some children have no toys."
(c) "There is some integer larger than all other integers."
(d) "All integers are divisible by 2."

10.4 Find the negation of the following statements. Decide whether the original statement or the negation is true.

(a) "There is an integer n such that $3n^2 + 1$ is even."
(b) "For all integers $n > 1$, the integer $n(n-1)$ is even."
(c) "There is an integer $n > 1$ such that n and $n + 2$ are both prime numbers."
(d) "There are no integers $n > 1$ such that n and $3n + 1$ are both prime numbers."

10.5 Prove the following conditional and biconditional statements about integers n:

(a) "If n is odd, then $n^2 + 3n$ is even."
(b) "If n is even, then $3n - 2$ is odd."
(c) "n is even if and only if $7n$ is even."
(d) "n is even if and only if n^3 is even."

10.6 (The Liar Paradox) Consider the statement

"This sentence is false"

Explain why this statement cannot be interpreted as either true or false, even though it appears to be formulated as a mathematical statement. What is the feature of this sentence that creates the paradox?

10.7 (Binary Trees) Flip a coin multiple times, and keep track of the results using a tree. The object we obtain is called a **binary tree**, a rooted tree such that each node that is not an end node branches to exactly two nodes (e.g., heads or tails). Here are some examples:

	Binary tree	Total number of nodes
B_1:		1
B_2:	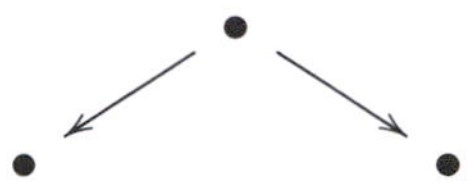	3
B_3:	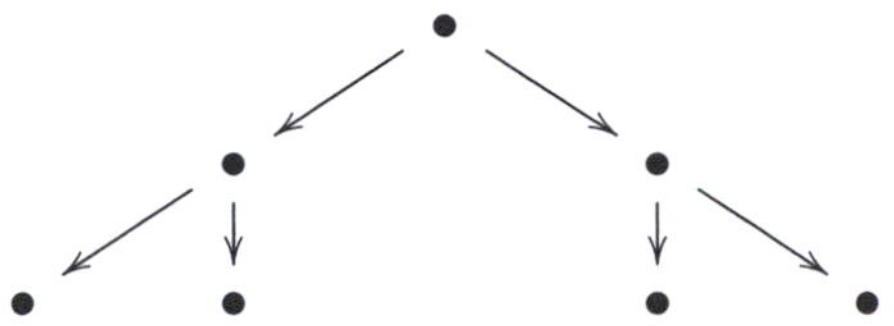	7
B_4:	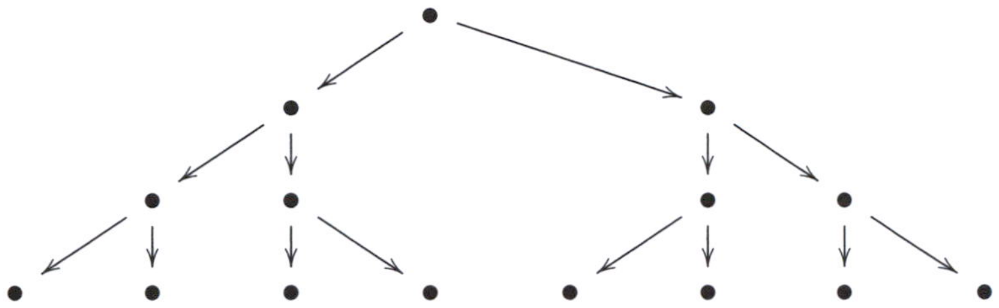	15

Use the Principle of Induction for Rooted Trees (10.2) to prove that the number of nodes in B_n is given by

$$\text{Number of nodes in } B_n = 2^n - 1$$

Hint: Observe that the binary tree B_{n+1} branches to two binary trees B_n. By the Induction Hypothesis, the picture looks like this:

The Binary Tree B_{n+1}

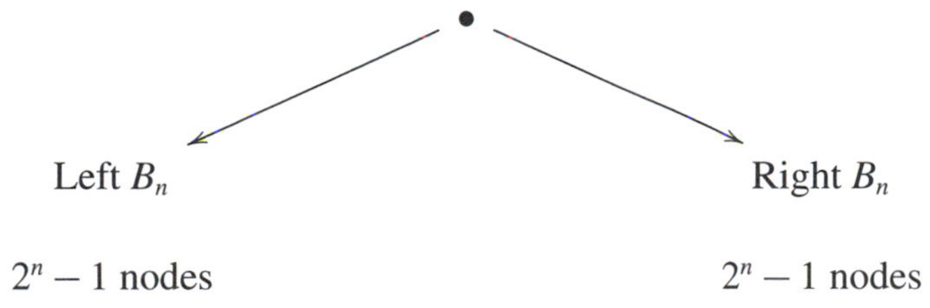

Left B_n — $2^n - 1$ nodes

Right B_n — $2^n - 1$ nodes

Now count the nodes in B_{n+1}.

10.8 A **ternary tree** is a rooted tree for which every node that is not an end node branches to exactly three nodes. Let T_n denote the ternary tree with $n-1$ stages. Use the Principle of Induction for Trees (10.2) to prove that

$$\text{Number of nodes in } T_n = \frac{3^n - 1}{2}$$

for $n = 1, 2, 3, \ldots$.

10.9 Follow these steps to prove the universal statement

"For all positive integers n, there are 2^n subsets of the set $\{1, 2, \ldots, n\}$"

First, we note that the empty set ∅ is considered a subset of any set. Here is a table of subsets of $\{1, 2, \ldots, n\}$ for small values of n:

n	**Subsets of $\{1,2,\ldots,n\}$**	**Number**
1	$\emptyset, \{1\}$	2
2	$\emptyset, \{1\}, \{2\}, \{1,2\}$	4
3	$\emptyset, \{1\}, \{2\}, \{3\}, \{1,2\}, \{1,3\}, \{2,3\}, \{1,2,3\}$	8

Fill in the table in the case $n = 4$. Can you see why there are twice as many subsets for $n = 4$ as for $n = 3$? Prove that

$$\text{Number of subsets of } \{1, 2, \ldots, n, n+1\} = 2 \cdot \text{number of subsets of } \{1, 2, \ldots, n\}$$

Use this fact to give a proof by induction.

10.10 Follow these steps to prove the following sum formula for all positive integers n:

$$1 + 2 + 3 + \cdots + (n-1) + n = \frac{n(n-1)}{2}$$

Explain why the statement is true for the Base Case, when $n = 1$. Assume that the (Induction Hypothesis) for n a fixed positive integer

$$1 + 2 + 3 + \cdots + (n-1) + n = \frac{n(n-1)}{2}$$

is true.

For the Induction Step, add the term $n+1$ to both sides of this equation. Regarding the new Right-hand Side, prove

$$\frac{n(n-1)}{2} + (n+1) = \frac{n(n+1)}{2}.$$

10.11 Use the Principle of Induction to prove the following number formulas:

(1) For all $n \geq 1$,

$$1 + 2^1 + 2^2 + \cdots + 2^n = 2^{n+1} - 1$$

This result is the same obtained in Exercise 10.7.

(2) Let $x > 1$ be a real number. For all $n \geq 1$,

$$1 + x + x^2 + \cdots + x^{n-1} + x^n = \frac{x^{n+1} - 1}{x - 1}.$$

Note that the case $x = 2$ corresponds to Part (1) while $x = 3$ corresponds to Exercise 10.8.

(3) For all $n \geq 1$,

$$1^2 + 2^2 + \cdots (n-1)^2 + n^2 = \frac{n(n+1)(2n+1)}{6}$$

Compare Exercise 10.10.

11 Laws of Probability

From a philosophical perspective, the idea of randomness is a somewhat mysterious concept. Mathematically, the study of random events leads to a calculable and widely applicable mathematical theory. Although we can never fully predict the outcome of a random process, we can nonetheless compute probabilities with great precision.

The foundations of probability theory closely parallel those of basic logic. An *event* plays the role of the mathematical statement. The first laws of probability theory are formulas for the probabilities that arise from applying connectives (and, or) and negation (not) to events. A conditional statement in mathematics translates, in turn, to the central notion of a *conditional probability*. The *Law of Conditional Probability* (Theorem 11.17) underlies a method for calculating the probabilities arising from sequential actions in games. We used this result extensively in Part I of this book in the form of the Law of the Probability Tree (Theorem 3.6).

We explore these foundational aspects of probability theory in this chapter. We begin with the idea of a *sample space*. This notion allows for the translation of the probability to arithmetic. We use the example of rolling a pair of dice to motivate the basic laws of connectives and negations. Drawing cards motivates a useful counting technique we call the *Combinations Formula* (Theorem 11.12). We next formulate the definition of a conditional probability and give the proof of Theorem 11.17. We conclude with a discussion of the proofs of the Law of the Probability Tree (Theorem 3.6) and the Linearity of Expectation (Theorem 3.13).

Sample Spaces

What does it mean to assign a probability to a random event? If the event involves human agency, perhaps the only reasonable approach is to gather empirical evidence. How else can we assess the likelihood that, say, a basketball player makes a free throw other than by consulting the player's past record? For such events, probabilities are not so much calculated as observed. A probability based on experimental evidence is called an **empirical probability**.

However, we may consider only physical experiments, wherein probabilities arise from features of the objects, such as the two sides of a coin or the six faces of a die. This point of view will allow us to proceed on firm mathematical footing but ultimately will prove too restrictive for many applications. For the purposes of this chapter, we will nonetheless restrict ourselves to this concrete setting. We make the following definition, which should be reminiscent of the definition of a game of pure chance.

Definition 11.1 A **probability experiment** is an action or sequence of actions whose execution results in the occurrence of exactly one outcome from a finite list of possible **simple outcomes**. We assume that the simple outcomes for a probability experiment are all equally likely to occur.

Examples of probability experiments include the basic activities in a game of chance: flipping a coin, rolling dice, or drawing a card. Notice, however, that the meaning of the term "simple outcome" for a probability experiment is much more restricted than the notion of an outcome of a game. We emphasize the requirement that the simple outcomes be *equally likely*. This assumption allows us to translate probability directly to arithmetic. The outcome of a game of chance generally will be a collection of simple outcomes of a given experiment. For example, when rolling a pair of dice, the game outcome "Roll=Doubles" corresponds to the subset {⚀⚀, ⚁⚁, ⚂⚂, ⚃⚃, ⚄⚄, ⚅⚅} of all 36 possible rolls. A collection of simple outcomes in a probability experiment is called an *event*.

Definition 11.2 Suppose that we are given a probability experiment. The **sample space** S is then defined to be the set of all simple outcomes. An **event** A is a subset of the sample space S of all simple outcomes. That is, A is a distinguished set of simple outcomes. We will often say that a simple outcome belonging to A is an outcome **favorable** to A.

We can picture the situation with sets:

Sample Space S and an Event A

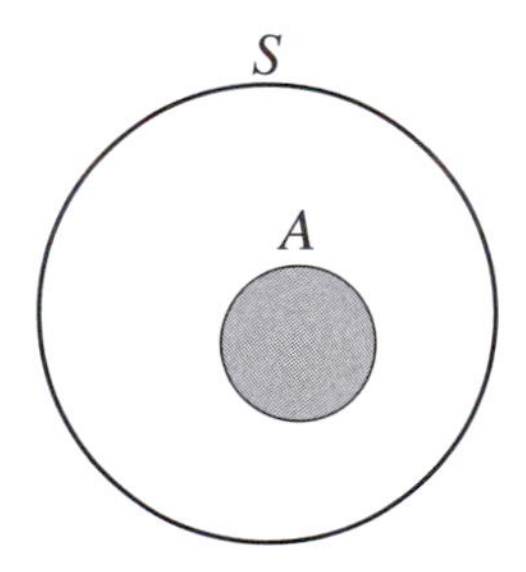

When S is a finite set, the **probability** $P(A)$ of A is defined to be the fraction of the total number of simple outcomes in S that are in A. We introduce the following notation for this count:

Notation 11.3 Given a finite set A, we write

$$n(A) = \text{the number of elements in } A$$

We then have the following definition:

efinition 11.4 Given a probability experiment with finite sample space S and an event A, we define the **probability of A** by the formula

$$P(A) = \frac{n(A)}{n(S)}$$

In practice, we will use the letter A to denote both the description of the event, for example, "Roll = Doubles," and the actual subset, here {⚀⚀, ⚁⚁, ⚂⚂, ⚃⚃, ⚄⚄, ⚅⚅}, of the sample space. The experiment of rolling a pair of dice offers a helpful illustration of these ideas. Here is a first example:

Example 11.5

Consider the probability experiment of rolling a pair of dice. The sample space S will be the set of 36 ordered pairs (Red die, Blue die) where both Red die and Blue die can each be ⚀ through ⚅. Consider the event

$$A = \text{"Sum of Dice} > 8\text{"}$$

We can picture the event A as a subset of the sample space consisting of the set of all 36 rolls:

The Event "Sum of Dice > 8"

Blue die

Red die	⚀	⚁	⚂	⚃	⚄	⚅
⚀						
⚁						
⚂						X
⚃					X	X
⚄				X	X	X
⚅			X	X	X	X

The probability of A is computed as

$$P(A) = \frac{n(A)}{n(S)} = \frac{10}{36}$$

Laws of Conjunctions and Negation

The conjunctions "and" and "or" and the negation "not" may be applied to events in a samples space allowing for the creation of new events from old. The definitions are as follows:

A **or** B = all simple outcomes favorable to either A, B, or both A and B
A **and** B = all simple outcomes favorable to both A and B simultaneously
not A = all simple outcomes unfavorable to A

We remark that each of these definitions may be viewed using Venn diagrams (see Exercise 11.2). We give formulas for the probabilities of these events in terms of the probabilities of A and B. A second example with dice will serve to illustrate the ideas.

Example 11.6

For the experiment of rolling a pair of dice, let

$$A = \text{“Roll is Doubles”} \quad \text{and} \quad B = \text{“Sum is 8”}$$

The sample space, with outcomes favorable to A marked with X's and to B marked with O's, is as follows:

"Roll is Doubles" (O) and "Sum is 8" (X)

Blue die

Red die

	⚀	⚁	⚂	⚃	⚄	⚅
⚀	O					
⚁		O				X
⚂			O		X	
⚃				⊗		
⚄			X		O	
⚅		X				O

Visibly, $n(A \text{ or } B) = 10$, whereas $n(A \text{ and } B) = 1$. Thus we have

$$P(A \text{ or } B) = \frac{10}{36} \quad \text{and} \quad P(A \text{ and } B) = \frac{1}{36}$$

Also, we can see that there are 30 outcomes not favorable to A. Thus

$$P(\text{not } A) = \frac{30}{36}$$

Notice that we do not count $n(A \text{ or } B)$ simply by adding X's and O's. This over counts by 1. The correct formula is

$$n(A \text{ or } B) = n(\text{X's}) + n(\text{O's}) - n(\otimes\text{'s}) = 6 + 5 - 1 = 10$$

That is, we first count the simple outcomes in A, then those in B, and then subtract 1 for the double-counted roll ⚃⚃. Generally, we have the counting identity

$$n(A \text{ or } B) = n(A) + n(B) - n(A \text{ and } B)$$

Similarly, we can count $n(\text{not } A)$ simply by subtracting the 6 simple outcomes favorable to A from the 36 total simple outcomes:

$$n(\text{not } A) = n(S) - n(A)$$

These simple observations lead directly to the first laws of probability:

Theorem 11.7 *Let A and B be events in a probability experiment. Then*

1. *Law of Conjunctions*: $P(A \text{ or } B) = P(A) + P(A) - P(A \text{ and } B)$
2. *Law of Negation*: $P(\text{not } A) = 1 - P(A)$

Proof The proof is just some arithmetic and the counting formulas we observed earlier. Let S be the sample space. For (1), we use the formula $n(A \text{ or } B) = n(A) + n(B) - n(A \text{ and } B)$. We have

$$\begin{aligned} P(A \text{ or } B) &= \frac{n(A \text{ or } B)}{n(S)} = \frac{n(A) + n(B) - n(A \text{ and } B)}{n(S)} \\ &= P(A) + P(B) - P(A \text{ and } B) \end{aligned}$$

Similarly, for (2), we have

$$P(\text{not } A) = \frac{n(\text{not } A)}{n(S)} = \frac{n(S) - n(A)}{n(S)} = 1 - P(A) \qquad \blacksquare$$

Theorem 11.7 (1) is especially nice if it so happens that $P(A \text{ and } B) = 0$. Note that $P(A \text{ and } B) = 0$ means that A and B cannot cannot happen simultaneously.

efinition 11.8 Two events A and B in a probability experiment are said to be **mutually exclusive** if $P(A \text{ and } B) = 0$.

For example, "Roll is Doubles" and "Sum is 9" are mutually exclusive events when rolling a pair of dice.

Corollary 11.9 The Law of Mutually Exclusive Events *Let A and B be mutually exclusive events. Then*

$$P(A \text{ or } B) = P(A) + P(B)$$

Proof This follows directly from Theorem 11.7 and Definition 11.8. $\blacksquare$

Counting Cards

The experiment of rolling dice provides a visual example of the basic concepts of probability. The experiment of drawing cards, in turn, highlights the central role of counting in probability theory. We introduce a useful formula for counting card hands that we will apply in various contexts. We begin with an example.

Example 11.10

Draw 2 cards from a deck of 52 without replacement. We consider the event

$$A = \text{“Pair of Jacks”}$$

Writing down the sample space for this experiment is impractical, but we can easily count the simple outcomes. However, we are faced with a basic decision: should we think of a two-card hand as an ordered pair, or should we ignore the order of the cards? We examine both possibilities.

Ordered Hands: It is easy enough and sometimes useful to keep track of the order in which we received the cards. In this case, it is convenient to write a simple outcome as an ordered pair (1st card, 2nd card) of two cards. With this view, we will distinguish, for example,

$$\text{Simple outcome 1} = (A\diamondsuit, J\spadesuit) \quad \text{from} \quad \text{Simple outcome 2} = (J\spadesuit, A\diamondsuit)$$

Counting the number of simple outcomes of ordered hands, we see that there are 52 possibilities for the 1st card and, having chosen this card, 51 possibilities remaining for the 2nd card, giving $52 \cdot 51 = 2{,}652$ possible ordered two-card hands. To get a pair of Jacks, first we must choose our 1st card to be a Jack: there are four choices. Then we must choose our 2nd card to be another Jack: there now are three choices left. We conclude that

$$P(\text{“Pair of Jacks”}) = \frac{12}{2{,}652}$$

Unordered Hands: In many games, the order in which the cards are received is not important; all that matters is the cards themselves. It is natural, then, to treat a simple outcome as an unordered hand of two cards. As far as notation goes, we may rely on the idea of a set. Recall that two sets are equal if they have the same elements, regardless of the order or manner in which the elements are described. Thus we write an unordered hand with set notation {1st card, 2nd card} and take the sample space to be the collection of all two-element sets of the 52 cards. For example, we have

$$\text{Simple outcome} = \{J\spadesuit, A\diamondsuit\}$$

which covers both the ordered two-card hands from before.

Let us compute the probability of $A =$ "Pair of Jacks" with this point of view. First, we count the number of unordered two-card hands. We could do so directly, but it is easier to observe that there are half as many unordered two-card hands as there are ordered two-card hands because each unordered set of two cards gives rise to exactly two ordered hands. Thus our sample space of unordered two-card hands consists of

$$n(\text{unordered two-card hands}) = \frac{52 \cdot 51}{2} = 1{,}326$$

different simple outcomes.

As for the event $A =$ "Pair of Jacks," observe that there are six unordered hands favorable to the event "Pair of Jacks." Here they are:

$$\{J\spadesuit, J\heartsuit\}, \{J\spadesuit, J\diamondsuit\}, \{J\spadesuit, J\clubsuit\}, \{J\heartsuit, J\diamondsuit\}, \{J\heartsuit, J\clubsuit\}, \{J\diamondsuit, J\clubsuit\}$$

We conclude that the probability computed with unordered hands as simple outcomes is

$$P(\text{"Pair of Jacks"}) = \frac{6}{1{,}326}$$

It is reassuring to note that both approaches for computing P("Pair of Jacks") yield the same probability. When computing probabilities, we generally are free to define the sample space according to our convenience as long as all the simple outcomes are equally likely. We next consider an example illustrating the advantage of using unordered card hands.

Example 11.11 Straight Poker

The pure-chance version of Poker, called **Straight Poker**, is played using five-card hands. We consider our two versions of the sample space for this experiment.

An ordered five-card hand is a 5-tuple of cards

$$(C_1, C_2, C_3, C_4, C_5)$$

A particular hand then might be

$$(A\spadesuit, 10\diamondsuit, 2\diamondsuit, Q\clubsuit, Q\heartsuit)$$

A different ordered hand can be made from the same cards. For example, the ordered hand $(Q\heartsuit, 10\diamondsuit, 2\diamondsuit, Q\clubsuit, A\spadesuit)$ uses the same cards but represents a different ordered hand.

We can directly compute the number of ordered five-card hands by counting the choices available for each one, in turn, and applying the Principle of Counting (Theorem 10.13). For the first card C_1, there are 52 choices. Having chosen C_1, there are then 51 choices for C_2, then 50 choices for C_3, 49 choices for C_4, and finally, 48 choices for C_5. Multiplying, we see that there are $52 \cdot 51 \cdot 50 \cdot 49 \cdot 48$ total ordered five-card hands. We introduce notation for this count, writing $P(52,5) = 52 \cdot 51 \cdot 50 \cdot 49 \cdot 48$ and, more generally,

$$P(n,k) = n \cdot (n-1) \cdots (n-k+2) \cdot (n-k+1)$$

for $n \geq k \geq 1$. The formula for $P(n,k)$ counts the number of ways to build an ordered list of k objects from a set of n objects. Note that the special case $n = k$ recovers the factorial function: $P(n,n) = n! = n \cdot (n-1) \cdots 3 \cdot 2 \cdot 1$. This is simply the number of ways to order a set of n objects or the number of **permutations** of n objects. Let us turn to the case of unordered five-card hands. Here is an example:

One Unordered Hand

$$\{A\spadesuit, 10\diamondsuit, 2\diamondsuit, Q\clubsuit, Q\heartsuit\} = \{2\diamondsuit, 10\diamondsuit, A\spadesuit, Q\heartsuit, Q\clubsuit\}$$

We introduce the notation

$$\binom{52}{5} = \text{the number of 5-element subsets of a set of 52 objects}$$

Recall that the set notation expresses the fact that we ignore the order of the objects when counting. The count is often referred to as the number of "5-combinations from 52" and written $C(52,5)$. We interpret this notation with the more suggestive name **52 choose 5**, the number of ways to **choose** 5 objects from 52. More generally, for $0 \leq k \leq n$, we have **n choose k**, written

$$\binom{n}{k} = \text{the number of } k\text{-element subsets of a set of } n \text{ objects}$$

We seek a formula for n choose k focusing first on the special case $\binom{52}{5}$. Here is the idea: we have already counted the ordered five-card hands with the formula $P(52,5) = 52 \cdot 51 \cdot 50 \cdot 49 \cdot 48$. We know that this significantly overcounts the unordered hands. The question is, by how much? An unordered hand, say, $\{A\spadesuit, 10\diamondsuit, 2\diamondsuit, Q\clubsuit, Q\heartsuit\}$, gives rise to $P(5,5) = 5!$ distinct orderings. Each ordering of these five cards gives a distinct ordered hand. Since each of the $\binom{52}{5}$ unordered hands gives rise to exactly $P(5,5)$ ordered hands, we have the equation

$$P(52,5) = P(5,5) \cdot \binom{52}{5}$$

Solving for $\binom{52}{5}$ now gives

$$\binom{52}{5} = \frac{P(52,5)}{P(5,5)} = \frac{52 \cdot 51 \cdot 50 \cdot 49 \cdot 48}{5 \cdot 4 \cdot 3 \cdot 2 \cdot 1} = 2{,}598{,}960$$

In general, we have the following result:

Theorem 11.12 Law of Combinations *Let $1 \leq k \leq n$. The number of k-element subsets of an n-element set is given by*

$$\binom{n}{k} = \frac{P(n,k)}{P(k,k)} = \frac{n \cdot (n-1) \cdots (n-k+2) \cdot (n-k+1)}{k \cdot (k-1) \cdots 2 \cdot 1}$$

Proof The proof follows the same argument as earlier. We observe that each of the $\binom{n}{k}$ unordered subsets gives rise to $P(k,k)$ ordered subsets, so we have the identity

$$P(n,k) = P(k,k) \cdot \binom{n}{k}$$

The result now follows from solving for $\binom{n}{k}$. ■

It is convenient to further define

$$\binom{n}{0} = 1$$

The numbers $\binom{n}{k}$ are often called **binomial coefficients** because of their role as the coefficients in the expansion of binomials of the form $(x+a)^n$. We continue to use the suggestive word "choose" when appropriate. Here is an example of the Combinations Formula in action:

Example 11.13

Twelve players meet in the fieldhouse to have a game of pickup basketball. How many ways can two teams of five be chosen?

First, we name the two teams "First" and "Second." We then *choose* five players for the "First" team. There are

$$\binom{12}{5} = \frac{12 \cdot 11 \cdot 10 \cdot 9 \cdot 8}{5 \cdot 4 \cdot 3 \cdot 2 \cdot 1} = 792 \text{ ways}$$

Next, we choose five players for the second team. There are

$$\binom{7}{5} = \frac{7 \cdot 6 \cdot 5 \cdot 4 \cdot 3}{5 \cdot 4 \cdot 3 \cdot 2 \cdot 1} = 21 \text{ ways}$$

Now we know that there are

$$792 \cdot 21 = 16{,}632 \text{ ways}$$

to pick the "First" and "Second" teams. Finally, note that we introduced an ordering of the teams that was not in the original question. Thus we should divide by two to get

$$8,316 \text{ ways to pick teams}$$

Conditional Probability

Conditional statements provide the framework for comparisons, deductions, and equivalences in mathematical theory. They are equally central figures in probability theory. A doctor makes a diagnosis based on his or her knowledge of preexisting conditions. A card sharp bases his or her bet on the status of the deck. Consideration of the underlying conditions and how they affect the outcome of an experiment is often the most difficult piece of a real-world probability calculation.

We motivate the definition of *conditional probability* with a gambling game.

Example 11.14 Fixing a Die

This is a two-player game of pure chance based on the roll of the dice. Each player puts \$5 in the pot. Player 1 wins the pot if the sum of the dice is greater than eight. Otherwise, Player 2 wins the pot. However, before the dice are rolled, Player 1 is given the option to pay an extra \$1 into the pot and fix the Red die on ⚃. In this case, only the Blue die is rolled, and the rules are the same as before.

Should Player 1 pay the extra \$1? It seems clear that setting the Red die to ⚃ is of some advantage to Player 1. But is it worth a \$1? To decide, we compare expected values.

We first compute the expected value of the game when Player 1 does not elect to pay to fix the die. We have

$$P(\text{“Sum} > 8\text{'}) = \frac{10}{36}$$

so the expected value for the game is

$$E = \$5 \cdot \frac{10}{36} - \$5 \cdot \frac{26}{36} = -\$2.22$$

When Player 1 pays the \$1, the Red die is fixed on ⚃. In this case, the pre-existing condition is the event "Red die = ⚃." We say that this event is a *given* condition.

Consider the sample space:

Two Events: "Sum > 8" (X) and "Red Die = ⚃" (O)

Blue die

Red die

	⚀	⚁	⚂	⚃	⚄	⚅
⚀						
⚁						
⚂						X
⚃	O	O	O	O	⊗	⊗
⚄				X	X	X
⚅			X	X	X	X

The given condition implies that the outcome is one of the boxes with O's. Of these six possibilities, two are favorable to the event "Sum > 8," namely, the simple outcomes marked ⊗. We have computed

$$P(\text{ “Sum} > 8 \text{ given Red die} = ⚃\text{”}) = \frac{n(\otimes\text{'s})}{n(\text{O's})} = \frac{2}{6}$$

The expected value, in this case, is

$$E = \$5 \cdot \frac{2}{6} - \$6 \cdot \frac{4}{6} = -\$2.33$$

Player 1 should not pay to set the Red die to ⚃.

The preceding calculation is an example of the following important definition:

finition 11.15 Let A and B be two events in a probability experiment. The **probability of A given B**, written $P(A \mid B)$, is the probability that A occurs given that B is assumed to have occurred. The probability $P(A \mid B)$ is called a **conditional probability.**

Just as we assume that the premise P of a conditional statement "If P, then Q" is true when giving a proof, we assume that the given condition B has definitely occurred when computing the probability of A given B, $P(A \mid B)$.

Earlier, we computed

$$P(\text{“Sum} > 8\text{”} \mid \text{“Red die} = ⚃\text{”}) = \frac{2}{6}$$

In general, to compute $P(A \mid B)$, we restrict our attention the outcomes favorable to B. In other words, B is taken as the sample space for the experiment because one of these outcomes certainly has occurred. Thus $n(B)$ is the denominator. Next, count the events in B that are favorable to A. In our notation, this is the number $n(A \text{ and } B)$. Dividing, gives the following formula:

Theorem 11.16 Conditional Probability Formula *Given events A and B in a probability experiment,*

$$P(A \mid B) = \frac{n(A \text{ and } B)}{n(B)}$$ ■

As a consequence, we obtain the following theorem:

Theorem 11.17 (Law of Conditional Probability) *Let A and B be events in a probability experiment. Then*

$$P(A \text{ and } B) = P(A \mid B) \cdot P(B)$$

Proof Begin with $P(A \mid B) = n(A \text{ and} B)/n(B)$, and multiply the numerator and denominator by $1/n(S)$. Thus,

$$P(A \mid B) = \frac{n(A \text{ and} B)}{n(B)}$$

$$= \frac{n(A \text{ and} B)}{n(B)} \cdot \frac{\frac{1}{n(s)}}{\frac{1}{n(s)}} = \frac{\frac{n(A \text{ and} B)}{n(S)}}{\frac{n(B)}{n(S)}} = \frac{P(A \text{ and} B)}{P(B)}$$

The result follows by multiplying both sides by $P(B)$. ■

Theorem 11.17 leads to a mathematical definition of *independent events.* Informally, two events are independent if the occurrence of one has no effect on the outcome of the other. Independent events arise naturally when two physically separate experiments are performed. In the context of a single experiment, the notion is more interesting. For the experiment of drawing a single card from a deck, consider the events

$$A = \text{"Jack"} \quad \text{and} \quad B = \text{"Red card"}$$

Notice that the probability of getting a Jack from a full deck is $P(A) = 4/52 = 1/13$. Now imagine that we repeat the experiment but this time ensure that B has occurred. That is, we remove all the Black cards from the deck. The probability of a Jack given a Red card is $P(A \mid B) = 2/26 = 1/13$. The probability has not changed. In our notation, we have $P(A|B) = P(A)$. It is also easy to check that $P(B|A) = P(B)$. The given information that one of the events A or B has occurred has no effect on the likelihood of the other event occurring.

As a consequence of the Law of Conditional Probability, we have the following result:

Theorem 11.18 *Let A and B be events in a probability experiment. Then the following statements are all equivalent:*

(a) $P(A \mid B) = P(A)$

(b) $P(B \mid A) = P(B)$
(c) $P(A \text{ and } B) = P(A) \cdot P(B)$

Proof The proof is left for you in Exercise 11.10. ■

Theorem 11.18 allows us to formulate the following definition:

finition 11.19 Let A and B be two events in a probability experiment. We say that A and B are **independent events** if the equivalent conditions (a), (b), and (c) in Theorem 11.18 hold for A and B.

Theorem 11.18 provides three equivalent conditions any one of which can be used to check for independence. Notice that (a) and (b) each express our intuitive notion of independence wherein one event does not influence outcome of the other. These are often easiest to check. The equivalence of (c) with these two conditions can be singled out as a useful corollary:

Corollary 11.20 The Law of Independent Events *Let A and B be independent events in a probability experiment. Then*

$$P(A \, and B) = P(A) \cdot P(B)$$ ■

The Law of the Probability Tree and the Linearity of Expectation

We conclude this chapter by explaining how Theorems 3.6 and 3.13 stated in Chapter 3 are consequences of our preceding work. Start with a two-stage game of chance with payoffs in which the first stage of the game is an experiment with an event A that governs the first branching. The second stage depends on the outcome (A or not A). In the case A occurs, a second experiment is performed with an event called B whereas if not A occurs, the second experiment performed has an event called C. The tree is as follows:

A Two-Stage Probability Tree

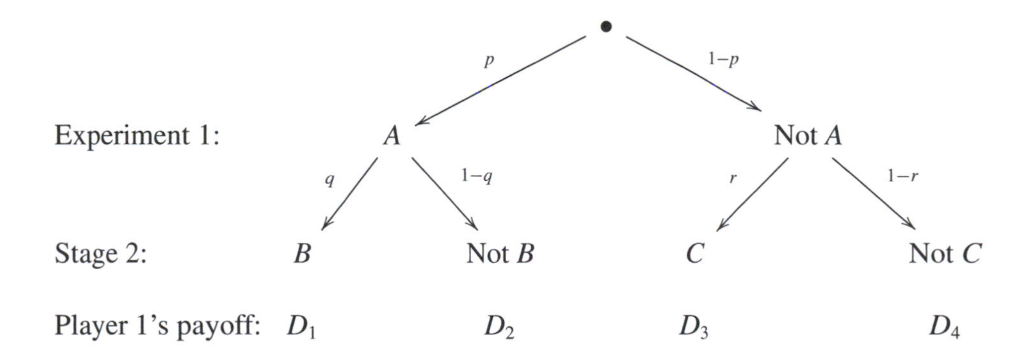

Reaching the end node B corresponds to the occurrence of both events A and B. Here p is the probability that A occurs, whereas q is the conditional probability of B *given* A occurred at the first stage. In our notation, this is the probability $q = P(B \mid A)$. Applying the Law of Conditional Probability (Theorem 11.17), we obtain

$$P(\text{reaching the node } B) = P(A \text{ and } B) = P(A) \cdot P(B \mid A) = pq$$

Next, note that the events corresponding to reaching two end nodes, say, B and C, are mutually exclusive because to reach B, we must have that A occurred, whereas to reach C, we must have that not A occurred. By Corollary 11.9, to compute the probability of reaching one or the other winning nodes, we add the probabilities. These arguments prove the Law of the Probability Tree for this example.

Let us now turn to the expected value E of the preceding game. Write E_A for the expected value of the subgame beginning when A occurs in Experiment 1 and $E_{\text{not A}}$ for the subgame beginning when not A occurs in Experiment 1. Then

$$E_A = qD_1 + (1-q)D_2 \quad \text{and} \quad E_{\text{not A}} = rD_3 + (1-r)D_4$$

Using the Law of the Probability Tree, we see that

$$\begin{aligned} E &= pqD_1 + p(1-q)D_2 + (1-p)rD_3 + (1-p)(1-r)D_4 \\ &= p(qD_1 + (1-q)D_2) + (1-p)(rD_3 + (1-r)D_4) \\ &= pE_A + (1-p)E_{\text{not A}} \end{aligned}$$

This proves the Linearity of Expectation (Theorem 3.13) for our example.

While we have only proved Theorems 3.6 and 3.13 for a very special case, the general results now follow by formal arguments using the Principle of Induction for Trees (10.2). We content ourselves with extending one of these arguments to give the idea.

Theorem 11.21 *Suppose that we are given a probability tree for a game of chance with branches labeled with branch probabilities. Then the probability of reaching any node in the tree is obtained by multiplying together all the branch probabilities arising in the unique path from the root node to the node in question.*

Proof We prove this result by induction on the depth of the probability tree.

Base Case: If the tree has depth $=1$, the individual branch probabilities give the probabilities of reaching the various end nodes. The result is trivial in this case.

Induction Hypothesis: Fix $n \geq 1$, and suppose that the probability of reaching a node in the tree is obtained multiplying by the branch probabilities from the root to this node for the every game tree of depth $\leq n$,

Induction Step: Now suppose that we are given a game of chance with a game tree with depth $= n+1$. Let B denote a node in the tree. If B is adjacent to the root node, then there is nothing to prove. Otherwise, let A be a node adjacent to the root node, such that B is a node in the A-rooted subtree. Here is a picture:

A Game Tree of Depth $= n+1$

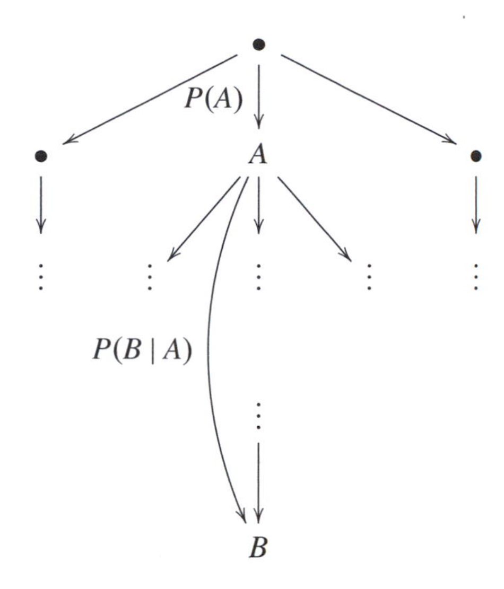

The curved branch is not a branch in the probability tree. It corresponds to following the path from A to B. By definition, the probability of taking this path is the conditional probability $P(B \mid A)$. Since the curved arrow represents a path in the A-rooted subtree that has depth $\leq n$, we may apply our Induction Hypothesis: $P(B \mid A)$ is the product of the branch probabilities in the path from A to B. Now, by the Law of Conditional Probability (Theorem 11.17), the probability of reaching B from the root node is

$$P(A \text{ and } B) = P(A) \cdot P(B \mid A)$$

Since $P(A)$ is a branch probability and $P(B \mid A)$ is a product of branch probabilities, $P(A \text{ and } B)$ is the product of the branch probabilities in the path from the root node to B, as needed. ■

This argument is another example, like Theorem 10.13, of how the method of Proof by Induction converts statements that are true for the case $n = 2$ to the general case.

Exercises

11.1 Compute the following probabilities for the experiment of rolling a pair of dice:

(a) P(“Doubles” and “Sum $\geq$ 8”) =
(b) P(“Sum odd” or “At least one die = ⚄”) =

(c) P("Sum $=6$" | "Doubles") =
(d) P("Sum ≥ 8" | "Doubles") =
(e) P("Sum ≥ 8" | "At least one die = ⚄") =
(f) P("Sum $\neq 7$" | "At least one die = ⚄") =

11.2 Let A, B, and C be events in a sample space S. A **Venn Diagram** represents these subsets as follows:

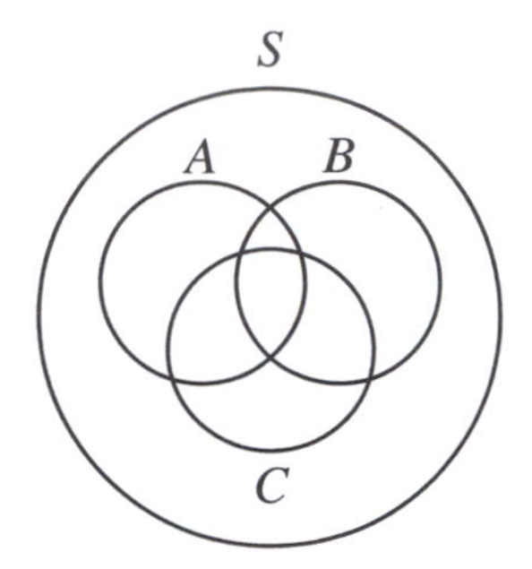

Shade the region corresponding to the following events in a Venn diagram:

(a) $(A \text{ and } B) \text{ or } C$
(b) $(\text{not } A) \text{ and } (B \text{ or } C)$
(c) $(\text{not } (A \text{ and } C)) \text{ or } (B)$
(d) $(A \text{ and } (\text{not } B)) \text{ or } (A \text{ and } (\text{not } C))$

11.3 Consider the following three events for drawing a single card from a deck of 52:

$A =$ "Card is a Jack" $\quad B =$ "Card is a Diamond" $\quad C =$ "Card is Red"

Decide which of the following pairs are Independent Events:

(a) A and B
(b) A and C
(c) B and C

11.4 Suppose that A and B are events for a probability experiment with $P(A) = 0.35$, $P(B) = 0.45$, and $P(A \text{ and } B) = 0.15$. Compute the following:

(a) $P(A \text{ or } B) =$
(b) $P(A \mid B) =$
(c) $P(B \mid A) =$
(d) $P(A \text{ or } B \mid A) =$
(e) $P(A \text{ or } B \mid A \text{ and } B) =$

11.5 Compute the following probabilities for the experiment of drawing two cards (without replacement) from a deck of 52. A **face card** is a Jack, Queen, King.

(a) P("A pair") =
(b) P("Both cards red") =

(c) $P($ "A pair" | "Both cards red"$) =$
(d) $P($ "Both cards red" | "A pair" $) =$
(e) $P($ "Two face cards" | "A pair" $) =$
(f) $P($ "Both cards red" | "Two face cards" $) =$

11.6 Consider the following variation on Fixing a Dice (Example 11.14). Each player puts $5 in the pot. Two dice are rolled. Player 1 wins the pot if the sum of the dice is 5, 6, or 7. Otherwise, Player 2 wins the pot. Before the roll, Player 1 is given the option to pay $2 and fix the Red die on a ⚂. Compute expected values to decide whether Player 1 should elect this option or not.

11.7 Compute the following probabilities for the experiment of drawing three cards (without replacement) from a deck of 52.

(a) $P($"Three Jacks"$) =$
(b) $P($"Three face cards"$) =$
(c) $P($"Three of the same value" | "Three face cards"$) =$
(d) $P($"Three face cards" | "Three of the same value" $) =$

11.8 Three cards are drawn from a deck of 52. All three cards are Hearts. What is the probability that the next two cards drawn are also Hearts?

11.9 Consider the experiment of drawing 4 cards from a deck of 52. Compute the following probabilities:

(a) $P($"Four of the same suit"$) =$
(b) $P($"Four face cards"$) =$
(c) $P($"Four face cards" | "Four of the same suit" $) =$
(d) $P($"Four of the same suit" | "Four face cards" $) =$

11.10 Use Theorem 11.17 to prove Theorem 11.18. Note that Theorem 11.18 asserts three biconditionals: "(a) if and only (b)," "(b) if and only if (c)," and "(c) if and only if (a)." However, it is sufficient to prove three conditionals, namely, "if (a), then (b)," "if (b), then (c)," and "if (c), then (a)."

11.11 The game Yahtzee is played with five dice. Compute the probability of rolling five dice and getting the following outcomes:

(a) All five dice have the same value.
(b) The five dice come up ⚀, ⚁, ⚂, ⚃, ⚄ in some order.
(c) The five dice come up ⚁, ⚁, ⚁, ⚁, ⚅ in some order.
(d) The five dice come up ⚀, ⚀, ⚀, ⚅, ⚅ in some order.

11.12 We say that a collection of events $E_1, E_2, \ldots, E_n$ in a probability experiment is **pairwise mutually exclusive** if E_i and E_j are mutually exclusive events for all $i \neq j$. Prove by induction that for all $n \geq 1$ and for any collection of n pairwise mutually exclusive events,

$$P(E_1 \text{ or } E_2 \text{ or } \cdots \text{ or } E_{n-1} \text{ or } E_n) = P(E_1) + P(E_2) + \cdots + P(E_{n-1}) + P(E_n)$$

11.13 Suppose that we are given a sequential-move game of chance with payoffs. Let \$$D$ be the payoff to Player 1 corresponding to an end node called B. Use Theorem 11.21 to prove that the contribution to the expected value E for Player 1 from node B is \$$D$ times the product of the branch probabilities in the unique path from the root of the probability tree to B. Use this result to prove the Linearity of Expectation (Theorem 3.13).

12 Fairness in Elections

The mathematical analysis of voting methods on the basis of fairness traces back to a debate between the Chevalier de Borda and the Marquis de Condorcet, two prominent eighteenth-century French intellectuals. Borda introduced the voting method we now call the Borda Count for elections with multiple candidates. He argued for the superiority of his method over the Plurality Method using a *fairness criterion*. Soon after Borda's work was published, Condorcet published a paper revealing a basic violation of fairness by the Borda Count. Condorcet observed that it is possible for a candidate to have the support of the majority of the voters and still lose the Borda Count, a *violation* of the *Majority Winner Criterion*. In this chapter, we introduce several fairness criteria: the *Majority Winner, Condorcet Winner, Majority Loser*, and *Condorcet Loser Criteria*. We examine the voting methods introduced in Chapter 5 using these criteria.

The study of voting methods by fairness criteria provides an excellent introduction to the dichotomy between proofs and counterexamples in mathematics, as discussed in Chapter 10. The fairness criteria that we present here are each universal statements about voting methods. The question of whether the voting method *satisfies* or *violates* the given fairness criterion becomes the prototypical mathematics question: is there a proof or is there a counterexample?

Fairness Criteria

To deserve the name, a fairness criterion should identify a property that corresponds to a "fair" outcome of an election. Devising fairness criteria is part art and part science. Fairness is something of a subjective question, and yet we need a precise, mathematical description of the criteria if we expect to use them to analyze voting methods. The fairness criteria we present will be structured as follows:

efinition 12.1

A **fairness criterion** is a universal statement of the following form: for all preference tables of a particular type, a voting method produces a social choice with a given property.

We say that a given voting method **satisfies** the fairness criterion if the universal statement is true for this voting method. We say that the voting method **violates** the fairness criterion if the universal statement is false for this voting method.

According to this formulation, a fairness criterion together with a given voting method presents the dichotomy discussed in Chapter 10: either there is a proof that the voting method always produces the desired result (satisfaction), or we can find a counterexample: a preference table of the given type such that the social choice is not as specified by the fairness criterion (a violation). Deciding which is true can be either an easy exercise or a challenging problem, as we will see.

We examine four fairness criteria in this chapter. Each singles out a special type of candidate, either one who should win or one who should lose. Most preference tables will not have such a candidate. But when they do, the fairness criteria described here require that this special candidate either be the social choice or definitely not be the social choice depending on the case. We begin with arguably the most natural criterion for fairness.

The Majority Winner Criterion

When the first-place votes are spread evenly among several strong candidates in an election, the social choice depends on the voting method used. But what if one candidate appears first on more than half the ballots? A **Majority Winner** is a candidate in an election who is the first preferred candidate for more than half the voters. It is reasonable to expect that a Majority Winner, when one exists, should be the social choice. This idea is the basis for our first fairness criterion.

Definition 12.2 The **Majority Winner Criterion** is the following universal statement: for every preference table having a candidate who is a Majority Winner, the Majority Winner is always the social choice.

To prove that a voting method satisfies the Majority Winner Criterion, we assume that we are given a preference table with a Majority Winner, say, candidate X. We then argue that X is definitely the social choice. To prove that a voting method violates the Majority Winner Criterion, we must construct a preference table with a Majority Winner who is not the social choice. The first result is tautological.

Theorem 12.3 *The Plurality Method satisfies the Majority Winner Criterion.*

Proof A Majority Winner in a given preference table automatically has a plurality of the first-place votes. Thus a Majority Winner is always the Plurality social choice. ■

Somewhat more surprising is the following fact, observed by Condorcet in his critique of the Borda Count:

Theorem 12.4 *The Borda Count violates the Majority Winner Criterion.*

Proof We need a counterexample, a preference table with a Majority Winner who somehow loses the Borda Count. Fortunately, this is easy to arrange. Here is an example:

A Majority Winner Loses the Borda Count

Number of voters	4	3
1st choice	X	A
2nd choice	A	B
3rd choice	B	C
4th choice	C	X

Notice that X is a Majority Winner, and yet X loses the Borda Count to candidate A, getting 12 Borda points to A's 17. This example then proves the theorem. ■

For the Hare and Coombs Methods, we have the following theorem:

Theorem 12.5 *The Hare and Coombs Methods satisfy the Majority Winner Criterion.*

Proof In both cases, a Majority Winner is elected on the first round by the definitions of the respective voting methods. ■

The Condorcet Winner Criterion

Our second fairness criterion is based on the principle that a candidate preferred by society to each other candidate should be the social choice. Given any preference table, we can use the Simple Majority Method (Definition 5.12) to run each pair of candidates head to head. A candidate who beats every other candidate head to head is called a **Condorcet Winner**. Notice that the Condorcet Winner, like the Majority Winner, is a feature of the preference table and is not a voting method. We have the following definition:

Definition 12.6 The **Condorcet Winner Criterion** is defined by the following universal statement: for every preference table having a Condorcet Winner, the Condorcet Winner is always the social choice.

Before turning to our voting methods, we observe a connection between the Majority Winner Criterion and the Condorcet Winner Criterion.

Theorem 12.7 *If a voting method violates the Majority Winner Criterion, then this voting method also violates the Condorcet Winner Criterion.*

Proof To prove this conditional statement, we assume that the voting method in question violates the Majority Winner Criterion. This means that there exists a preference table with a Majority Winner candidate, say, X, such that X is not the social choice. Now observe that since X has a majority of the first-place votes in this preference table, X beats every other candidate head to head. Thus the same preference table provides an example of a Condorcet Winner who is not the social choice. This is the needed violation of the Condorcet Winner Criterion. ■

As an immediate consequence, we obtain the following corollary:

Corollary 12.8 *The Borda Count violates the Condorcet Winner Criterion.*

Proof Use Theorem 12.4 and then apply Theorem 12.7. ■

Next, we prove the following theorem:

Theorem 12.9 *The Plurality and Hare Methods both violate the Condorcet Winner Criterion.*

Proof We give one example that produces both violations:

A Condorcet Winner Loses the Plurality and Hare Methods

Number of voters	4	3	2
1st choice	A	B	X
2nd choice	X	X	B
3rd choice	B	A	A

Notice that X is the Condorcet Winner, and yet X loses the Plurality election to A. For the Hare Method, X is the first candidate eliminated, and B is the Hare social choice. ■

A related example is used to prove the following theorem:

Theorem 12.10 *The Coombs Method violates the Condorcet Winner Criterion.*

Proof For this violation, we need four candidates:

A Condorcet Winner Loses the Coombs Method

Number of voters	9	6	4	1	1
1st choice	A	X	X	B	C
2nd choice	B	B	C	X	X
3rd choice	C	C	A	A	B
4th choice	X	A	B	C	A

Again, X is the Condorcet Winner, and yet X is eliminated in the first round of the Coombs Method, having the most last-place votes. ■

We next consider fairness criteria identifying candidates who are particularly unsuitable to be the social choice.

The Majority Loser Criterion

While it may be difficult to decide on the most worthy candidate in an election, sometimes there is a candidate who can be easily ruled out. Given a preference table, we define a **Majority Loser** to be a candidate who receives more than half the total last-place votes.

finition 12.11 The **Majority Loser Criterion** is defined by the universal statement: for every preference table having a Majority Loser, the Majority Loser is never the social choice.

We analyze our voting methods using this fairness criterion. First, we have the following theorem:

Theorem 12.12 *The Plurality Method violates the Majority Loser Criterion.*

Proof A violation is given by the following example:

A Majority Loser Wins the Plurality Method

Number of voters	4	3	2
1st choice	X	A	B
2nd choice	A	B	A
3rd choice	B	X	X

Notice that X has a majority of the last-place votes and a plurality of the first-place votes. ■

Next, we address the question of whether a majority loser could ever win the Borda Count. Let us consider a possible scenario: for definiteness, suppose that we have a violation with, say, 11 total voters, 4 candidates, and a majority loser X. Since a majority of voters have X ranked last, X will receive 0 points from at least 6 of the voters. This means that the maximum Borda points X can achieve occurs when 5 voters rank X first. In this case, X receives

$$6 \cdot 0 + 5 \cdot (4 - 1) = 15 \quad \text{Borda points}$$

Now suppose that X is actually the Borda Count social choice. Then the other 3 candidates must receive < 15 Borda points. Let us assume that they each receive 14 Borda points. Then we conclude that

$$\text{Total Borda points} = 14 + 14 + 14 + 15 = 57 \quad \text{Borda points}$$

However, we can count the total Borda points directly. Each of the 11 voters gives out $3 + 2 + 1 + 0 = 6$ Borda points. Thus

$$\text{Total Borda points} = 6 \cdot 11 = 66 \quad \text{Borda points}$$

This contradiction means that X cannot be the Borda Count social choice.

We would like to convert this scenario into a general proof. As a preliminary, we compute the total Borda points for an election. The result is a consequence of a famous identity.

Theorem 12.13 *Let $n \geq 1$ be an integer. Then*

$$1+2+\cdots+(n-1)+n=\frac{n(n+1)}{2}$$

Proof This result can be proved in several ways (compare Exercises 10.10 and 12.17). The proof we give here is attributed to the great mathematician Carl Gauss when he was an elementary school student. Let $S = 1+2+3+\cdots+(n-2)+(n-1)+n$, the sum in question. We may just as well compute S by adding in the reverse order: $S = n+(n-1)+(n-2)+\cdots+3+2+1$. We now add these two representations of the sum S as follows:

$$\begin{array}{rcccccccccc} S & = & 1 & + & 2 & + \cdots + & (n-1) & + & n \\ +\ S & = & n & + & (n-1) & + \cdots + & 2 & + & 1 \\ \hline 2S & = & (n+1) & + & (n+1) & + \cdots + & (n+1) & + & (n+1) \end{array}$$

Notice that on the right side we have n summands each equal to $n+1$. We conclude that $2S = n(n+1)$, as needed. ∎

As a consequence, we deduce the following corollary:

Corollary 12.14 *Given an election with k voters and m candidates, the total Borda points B_{total} for all candidates is*

$$B_{\text{total}} = \frac{km(m-1)}{2}$$

Proof With our convention, each voter assigns $m-1$ Borda points to his or her first choice, $m-2$ Borda points to his or her second choice, on down to 0 points for his or her least preferred candidate. Using Theorem 12.13 for the case $n = m-1$, each voter awards a total of

$$0+1+2+\cdots+(m-1)=\frac{(m-1+1)(m-1)}{2}=\frac{m(m-1)}{2}$$

Borda points by Theorem 12.13. The result now follows. ∎

With this calculation in hand, we prove the following theorem:

Theorem 12.15 *The Borda Count satisfies the Majority Loser Criterion.*

Proof Let k be the number of voters and m the number of candidates. Let us first assume that k is odd and write $k = 2n+1$ for some integer n.

Suppose that X is a Majority Loser. We obtain an upper bound on X's Borda points, as we did earlier. We know at least $n+1$ voters rank X last. Thus, at most

n voters give X any points at all. At best, these n voters rank X first. In this case, each awards X a total of $m-1$ Borda points. We conclude that

$$X\text{'s Borda points} \leq n(m-1)$$

Now suppose that X is the Borda Count social choice. Then other candidates must receive fewer Borda points than X. We obtain an upper bound for B_{total}, the total Borda points

$$\begin{aligned} B_{\text{total}} &\leq (\text{number of candidates}) \cdot (X\text{'s Borda points}) \\ &\leq mn(m-1) \end{aligned}$$

However, applying Corollary 12.14, we have

$$\begin{aligned} B_{\text{total}} &= \frac{km(m-1)}{2} \\ &= \frac{(2n+1)m(m-1)}{2} \\ &> \frac{(2n)m(m-1)}{2} \\ &= mn(m-1) \end{aligned}$$

We now have two contradictory inequalities involving B_{total}. The contradiction implies that X cannot be both the Majority Loser and the Borda Count social choice.

It remains for us to handle the case where the number k of voters is even. The proof is similar and left for you in Exercise 12.18. ■

The proof of Theorem 12.15 is an example of a **proof by contradiction.** In our proof, we assumed a violation of the Majority Loser Criterion and deduced, as a direct consequence, a numeric contradiction. The contradiction implies that such a violation is impossible.

It is easier to settle the question for the Hare and Coombs Methods. We have the following theorem:

Theorem 12.16 *The Hare and Coombs Methods satisfy the Majority Loser Criterion.*

Proof For the Hare Method, notice that a Majority Loser X cannot win a head-to-head election by the Simple Majority Method. To become the Hare social choice, a candidate must, in particular, beat the other last surviving candidate head to head. Thus X cannot win a Hare Method election.

For the Coombs Method, we simply observe that a Majority Loser is the first candidate eliminated. ■

The Condorcet Loser Criterion

We identify another unsuitable candidate in a preference table, this time using head-to-head elections. Define a **Condorcet Loser** to be a candidate who loses head to head to every other candidate in the election.

Definition 12.17 The **Condorcet Loser Criterion** is defined by the universal statement: for every preference table with a Condorcet Loser, the Condorcet Loser is never the social choice.

We begin our analysis with the following theorem:

Theorem 12.18 *The Plurality Method violates the Condorcet Loser Criterion.*

Proof In the preference table given in the proof of Theorem 12.12, candidate X has a majority of the last-place votes and yet wins the Plurality Method. Since a Majority Loser is easily seen to be a Condorcet Loser, the result follows. ■

Notice that we have proven the following theorem:

Theorem 12.19 *A voting method that violates the Majority Loser Criterion also violates the Condorcet Loser Criterion.* ■

Again, the analyses for the Hare and Coombs Methods are easy.

Theorem 12.20 *The Hare and Coombs Methods satisfy the Condorcet Loser Criterion.*

Proof The last possible stage of either a Hare or a Coombs Method election pits the last two surviving candidates in a head-to-head election. If a Condorcet Loser happens to survive to the last round, then this candidate will lose this head-to-head election. Thus we see that a Condorcet Loser can never be the Hare or Coombs social choice. ■

We remark that the proof of Theorem 12.20 depends on our convention of eliminating exactly one candidate at each stage in the Hare and Coombs Methods, breaking ties if necessary. If all tied candidates are eliminated at each stage, then, in fact, both methods violate the Condorcet Loser Criterion (see Exercises 12.19 and 12.20).

We summarize our work so far in the following table:

Summary of Fairness Critera

Method	Majority Winner	Condorcet Winner	Majority Loser	Condorcet Loser
Plurality	Satisfies	Violates	Violates	Violates
Borda	Violates	Violates	Satisfies	?
Hare	Satisfies	Violates	Satisfies	Satisfies
Coombs	Satisfies	Violates	Satisfies	Satisfies

This leaves us with the question of the Borda Count: does it satisfy or violate the Condorcet Loser Criterion? We have seen that a Majority Loser cannot win the Borda Count. But what about a Condorcet Loser? We leave this question for you to ponder and return to it in Chapter 21.

Exercises

12.1 Decide whether a Dictatorship satisfies or violates the Majority Winner, Condorcet Winner, Majority Loser, and Condorcet Loser fairness criteria. Justify your answer with a proof (satisfies) or an example (violates).

12.2 Decide whether the following preference table gives an example of a violation of the Condorcet Winner Criterion by the Borda Count:

Number of voters	9	8	14	11
1st choice	A	B	C	A
2nd choice	B	A	B	B
3rd choice	D	C	A	C
4th choice	C	D	D	D

12.3 Decide whether the following preference table gives an example of a violation of the Condorcet Loser Criterion by the Plurality Method:

Number of voters	8	7	3	3
1st choice	A	B	C	D
2nd choice	B	D	B	B
3rd choice	D	C	A	C
4th choice	C	A	D	A

12.4 The Pairwise Comparison with an Agenda voting method was defined in Definition 5.16. Using the reverse alphabetical agenda: D, C, B, A, and so on, decide whether the Pairwise Comparison Method satisfies or violates the Majority Winner, Condorcet Winner, Majority Loser, and Condorcet Loser fairness criteria. Justify your answer with a proof (satisfies) or an example (violates).

12.5 Give an example of an election with a candidate X such that (a) X is the Condorcet Winner, (b) X is not a Majority Winner, and (c) X loses the Borda Count.

12.6 Give an example of an election with a candidate X such that (a) X has a plurality of last-place votes and (b) X wins the Hare Method.

12.7 Give an example of an election with a candidate X such that (a) X is the Condorcet Winner, (b) X is the Plurality social choice, and (c) X loses the Coombs Method.

12.8 The **Double Plurality Method** is the voting method defined as follows: if there is a Majority Winner, that candidate is the Double Plurality social choice. Otherwise, the two candidates with the most first-place votes are run head to head using the Simple Majority Method; the winner is the Double Plurality social choice.

(a) Determine the Double Plurality social choice for the following preference table:

Number of voters	**22**	**17**	**25**	**18**	**11**
1st choice	A	B	C	D	D
2nd choice	B	C	A	C	C
3rd choice	C	A	D	B	A
4th choice	D	D	B	A	B

(b) Decide whether the Double Plurality Method satisfies or violates the Majority Winner, Condorcet Winner, Majority Loser, and Condorcet Loser fairness criteria. Justify your answer with a proof (satisfies) or an example (violates).

12.9 Prove that the Double Plurality Method, defined in Exercise 12.8, is manipulable (see Definition 6.3).

12.10 Prove that the Double Plurality Method exhibits the No-Show Paradox (Definition 6.11).

12.11 The **Bucklin Method** is the voting method defined as follows: suppose that there are k voters. Count the first-place votes for each candidate. If any candidate has a majority ($>k/2$) of first-place votes, that candidate is the Bucklin social choice. Otherwise, count the first- and second-place votes for each candidate. If any candidate has more than $k/2$ first- and second-place votes, that candidate is the Bucklin social choice. Continue in this way, counting next the first-, second- and third-place votes until some candidate has more than $k/2$ total. In case more than one candidate receives more than $k/2$ top votes at some stage, the Bucklin social choice is the candidate with the largest number of such votes.

(a) Determine the Bucklin social choice for the following preference table:

Number of voters	**10**	**9**	**8**	**7**	**6**
1st choice	A	D	B	C	D
2nd choice	B	C	A	B	C
3rd choice	C	A	C	D	A
4th choice	D	B	D	A	B

(b) Decide whether the Bucklin Method satisfies or violates the Majority Winner, Condorcet Winner, Majority Loser, and Condorcet Loser fairness criteria. Justify your answer with a proof (satisfies) or an example (violates).

12.12 Prove that the Bucklin Method is manipulable.

12.13 Prove that the Bucklin Method exhibits the No-Show Paradox (Definition 6.11).

12.14 Define the **Individual Choice Criterion** as follows: a voting method satisfies the Individual Choice Criterion if, for every preference table, the social choice is the first choice of at least one individual voter. Decide whether each of our voting methods (Plurality, Borda, Hare, and Coombs) satisfies or violates the Individual Choice Criterion. Prove your answer in each case.

12.15 Define the **Plurality Loser Criterion** as follows: a voting method satisfies the Plurality Loser Criterion if, for every preference table, when candidate X has a plurality of last-place votes, then candidate X is never the social choice. Decide whether each of our voting methods (Plurality, Borda, Hare, and Coombs) satisfies or violates the Plurality Loser Criterion. Prove your answer in each case.

12.16 Prove that a Condorcet Winner and a Condorcet Loser, if either exist, are unique.

12.17 Give an alternate proof of Theorem 12.13 as follows: recall that Theorem 11.12 gives a formula for the number of ways to choose two objects from a set of $n+1$, that is,

$$\binom{n+1}{2} = \frac{n(n+1)}{2}$$

Suppose that the $n+1$ element set is $\{1,2,\ldots,n,n+1\}$. How many pairs involve the number 1? How many pairs not counted yet use the number 2? Add these counts to prove Theorem 12.13.

12.18 Give the proof of Theorem 12.15 when the number of voters in the Borda Count election is even.

12.19 Suppose in the Hare Method that all candidates who tied for the least first-place place votes in each round are eliminated at once. Prove that this version of the Hare Method violates the Condorcet Loser Criterion.

12.20 Suppose in the Coombs Method that all candidates who tied for the most last-place votes in each round are eliminated at once. Prove that this version of the Coombs Method violates the Condorcet Loser Criterion.

13 Weighted Voting

Arguably, the first principle of fairness for a voting method is that each vote should count equally in the final outcome. In this chapter, we study voting systems that violate this basic principle. These are voting systems in which differences between voters are purposely introduced to reflect differences in the rank of the player or, in the case of regions, the size of the population. Examples of such systems include the U.S. Electoral College, the U.N. Security Council, and countless corporate and legislative committees. As is customary, we study such voting systems under the simplifying assumption that there are only two possible outcomes of the election, "yes" and "no," a yes-no voting system as in Definition 7.4. In this setting, the relative power of the individual voters is the central issue. We apply the counting techniques, notably the Combinations Formula (Theorem 11.12), developed in Chapter 11 to study the power of voters in weighted systems. We also consider the question of assigning *weights* to voters in a yes-no voting system. This problem leads us to the notion of a *mathematical invariant*. Finally, we show that Question 1.6 regarding the U.S. Senate introduced in Chapter 1 leads to a *combinatorial identity* called the *recursive law* for combinations.

Yes-No Voting Systems and Power

Much of the controversy over voting methods stems from issues arising from the presence of multiple candidates. Even the most democratic methods can appear "unfair" in the attempt to untangle social preferences when their are several competing alternatives. We remove this issue here and focus on elections with only two candidates. Recall that a yes-no voting system is a collection of voters who each vote "yes" or "no" and a rule for determining the outcome of the vote that is either to approve the measure ("yes") or reject it ("no") (Definition 7.4).

One voting method available for a yes-no voting system is the Simple Majority Method. Just count the votes for "yes" and those for "no"; the majority wins. Of course, in this case, all the voters have the same amount of power. Here is a real-world example with power differentials:

Example 13.1 U.N. Security Council

The U.N. Security Council consists of representatives from 15 countries. The voters are split into two classes: the 5 permanent members (China, France, the Russian Federation, the United Kingdom, and the United States) and the 10 nonpermanent members (in 2014, these were Argentina, Austria, Chad, Chile, Jordan, Lithuania, Luxembourg, Nigeria, Korea, and Rwanda). A measure passes the U.N. Security Council if (1) it receives a "yes" vote from all 5 permanent members and (2) it receives a "yes" vote from at least 4 nonpermanent members.

In the Chapter 7, we introduced the power index (Definition 7.6) for quantifying the power of voters in a yes-no voting systems. Recall that given voters $P_1, P_2, \ldots, P_n$, we define C_i to be the number of winning coalitions for which P_i is a critical voter (see Definition 7.5). The power index is then the list of fractions

$$\left[\frac{C_1}{C}, \frac{C_2}{C}, \ldots, \frac{C_n}{C}\right]$$

where $C = C_1 + C_2 + \cdots + C_n$. We compute the power index for the U.N. Security Council using counting techniques.

We take advantage of the fact that there are only two types of voters. Let $P_1, \ldots, P_5$ denote the 5 permanent members, and let $N_1, \ldots, N_{10}$ denote the 10 nonpermanent members. Suppose that we compute

$$P = \text{number of winning coalitions in which } P_1 \text{ is a critical voter}$$

and

$$N = \text{number of winning coalitions in which } N_1 \text{ is a critical voter}$$

Then the total number of critical voters will be $5P + 10N$, and the power for a permanent member will be

$$\frac{P}{5P + 10N}$$

whereas the power for a nonpermanent member will be

$$\frac{N}{5P + 10N}$$

Thus we are reduced to making only two calculations.

Let us compute P. We identify the winning coalitions in which P_1, the first permanent member, is a critical voter. Note that P_1 is a critical voter in every winning coalition. Thus we must count the winning coalitions. A winning coalition consists of all five permanent members plus at least four nonpermanent members.

To count these coalitions, we use the "choose." We see that

$$\begin{aligned} P &= \text{number of winning coalitions} \\ &= \text{number of subsets of the nonpermanent members with 4 or more elements} \\ &= \binom{10}{4} + \binom{10}{5} + \binom{10}{6} + \binom{10}{7} + \binom{10}{8} + \binom{10}{9} + \binom{10}{10} \\ &= 848 \end{aligned}$$

Next, we focus on N. Observe that N_1 is a critical voter in a winning coalition precisely when that coalition contains (1) all five permanent members plus (2) exactly three other nonpermanent members. Thus

$$N = \binom{9}{3} = 84$$

Here we choose the three nonpermanent members from $N_2, \ldots, N_{10}$ because we are assuming that N_1 is already chosen.

In conclusion, we computed the power index for the U.N. Security Council to be

Power Index for the U.N. Security Council

$$\left[\frac{848}{5{,}080}, \frac{848}{5{,}080}, \frac{848}{5{,}080}, \frac{848}{5{,}080}, \frac{848}{5{,}080}, \frac{84}{5{,}080}, \frac{84}{5{,}080}, \frac{84}{5{,}080}, \frac{84}{5{,}080}, \frac{84}{5{,}080}, \frac{84}{5{,}080}, \frac{84}{5{,}080}, \frac{84}{5{,}080}, \frac{84}{5{,}080}, \frac{84}{5{,}080}\right]$$

The power index turns any yes-no voting system into a counting problem. For voting systems such as the U.S. Electoral College, the calculations are too extensive to be done by hand. For smaller voting systems in which there are only a few different roles for voters, the problem is an exercise in use of the Combinations Formula, as we have just seen. We give one further example along the same lines:

Example 13.2 Academic Committee

A committee consisting of three deans (D_1, D_2, D_3) and four professors (P_1, P_2, P_3, P_4) convenes to decide on the passage of mandates proposed by members of the university. The deans are each awarded 10 votes, and the professors are awarded 5 votes. A mandate passes the academic committee if it receives at least 30 total votes.

The academic committee is a weighted voting system as in Definition 7.3. Here the quota is $q = 30$. Recall that we write such a system as

$$[30\colon\ 10, 10, 10, 5, 5, 5, 5]$$

We compute the power index for this voting system.

As in the case of the U.N. Security Council, it suffices to compute

$$D = \text{number of winning coalitions in which } D_1 \text{ is critical}$$

and

$$P = \text{number of winning coalitions in which } P_1 \text{ is critical}$$

We make this calculations using the following tables counting winning coalitions. The distinguished player (D_1 and P_1) is included in the first column. In the second column, we decrease the number of deans included. The third column adds in professors as needed to make the coalition winning and the distinguished player a critical player. The counts are then made by multiplying the "choose" values or sum thereof from each column. Recall that we have defined $\binom{n}{0} = 1$.

D_1 Is a Critical Player

D_1	D_2,D_3	P_1,P_2,P_3,P_4	Count	Total
D_1	2	0, 1	$\binom{2}{2}\left[\binom{4}{0}+\binom{4}{1}\right]$	5
D_1	1	2, 3	$\binom{2}{1}\left[\binom{4}{2}+\binom{4}{3}\right]$	20
D_1	0	4	$\binom{2}{0}\binom{4}{4}$	1

P_1 Is a Critical Player

P_1	D_1,D_2,D_3	P_2,P_3,P_4	Count	Total
P_1	2	1	$\binom{3}{2}\cdot\binom{3}{1}$	9
P_1	1	3	$\binom{3}{1}\cdot\binom{3}{3}$	3

We conclude that $D = 26$ and $P = 12$. The power index for the academic committee is thus

$$\left[\frac{26}{126}, \frac{26}{126}, \frac{26}{126}, \frac{12}{126}, \frac{12}{126}, \frac{12}{126}, \frac{12}{126}\right]$$

Weights and Equivalences

Weighted voting systems are often used for representative committees wherein the weight distribution is chosen to reflect the population differences of the regions represented. The most famous example is, of course, the U.S. Electoral College. The assignment of weights to representative voters is called **apportionment.** Apportionment can be a delicate problem; the translation of large populations to small weights can have strange side effects. Indeed, the power index we study here came to fame in a legal paper written by John Banzhaf, who applied the index to argue the unfairness of a Nassau County Board of Supervisors in the 1960s (see Exercise 7.3). The weighted voting system used by Nassau County took the form

$$[16\colon 9,9,7,3,1,1]$$

with the three largest weights assigned to regions of the city of Hempstead and the three smaller values to the towns of Oyster Bay, Glen Cove, and Long Beach. Banzhaf observed that the smaller towns actually have no power in this weighted voting system. A voter in a yes-no system who is never a critical voter, that is, who has no power, is called a **dummy**. For all intents and purposes, the Nassau County Board of Supervisors could have been using the weighted system

$$[2\colon 1,1,1,0,0,0].$$

We call the weighted voting systems $[16\colon 9,9,7,3,1,1]$ and $[2\colon 1,1,1,0,0,0]$ *equivalent* yes-no voting systems. The precise definition is as follows:

efinition 13.3 We say that two yes-no voting systems are **equivalent** if the two systems have the same voters and exactly the same winning coalitions.

Definition 13.3 highlights the fact that the winning coalitions determine a yes-no system regardless of the particular description given. In fact, to describe a yes-no system, we could just list the winning coalitions if time and space were no object. A particularly interesting type of equivalence is one between a yes-no system, described without explicitly mentioning weights and a quota, and a bona fide weighted voting system. Here is an example:

Example 13.4 Finding Weights for the U.N. Security Council

Recall from Example 13.1 that the U.N. Security Council consists of five permanent members, $P_1,\ldots,P_5$, and ten nonpermanent members, $N_1,\ldots,N_{10}$. A measure passes if it receives a yes vote from (1) all five permanent members and (2) at least four nonpermanent members.

Let us write a for the weight of each permanent member and b for the weight of each nonpermanent member. If q is the quota, then we must have

$$5a+4b \geq q \quad \text{and} \quad 4a+9b < q$$

We use these inequalities to determine working values of $a, b,$ and q. For example, we might try making $a = 5$ and $b = 1$. We would then need $q \leq 29$ so that $29 = 5a + 4b \geq q$. Unfortunately, we now have $4a + 9b = q$, so these values will not work. We need to use larger values of a and b. Let's try $a = 10$ and $b = 1$. We then need $q \leq 54 = 5a + 4b$. Taking $q = 54$, we see that $4a + 9b = 49 < 54$. These values of a, b, and q do the job. We conclude that the U.N. Security Council yes-no voting system is equivalent to the weighted voting system

U.N. Security Council as Weighted System

$$[54: 10, 10, 10, 10, 10, 1, 1, 1, 1, 1, 1, 1, 1, 1, 1]$$

Example 13.4 suggests an extension of the notion of a weighted voting system.

Definition 13.5 We say that a yes-no voting system is **weighted** if it is equivalent to a voting system presented with weights and quota as in Definition 7.3.

In this terminology, the U.N. Security Council is a weighted voting system, even though it is not presented with weights. Definition 13.5, in turn, suggests the following question:

Is every yes-no voting system weighted?

This question goes to the basic dichotomy between proofs and counterexamples presented by a universal statement, as discussed in Chapter 10. Suppose that there is a proof that every yes-no system is weighted. Such a proof would imply a general approach to finding weights that is guaranteed to work for every yes-no voting system. However, the answer is "No" if there is even just one example of a yes-no voting system that is not weighted. A proof for "No" requires just one example.

In fact, the answer is "No." The following modification of the U.N. Security Council will serve the purpose of a counterexample:

Example 13.6 Modified U.N. Security Council

The U.N. Security Council voting system is modified as follows: the players are the same. There are five permanent members, P_1, P_2, P_3, P_4, P_5, and ten nonpermanent members, $N_1, N_2, N_3, N_4, N_5, N_6, N_7, N_8, N_9, N_{10}$. A measure passes the modified council if at has the support of at least four permanent members and at least five nonpermanent members.

We claim that the Modified U.N. Security Council is not a weighted system. How do we prove such a claim? Clearly, it is not enough to say that we have tried and failed to find weights and quota that work! We must somehow *prove* that finding such weights is impossible. Our proof will make use of a venerable mathematical technique, the idea of an *invariant*. We would like to find a property that is enjoyed by every weighted voting system but that does not hold true for the Modified U.N. Security Council. Our property should be defined in terms of winning coalitions in view of Definition 13.3.

The property we use was introduced by Alan Taylor and William Zwicker, who gave a complete answer to the question of which yes-no voting systems are weighted (see Theorem 13.11). Suppose that we are given a yes-no voting system and two collections X and Y of voters. Define a **trade** between X and Y to be some exchange of voters between the sets X and Y. We do not place any restrictions on the number of voters exchanged. For instance, we can just move some voters from X to Y or vice versa. We do require that every trade result in a genuine change in both sets of voters. We will write X' and Y' for the new collections of voters, after the trade. Here is an example of a trade:

A Trade of Voters

X	**Y**
V_1	V_2
V_2	V_4
V_3	V_6
V_4	V_8
	V_{10}

$\Longrightarrow$

Trade

$X: V_1, V_3$

$Y: V_6, V_8, V_{10}$

X′	**Y′**
V_2	V_1
V_4	V_2
V_6	V_3
V_8	V_4
V_{10}	

The coalitions may share voters such as V_4 here. A trade requires that unshared voters be exchanged.

The idea of a trade extends naturally to a list of n collections of voters $X_1, \ldots, X_n$. Here we allow any number of exchanges between the various collections but again require that some genuine change in at least two of the collections occurs. We again write the new collections as $X'_1, \ldots, X'_n$.

We are now prepared to introduce the key notion for proving that the Modified U.N. Security Council is not a weighted system.

Definition 13.7 We say that a yes-no voting system is **trade robust** if given any trade among winning coalitions $X_1, \ldots, X_n$, at least one of the resulting coalitions $X'_1, \ldots, X'_n$ is still a winning coalition.

We first observe that two equivalent yes-no voting systems as in Definition 13.3 are either both trade robust or else both fail to be trade robust. This fact is an immediate consequence of the definition of trade robustness, expressed, as it is, in terms of winning coalitions. In mathematics, a property (such as trade robustness) that is preserved by a notion of equivalence is called a **mathematical invariant.** Invariants play a special role in mathematics, especially as devices for distinguishing nonequivalent mathematical objects. In our case, the logic goes as follows: suppose that we show that one voting system is not trade robust and another is trade robust. Then we can immediately conclude that the two yes-no systems are not equivalent. This will be our method for showing that the Modified U.N. Security Council voting system is not weighted. To begin we prove the following theorem:

Theorem 13.8 *The Modified U.N. Security Council is not trade robust.*

Proof As usual, we write $P_1, \ldots, P_5$ for the permanent members and $N_1, \ldots, N_{10}$ for the nonpermanent members. We then define two winning coalitions X and Y as follows:

A Trade

X	**Y**		**X′**	**Y′**
P_1	P_1	$\Longrightarrow$	P_1	P_1
P_2	P_2		P_2	P_2
P_3	P_3		P_3	P_3
P_4	P_5		P_4	N_1
N_1	N_1		P_5	N_2
N_2	N_2	**Trade**	N_1	N_3
N_3	N_3	$X: N_5$	N_2	N_4
N_4	N_4	$Y: P_5$	N_3	N_5
N_5	N_{10}		N_4	N_{10}

Notice that neither X' nor Y' is a winning coalition. Thus the Modified U.N. Security Council is not trade robust. ∎

We next prove the following general result:

Theorem 13.9 *Every weighted voting system is trade robust.*

Proof We suppose that we are given a yes-no system that is weighted. Recall that this means that our voting system is equivalent to a weighted voting system described in terms of weights and a quota, say, q. Since trade robustness is preserved by equivalence, we may assume that our yes-no system is, in fact, this weighted voting system.

Given a coalition X of voters, we write $w(X)$ for the total weight of voters in X, that is,

$$w(X) = \text{sum of weights of voters in the coalition } X$$

With this notation, we can identify winning coalitions by the following biconditional:

$$X \text{ is a winning coalition} \quad \text{if and only if} \quad w(X) \geq q$$

Now suppose that we are given two winning coalitions X and Y. Then $w(X) \geq q$ and $w(Y) \geq q$. Consequently, we see that

$$w(X) + w(Y) \geq 2q$$

Next, suppose that we perform a trade of voters between the coalitions X and Y yielding new coalitions X' and Y'. We claim that

$$w(X)+w(Y)=w(X')+w(Y')$$

To see this equality, observe that $w(X)+w(Y)$ is the sum of the weights of all voters listed in X and Y, with the weights of voters appearing in both coalitions, the duplicates, added twice. The trade concerns the nonduplicate voters in X and Y, some of whom are moved from the first list to the second and vice versa. Since the duplicate voters are not moved and the nonduplicate voters are just rearranged, adding all the weights according to the lists X' and Y' is thus the same as the original.

Now we play devil's advocate: suppose that both $w(X') < q$ and $w(Y') < q$. Then

$$w(X')+w(Y') < 2q$$

This leads to the contradiction

$$2q \leq w(X)+w(Y)=w(X')+w(Y') < 2q$$

The contradiction implies either $w(X') \geq q$ or $w(Y') \geq q$, so at least one of the new coalitions is still a winning coalitions. We have proven a weighted system is trade robust when the trade involves two winning coalitions. The same argument can be used for multiple winning coalitions; we leave the details for Exercise 13.15. ■

We apply this result to prove that our example is not a weighted system.

Corollary 13.10 *The Modified U.N. Security Council is not a weighted voting system.*

Proof The Modified U.N. Security Council is not trade robust by Theorem 13.8. Every weighted voting system is trade robust by Theorem 13.9. We conclude that the Modified U.N. Security Council is not equivalent to a weighted voting system because the property of trade robustness is an invariant of equivalence of yes-no systems. ■

The argument given in Corollary 13.10 is an effective method for proving that yes-no systems are not weighted. We encourage you to pursue some other examples in the exercises. We mention a further question that arises now from Theorem 13.9. We have proved the conditional statement

If a yes-no system is weighted, then it is trade robust.

It is natural to ask whether the converse statement is true, that is, whether the conditional statement

If a yes-no system is trade robust, is it weighted?

is a true statement. The answer to this question is contained in the following remarkable theorem:

Theorem 13.11 Taylor-Zwicker *A yes-no voting system is weighted if and only if it is trade robust.* ■

Theorem 13.11 is an excellent example of a deep and surprising equivalence between two mathematical definitions.

The U.S. Senate

We return to the U.S. Senate (Example 1.5) and our question from Chapter 1.

Question 1.6 *Who has more power in the U.S. Senate, an individual senator or the vice president?*

We view the U.S. Senate as a yes-no voting system with 100 Senators $S_1,\ldots,S_{100}$ and the vice president V. A measure put before the Senate is voted on by the senators. If the Senate vote is 50–50, then the vice president votes the measure up or down. We begin with the vice president. Let

$$v = \text{number of winning coalitions for which } V \text{ is a critical voter}$$

Observe that V is critical in any winning coalition of which he or she is a member because he or she only votes to break ties. Of course, these winning coalitions always include exactly 50 senators. We conclude that

$$v = \binom{100}{50}$$

Next, let us focus on any given senator, say, S_1. Let

$$s = \text{number of winning coalitions for which } S_1 \text{ is a critical voter}$$

We have two types of winning coalitions for which S_1 is a critical voter. First, we have those coalitions consisting of S_1 plus exactly 50 other senators from the remaining 99. The vice president does not vote in this scenario. We also have the coalitions in which S_1 plus 49 senators plus the vice president support the measure.

S_1 Is a Critical Voter

S_1	$S_2,\ldots,S_{100}$	V	Count
S_1	50	0	$\binom{99}{50}$
S_1	49	1	$\binom{99}{49}$

We conclude that

$$s = \binom{99}{50} + \binom{99}{49}$$

Question 1.6 now reduces to

Question 13.12 *Which is larger,*

$$\binom{100}{50} \quad \text{or} \quad \binom{99}{50} + \binom{99}{49}$$

We address this question in what follows by proving a *combinatorial identity*.

Pascal's Triangle

Our reformulation of the U.S. Senate question (Question 1.6) as Question 13.12 provides an example of a problem in **combinatorics**, the mathematical study of counting problems. Combinatorics and probability share roots in the work of Blaise de Pascal, who pioneered counting techniques using combinations. Pascal assembled the counts for the numbers of subsets of a set of n elements into a triangle, now called **Pascal's Triangle**, as follows:

Pascal's Triangle

$$\binom{0}{0}$$

$$\binom{1}{0} \quad \binom{1}{1}$$

$$\binom{2}{0} \quad \binom{2}{1} \quad \binom{2}{2}$$

$$\binom{3}{0} \quad \binom{3}{1} \quad \binom{3}{2} \quad \binom{3}{3}$$

$$\binom{4}{0} \quad \binom{4}{1} \quad \binom{4}{2} \quad \binom{4}{3} \quad \binom{4}{4}$$

$$\vdots \quad \vdots \quad \vdots \quad \vdots \quad \vdots$$

Filling in the numbers reveals a wealth of patterns:

Pascal's Triangle for $n = 6$

						1						
					1		1					
				1		2		1				
			1		3		3		1			
		1		4		6		4		1		
	1		5		10		10		5		1	
1		6		15		20		15		6		1

Perhaps the first thing we notice is the basic symmetry present, how the numbers repeat in each row. This symmetry is an example **combinatorial identity**, a numeric equation arising from two different ways to count the same collection of objects.

Theorem 13.13 Symmetry Identity *Let* $1 \leq k \leq n$. *Then*

$$\binom{n}{k} = \binom{n}{n-k}$$

Proof We use the language of elections. We assume that there are n candidates total. The left-hand side $\binom{n}{k}$ counts the number of ways to choose a set of k candidates. We call these chosen candidates *winners*. Now each choice of k winners produces, by default, a set of $n - k$ *losers*, the candidates that were not chosen. Similarly, choosing $n - k$ winners produces, by default, a choice of k losers. The Symmetry Identity follows. ■

The combinatorial identity we are interested in is not as evident as the Symmetry Identity. Imagine writing down the next row for $n = 7$. If we have a calculator, we can just compute the numbers $\binom{7}{k}$ for $k = 0, \ldots, 7$. However, a calculator is not necessary. Notice that each entry in the row for $n = 6$ is the sum of two entries in the row for $n = 5$. For example,

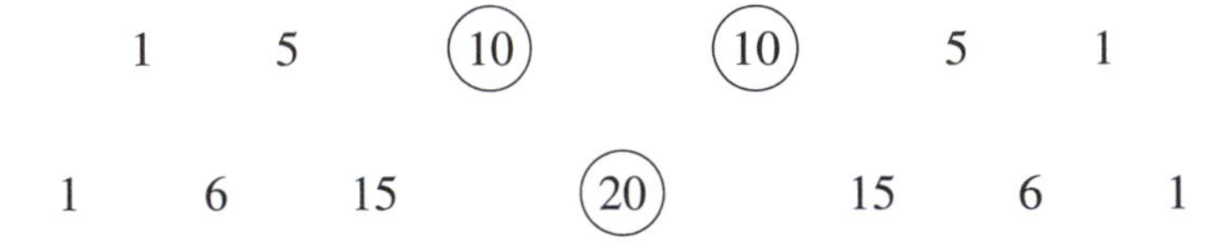

This pattern is called a **recursive identity** because it allows us to compute combinations in the case $n = m + 1$ if we know the values for $n = m$. We prove the general case with the following theorem:

Theorem 13.14 Recursive Identity for Combinations *Let* $1 \le k \le n$*. Then*

$$\binom{n+1}{k} = \binom{n}{k} + \binom{n}{k-1}$$

Proof Our proof is motivated by the U.S. Senate example. Recall that

$$\binom{n+1}{k} = \text{number of } k\text{-element subsets of a set of } n+1 \text{ elements}$$

We write the set A of $n+1$ elements in the suggestive form

$$A = \{S_1, S_2, \ldots, S_n, V\}$$

The subsets of A with k elements can be separated into two mutually exclusive groups:

Group 1: k senators chosen from $S_1, \ldots S_n$ without V:

$$\binom{n}{k} \text{ subsets}$$

Group 2: $k-1$ senators chosen from $S_1, \ldots S_n$ plus V:

$$\binom{n}{k-1} \text{ subsets}$$

Since these groups exhaust the possible k-element subgroups of A, we conclude that

$$\binom{n+1}{k} = \binom{n}{k} + \binom{n}{k-1}$$

■

Theorem 13.14 resolves Question 1.6: each senator has the same voting power as the vice president; the U.S. Senate, as yes-no system, is a Simple Majority voting system. Alternately, we could say that the fact that the U.S. Senate is actually a Simple Majority system despite its deceptive description is basically the proof of the Recursion Identity!

We mention one last combinatorial identity that can be useful for summing winning coalitions.

Theorem 13.15 Sum Identity *Let n be a positive integer. Then*

$$\binom{n}{0} + \binom{n}{1} + \binom{n}{2} + \cdots + \binom{n}{n-1} + \binom{n}{n} = 2^n$$

Proof We leave the proof of this identity for you in Exercise 13.14. ■

Exercises

13.1 Use combinations to compute the power index for the following weighted voting systems:

(a) $[35\colon\ 10,10,10,5,5,5,5,5,5]$
(b) $[120\colon\ 50,50,20,20,20,20,20]$
(c) $[40\colon\ 10,10,5,5,5,5,5,5,5,5]$

13.2 Consider the following yes-no voting system with voters P_1,P_2,P_3,P_4 and winning coalitions:

$$P_1,P_2 \quad P_1,P_2,P_3 \quad P_1,P_2,P_4 \quad P_1,P_3,P_4 \quad P_1,P_2,P_3,P_4$$

Prove that this is a weighted voting system by finding quota and weights. Compute the power index for this voting system. *Hint*: Compute the power index!

13.3 Count the winning coalitions in the following weighted voting systems:

(a) [24 : 4, 4, 4, 4, 4, 2, 2, 2, 2, 2, 2, 2, 2]
(b) [100: 20, 20, 20, 20, 20, 10, 10, 10, 10, 10, 10]
(c) [70: 10, 10, 10, 10, 10, 10, 10, 5, 5, 5, 5, 5, 5]

13.4 Compute the power index for the following weighted voting systems:

(a) $[80\colon\ 20,20,20,10,10,10,10,10,10,10,10]$
(b) $[20\colon\ 10,5,5,5,5,1,1,1,1,1]$
(c) $[100\colon\ 50,50,10,10,10,10,5,5,5,5,5,5,5,5,5,5]$

13.5 Consider the following yes-no voting system with voters P_1,P_2,P_3,P_4,P_5 and winning coalitions:

P_1,P_2	P_1,P_3	P_1,P_2,P_3	P_1,P_2,P_4	P_1,P_2,P_5
P_1,P_3,P_4	P_1,P_3,P_5	$P_1,P_2,P_3,P_4,$	P_1,P_2,P_3,P_5	
P_1,P_2,P_4,P_5	P_2,P_3,P_4,P_5	P_1,P_3,P_4,P_5	P_1,P_2,P_3,P_4,P_5	

Prove that this is a weighted voting system by finding quota and weights. Compute the power index for this voting system.

13.6 Consider the following yes-no voting system with voters P_1,P_2,P_3,P_4,P_5 and winning coalitions:

P_1,P_2	P_1,P_3	P_2,P_3	P_1,P_2,P_3	P_1,P_2,P_4
P_1,P_2,P_5	P_1,P_3,P_4	P_1,P_3,P_5	$P_2,P_3,P_4,$	P_2,P_3,P_5
P_1,P_4,P_5	P_1,P_2,P_3,P_4	P_1,P_2,P_4,P_5	P_1,P_3,P_4,P_5	
P_2,P_3,P_4,P_5	P_1,P_2,P_3,P_4,P_5			

Prove that this is a weighted voting system by finding quota and weights. Compute the power index for this voting system.

13.7 Before 1966, the U.N. Security Council consisted of 11 members, 5 permanent and 6 nonpermanent. A total of 9 yes votes including all 5 permanent members was required for passage of a measure. Find an equivalent weighted voting system for the original U.N. Security Council. Did a given permanent member have more or less power after 1966?

13.8 Define a yes-no voting system with four senators, $S_1,\ldots,S_4$, and eight congressmen, $C_1,\ldots,C_8$. To pass, a measure must have the support of all four senators and at least four congressmen. Prove that this is a weighted voting system by finding quota and weights.

13.9 The bicameral U.S. Congress is a yes-no voting system with voters of two types: The House of Representatives has 435 members, $C_1,\ldots C_{435}$. The Senate has 101 members, $S_1,\ldots,S_{101}$ (including the vice president, counted here as S_{101}). A measure passes the Congress if it receives a simple majority of votes in both the House and the Senate. Prove that the bicameral U.S. Congress is not trade robust and so not a weighted voting system.

13.10 Consider a yes-no voting system with six voters named A,B,C,D,E, and F. The voters A,B,C, and D are assigned the weights $4,3,2$, and 1, respectively. To pass, a measure must reach a quota of $q=6$ from voters A,B,C, and D and must also have either E or F's support. Decide whether or not this is a weighted voting system. Prove your answer.

13.11 Consider a yes-no voting system with five voters named A,B,C,D, and E. The voters A,B,C, and D are assigned the weights $4,3,2$, and 1, respectively. To pass, a measure must reach a quota of $q=6$ from voters A,B,C, and D and must also have E's support (E has veto power). Decide whether or not this is a weighted voting system. Prove your answer.

13.12 Consider a yes-no system with 10 voters, $P_1,\ldots,P_{10}$. A measure passes if it has the support of at least 4 even-numbered voters (i.e., P_2,P_4,P_6,P_8, and P_{10}) and at least 3 odd-numbered voters. Decide whether or not this is a weighted voting system. Prove your answer.

13.13 A bicameral voting system is defined as follows: the first body has five members and approves measures by the weighted voting system

$$[10\colon 6,5,3,2,1]$$

The second body also has five members and approves measures by the weighted system

$$[12\colon 6,6,4,4,2]$$

A measure is approved by the bicameral system if it is approved by both bodies. Prove that this system is not trade robust and hence not weighted.

13.14 Prove Theorem 13.15:

$$\binom{n}{0}+\binom{n}{1}+\binom{n}{2}+\cdots+\binom{n}{n-2}+\binom{n}{n-1}+\binom{n}{n}=2^n$$

Hint: Prove that there are 2^n subsets of a set of n elements. See Exercise 10.9.

13.15 Extend the argument given in Theorem 3.9 to prove a trade among n winning coalitions in a weighted system results in a least one winning coalition.

14 Gambling Games

Gambling games have inspired two major mathematical developments related to our work. In the seventeenth century, a writer named the Chevalier de Méré posed a question about probabilities with repeated rolls of the dice to the eminent mathematician and philosopher Blaise de Pascal. The problem led Pascal and another legend of mathematics, Pierre de Fermat, to develop the foundations for a modern theory of probability. In the twentieth century, the game of Poker with its complex mix of chance and strategy became a central problem for the development of game theory. Both Emilé Borel and John von Neumann published early work on the game in restricted settings. The psychology of betting and bluffing in Poker became a useful metaphor for the variables at play in economic and political interactions. While the complexity of Poker has proven too great to admit anything like a complete mathematical theory, working algorithms for playing the game can be extremely lucrative.

In this chapter, we discuss two popular gambling games: *Craps* and *Poker*. Our analysis of these games will motivate several new ideas. For the game of Craps, we explain the technique of "introducing a conditional," which allows us to compute a probability in an infinite game. Straight Poker, in turn, presents the problem of counting card hands. We apply the Combinations Formula (Theorem 11.12) to compute the probability of Poker hands. We then return to the game Five-Card Stud and in particular address Question 1.2 from Chapter 1. Our resolution of this question will motivate Bayes' Theorem (Theorem 14.5).

Craps

The dice game Craps creates excitement from the simple of act of repeated rolling of a pair of dice. We introduce this casino game now.

Example 14.1 Craps

The game is played on a long velvet table with one dice roller (the **shooter**) and many side bettors and spectators. Craps can go on indefinitely, with the tension and side bets mounting with each throw. The rules are as follows: the shooter throws the dice. If he or she rolls a sum of 7 or 11, he or she wins. If the shooter throws a sum of 2, 3, or 12 (**craps**), he or she loses. Otherwise, the sum rolled is called the **point**. We denote this sum by P. The shooter continually rolls until either (1) he or she rolls a sum of P again or (2) he or she rolls a sum of 7. In the first case, the shooter wins the bet, and in the second case, the shooter loses.

Craps is not a finite game, although it ends fairly quickly in practice. In a casino, a shooter keeps the dice as long as he or she wins. The record for the longest run of wins was set in Atlantic City in 2009. Pat DeMauro threw 154 straight wins in craps, holding the dice for 4 hours and 18 minutes.

We consider the problem of computing the probability

$$p = P(\text{“Shooter Wins Craps”})$$

The main difficulty we confront here is the indefinite length of the game. To overcome this difficulty, we imagine summarizing this game with a tree. It is then unnecessary to record all the rolls that result in no final outcome of the game. We can safely exclude these branchings from the game tree. Recall that we introduced the notion of an abridged game tree in Chapter 4 as a game tree with some nodes and branches omitted. The following abridged game tree conveniently conceals what could be a large number of rolls:

Abridged Game Tree for Craps

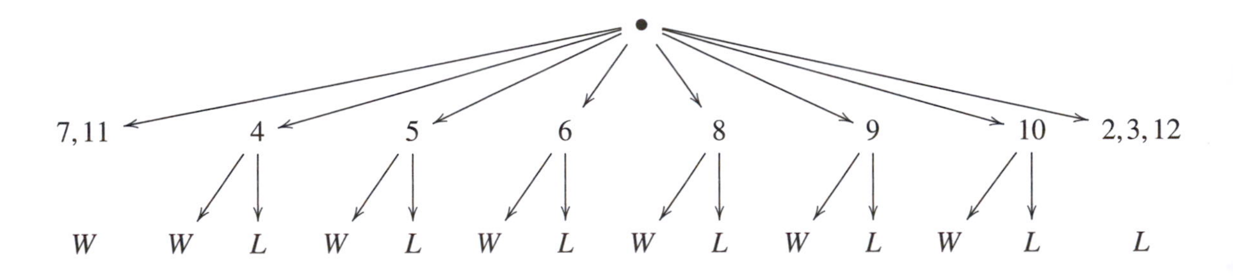

We have labeled the end nodes with the outcomes for the shooter: W = “Shooter wins” and L = “Shooter loses.”

The first stage represented in the preceding tree is just the first roll of the game. These probabilities are directly computed using a dice table. The more interesting branch probabilities are those at the second stage. These occur when a point is rolled. We seek, in each case, the probability of winning (or losing) given that the point is P. Let us focus on a particular case, say, $P = 10$. Here is the picture:

Point Is 10

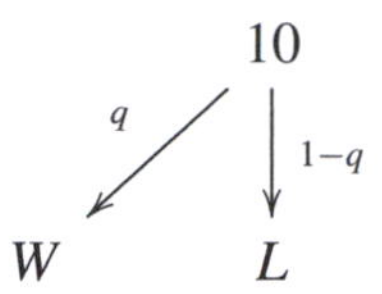

The branch probability q is a conditional probability:

$$q = P(\text{“Shooter wins given point is 10”}) = P(\text{W} \mid \text{“}P = 10\text{"})$$

To compute q, we use the abridged game tree as a guide and introduce a new conditional. At some roll in the future, the game of craps will end – or at least we assume that this is so! We imagine that we have just rolled the last roll. Since the point is 10, the last roll must be either a sum of 10 or a sum of 7. We introduce the following given condition: “Sum = 7 or 10.” We then observe that

$$q = P(\text{“Sum = 10”} \mid \text{“Sum = 7 or 10”})$$

We can compute this probability easily using the following dice table:

Last Roll in Craps with Point $P = 10$

		Blue Die					
		1	2	3	4	5	6
	1						O
	2					O	
Red Die	3				O		
	4			O			⊗
	5		O			⊗	
	6	O			⊗		

Our conditional assures us that the roll corresponds to one of the nine boxes with circles. The shooter wins if the box is labeled with an X. Thus

$$q = \frac{n(\otimes\text{’s})}{n(\text{O’s})} = \frac{3}{9} = \frac{1}{3}$$

The conditional probabilities $P(W \mid P = 4), P(W \mid P = 5), \ldots$ may be calculated in a similar fashion. The probability

$$p = P(\text{“Shooter wins craps”})$$

is then computed, as usual, using the Law of the Probability Tree. We leave the details for you in Exercise 14.1. We remark that betting on the shooter to win Craps represents one of the best straight bets in the casino.

Straight Poker

The game of Poker has countless variations. Generally, variations on the basic game alter the manner with which a player's hand is revealed. Although a Poker hand always consists of five cards, many versions of the game involve more than five cards per player, with each player selecting his or her best five-card hand. These innovations add complexity and suspense to the game. Perhaps most important, they create more opportunities for betting. We here consider Poker to be a game of pure chance. We focus on the probabilities that arise from dealing a five-card hand of cards.

Example 14.2 Straight Poker

This is a multiplayer game of pure chance. Start with a shuffled deck of cards. Recall that a standard deck has 52 cards, with 13 card values (A, 2, 3, 4, 5, 6, 7, 8, 9, 10, J, Q, K) and four suits ($\clubsuit, \diamondsuit, \heartsuit, \spadesuit$). Each player receives five cards from the deck. The player with the highest ranked hand wins, where the ranking of hands is determined by the following hierarchy:

(14.1) **The Hierarchy of Poker Hands**

1. **Straight Flush**: Five consecutive card values all of the same suit. Aces can be high or low.

   Examples:

   $$A\clubsuit 2\clubsuit 3\clubsuit 4\clubsuit 5\clubsuit \quad \text{or} \quad 10\heartsuit J\heartsuit Q\heartsuit K\heartsuit A\heartsuit$$

2. **Four of a Kind**: Four of the five cards have the same value.

   Examples:

   $$Q\clubsuit Q\diamondsuit Q\heartsuit Q\spadesuit 7\diamondsuit \quad \text{or} \quad 2\clubsuit 2\diamondsuit 2\heartsuit 2\spadesuit J\clubsuit$$

3. **Full House**: Three of the five cards have one value, and the remaining two have a second value.

   Examples:

   $$J\clubsuit J\diamondsuit J\heartsuit Q\heartsuit Q\diamondsuit \quad \text{or} \quad 7\diamondsuit 7\heartsuit 7\spadesuit A\clubsuit A\heartsuit$$

4. **Flush**: All five cards are the same suit, but the card values are not consecutive as in a Straight Flush.

   Examples:

   $$4\heartsuit 6\heartsuit J\heartsuit Q\heartsuit K\heartsuit \quad \text{or} \quad 3\clubsuit 7\clubsuit 8\clubsuit 9\clubsuit 10\clubsuit$$

5. **Straight**: Five consecutive card values, not all five cards of the same suit. Again, Aces can be high or low.

   Examples:

   $$4\clubsuit 5\heartsuit 6\heartsuit 7\spadesuit 8\heartsuit \qquad \text{or} \qquad A\diamondsuit 2\spadesuit 3\diamondsuit 4\diamondsuit 5\heartsuit$$

6. **Three of a Kind**: Three cards have the same value, the remaining two do not have this value, nor do these two have the same value.

   Examples:

   $$6\clubsuit 6\heartsuit 6\spadesuit Q\heartsuit K\spadesuit \qquad \text{or} \qquad A\diamondsuit A\heartsuit A\spadesuit J\diamondsuit 2\spadesuit$$

7. **Two Pairs**: Two pairs of cards have the same values; the fifth card does not match either of these two values.

   Examples:

   $$10\heartsuit 10\spadesuit J\clubsuit J\heartsuit K\heartsuit \qquad \text{or} \qquad 8\diamondsuit 8\heartsuit 9\diamondsuit 9\spadesuit Q\diamondsuit$$

8. **Pair**: One pair of cards has the same value; the other three cards have three different values.

   Examples:

   $$2\heartsuit 2\spadesuit 9\spadesuit Q\heartsuit A\clubsuit \qquad \text{or} \qquad J\diamondsuit J\spadesuit 8\diamondsuit 9\spadesuit 10\diamondsuit$$

9. **Nothing**: None of the preceding apply.

   Examples:

   $$4\heartsuit 7\spadesuit J\heartsuit Q\diamondsuit K\heartsuit \qquad \text{or} \qquad 5\diamondsuit 8\spadesuit 9\spadesuit J\diamondsuit A\diamondsuit$$

Straight Flush is the best hand, and Nothing is the worst. Note that if two players have the same hand, as in the preceding list, the winner is based on high cards. For example,

$$J\clubsuit J\heartsuit A\diamondsuit Q\clubsuit 10\diamondsuit \qquad \text{beats} \qquad J\diamondsuit J\spadesuit A\heartsuit 9\diamondsuit 7\spadesuit$$

because after the tie of pairs and high cards (aces), Player 1 has the queen over the nine.

The problem we address here concerns the preceding hierarchy. It is natural to ask why this particular ordering is used. For example, why is Three of a Kind ranked higher than Two Pairs? Why does a Flush beat a Straight? Presumably, this ordering has to do with the likelihood of the various hands. In fact, we have the following theorem:

Theorem 14.3 *The Poker hands are ranked in increasing order of their probabilities.* ■

The proof of this theorem is pure computation. Specifically, each probability from P(Straight Flush) down to P(Nothing) must be calculated and compared. We make some of these calculations here, leaving the rest for you in the exercises. We use unordered five-card hands for our sample space and make extensive use of the Combinations Formula (Theorem 11.12). In particular, recall that the total number of simple outcomes (unordered five-card hands) is given by

$$\binom{52}{5} = \frac{52 \cdot 51 \cdot 50 \cdot 49 \cdot 48}{5 \cdot 4 \cdot 3 \cdot 2 \cdot 1} = 2{,}598{,}960$$

We start at the top of the hierarchy.

Straight Flush

This one is the easiest of all. A Straight Flush is determined by two choices: a suit (4 choices) and the lowest card in the straight (Ace through Ten, giving 10 choices). Thus there are 40 total straight flushes, and

$$P(\text{“Straight Flush”}) = \frac{40}{\binom{52}{5}} = 0.000015$$

Flush

To describe a Flush, we first specify a suit (4 choices) and then specify the 5 values in this suit. Once we fix a suit, we are restricted to 13 card values. Thus there are $\binom{13}{5}$ possibilities in each fixed suit and so $4 \cdot \binom{13}{5}$ total. Now we must be careful because we cannot allow our hand to be a Straight Flush. However, we know that there are 40 of these, so we can subtract these from our count. We conclude that

$$P(\text{“Flush”}) = \frac{4 \cdot \binom{13}{5} - 40}{\binom{52}{5}} = 0.00197$$

Three of a Kind

Again, we imagine building the hand, in this case a Three of a Kind, in several steps and count the number of choices at each step. We give two methods for counting Three-of-a-Kind hands.

Direct Method. Begin by specifying the card value (Two through Ace) we have for our Three-of-a-Kind hand: there are 13 possibilities here. Next, specify the suits for these 3 cards: there are 4 suits, and we need to choose 3, so there are $\binom{4}{3} = 4$ possibilities. Finally, we will pick our last two cards. Here we must be careful. We do not want either of the last two cards to be the fourth remaining card of our Three of a Kind. We remove this card and reduce our choices to the 48 cards remaining. We also don't want to get a Full House. So we want to draw 2 cards from 48 and not get a pair. It is easier to determine how many two-card hands are pairs: there are 12 values, so choose one and then choose the two suits, that is, 4 suits, choose 2, or $\binom{4}{2} = 6$ choices. Thus there are $\binom{48}{2} - 12 \cdot 6 = 1{,}056$ ways to choose the last two cards. We conclude that

$$P(\text{“Three of a Kind”}) = \frac{4 \cdot 13 \cdot 1056}{\binom{52}{5}} = 0.02113$$

Values-First Method. We avoid the worry about getting better hands (such as Four of a Kind or Full House) and having to subtract by introducing a method we call the *values-first method.* This method works well for hands featuring matching values. First, we note that a Three-of-a-Kind hand has a total of 3 card values: two singletons and one value that is the Three of the Kind. As a first step, we select these 3 values. There are 13 values, and we need 3, so there are $\binom{13}{3}$ choices. For concreteness, let us suppose that we have chosen 2, Jack, and King for our three values. We construct an unordered Three-of-a-Kind hand from this choice.

For the second step, we distinguish one of the three values 2, Jack, or King to be the Three-of-a-Kind value. By default, the other two values will be singletons. There are 3 choices for the Three-of-a-Kind value. Let us suppose that King has been chosen to be the Three-of-a-Kind value. Notice that, equivalently, we could choose the two singleton values (2, Jack). This equivalence is reflected in the fact that

$$\binom{3}{1} = \binom{3}{2} = 3$$

Finally, we pick the suits. For the Three-of-a-Kind value King, we need 3 cards and so must pick 3 suits for these. There are $\binom{4}{3}$ choices for these suits. For, the other two values (2 and Jack), we just need one suit and so have 4 choices each. We summarize this approach as follows:

Values-First Count for Three of a Kind

Step 1:	Choose 3 values	$\binom{13}{3}$ choices
Step 2:	Distinguish one value for “Three of a Kind”	3 choices
Step 3:	Choose suits:	
	Choose suits for Three of a Kind:	$\binom{4}{3}$ choices
	Choose suits for singletons:	$4 \cdot 4$ choices
		$\binom{13}{3} \cdot 3 \cdot \binom{4}{3} \cdot 4 \cdot 4 = 54{,}912$ choices

The values-first approach is especially useful for lower-ranked hands such as Pair and Two Pairs.

Five-Card Stud

We now address Question 1.2 from Chapter 1. Recall the scenario from Example 1.1, summarized as follows:

A Hand of Five-Card Stud

	Up cards			Down cards	
Us	A♠	A♢	A♣	3♢	Q♡
Opponent	2♣	5♣	7♣	?	?

The current pot is \$100,000, and our opponent has just bet \$50,000 more. Our question is as follows:

Question 1.2 *Should we call or fold in A Hand of Five-Card Stud?*

Again, we see that the only hand that beats our Three of a Kind is a Flush. Let

$$p = P(\text{“Opponent has Flush”})$$

so p is the probability the two unseen cards are two clubs. The calculation of p is not difficult. Remove the 8 cards showing from consideration, leaving 44 cards. Of these 44 cards, there are 9 clubs. Thus

$$p = \frac{\binom{9}{2}}{\binom{44}{2}} \approx 0.04$$

If our opponent's cards were not dealt yet and we were simply betting against him or her getting the Flush, we would have a great bet with a 96 percent chance of winning. However, more information is available in the given scenario. Our opponent has seen the cards, and he or she has bet \$50,000 on the hand! The fact of the bet represents a very important given condition. The relevant probability for our question is the conditional probability

$$P(\text{“Opponent has Flush”} \mid \text{“Opponent has Bet”})$$

Recall that we introduced a key simplification. We have placed the odds of our opponent bluffing at only a 3 percent chance. We interpret this to mean that our opponent bets without a Flush with probability $b = 0.03$. The probability tree is now reminiscent of the tree for Betting on Diamonds (Example 8.1). We keep track of our opponent's events as follows:

Probability Tree for A Hand of Five-Card Stud

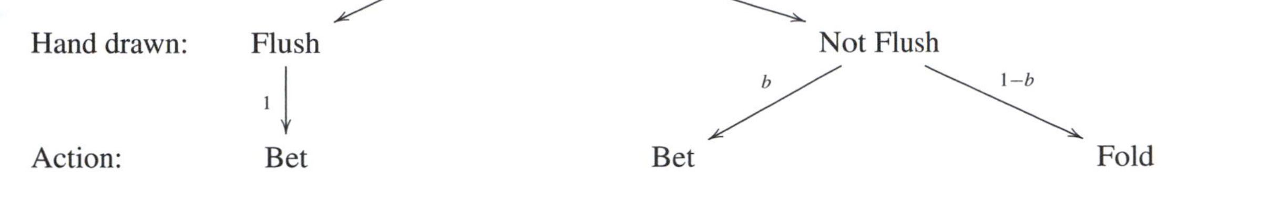

Probability: p $\quad (1-p)b \quad (1-p)(1-b)$

We compute the conditional probability P(“Has Flush” | “Bet”) using this tree and the Law of Conditional Probability (Theorem 11.17). Thus,

$$
\begin{aligned}
P(\text{“Has Flush”} \mid \text{“Bet”}) &= \frac{P(\text{“Has Flush” and “Bet”})}{P(\text{“Bet”})} \\
&= \frac{p}{b(1-p)+p}
\end{aligned}
$$

Before turning to our particular values of p and b, let us consider the formula. If our opponent never bluffs ($b = 0$), then

$$P(\text{“Has Flush”} \mid \text{“Bet”}) = 1$$

Similarly, if our opponent always bluffs ($b = 1$), then

$$P(\text{“Has Flush”} \mid \text{“Bet”}) = p$$

Notice that if our opponent always bluffs, then he or she always bets. The given condition is no information in this case. In general, the larger b is, the less likely our opponent has the cards. This is to be expected. If we perceive that our opponent is a bluffer, we are more likely to call. We emphasize that fixing the bluffing percentage for our opponent represents a significant simplification of the game.

For our scenario, we compute

$$P(\text{“Has Flush”} \mid \text{“Bet”}) = \frac{(0.04)}{(0.04)+(0.03)(0.96)} = 0.58$$

There is a 58 percent chance that we will lose this hand if we call.

Although the cautious player may be inclined to fold based on our analysis so far, the veteran gambler will consider still one further piece of information. The probability calculation alone does not take into the account the significance of the size of the bet or, for that matter, the size of the pot. When there is money involved in a game of chance, we should consult the expected value to truly evaluate what is at stake.

For our scenario, we make an important observation: regardless of how much we put into it, the pot is not our money now. We count our gains and losses starting at the moment our opponent bets. With this view, the pot of \$100,000 is money to be won. If we fold, we get nothing. If we call and win, we gain the original pot plus our opponent's bet for a total of \$150,000. If we call and lose, we are out \$50,000. The expected value of calling is then

$$E = \$150,000(0.42) - \$50,000(0.58) = \$34,000$$

Since the expected value of calling is substantially more than that of folding ($E = 0$), our analysis suggests that we should call the bet. A bet for which the probability of winning is less than even but for which the expected value is positive is called a **value bet**.

Bayes' Law

The calculation of the conditional probability $P(\text{"Has Flush"} \mid \text{"Bet"})$ is a special case of Bayes' Law. We provide the following definition:

Definition 14.4 Given a collection of events $B_1, \ldots, B_n$ for a probability experiment, we say that the collection is a **partition of the sample space** if

(a) Each pair of events B_i and B_j are mutually exclusive for $i \neq j$, and
(b) The sum of the probabilities $P(B_1) + \cdots + P(B_n) = 1$.

In our analysis of the Hand of Five-Card Stud, the events "Bet" and "Fold" represent the relevant partition. We prove the following theorem:

Theorem 14.5 Bayes' Law *Let $B_1, \ldots, B_n$ be events in a probability experiment that partition the sample space. Let W be another event. Then, for each $i = 1, \ldots, n$, we have*

$$P(W \mid B_i) = \frac{P(W \, and \, B_i)}{P(W \, and \, B_1) + P(W \, and \, B_2) + \cdots + P(W \, and \, B_n)}$$

Proof Since the collection of $B_1, \ldots, B_n$ forms a partition, we see that the events "W and B_j" are pairwise mutual exclusive events. Further, every outcome of A is favorable to some B_j by condition (b) in the definition. It follows from the Law of Mutually Exclusive Events that

$$P(W) = P(W \, \text{and} \, B_1) + \cdots + P(W \, \text{and} \, B_n)$$

(See Exercise 14.13.) The result is now obtained by solving the Law of Conditional Probability equation $P(W \, \text{and} \, B_i) = P(W) \cdot P(W \mid B_i)$ for $P(W \mid B_i)$. ■

Bayes' Law introduces an important point of view. Suppose that we reach a particular outcome after a series of probability experiments. Bayes' Law allows us to reverse the chronology and determine the probability that an event occurred earlier in the process based on our current position. This point of view has significant applications in several areas of applied mathematics. We give one example here from the realm of politics.

Example 14.6 Interviewing a Voter

A community with three political parties (Democrat, Republican, and Independent) is preparing to vote on a tax reform bill. While 90 percent of the Independent voters favor the bill, the bill is less popular among the Democratic or Republican voters: only 30 percent of the Democratic voters and 20 percent of the Republican voters favor the bill. The Democratic voters make up 55 percent of the population, the Republican voters 40 percent, and the Independent voters only 5 percent. A news reporter interviews a voter chosen at random and discovers that the voter favors the bill. What is the probability that the voter is an Independent?

We solve this problem using the following probability tree:

Probability Tree for Interviewing a Voter

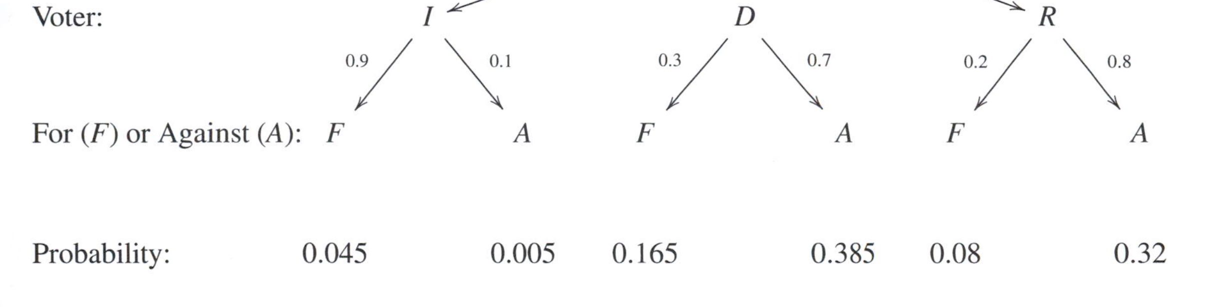

By Bayes' Law, we compute

$$P(I \mid F) = \frac{0.045}{0.045 + 0.165 + 0.08} = 0.155$$

There is a 15.5 percent chance that the voter is an Independent.

Exercises

14.1 Complete the labeling of the following abridged probability tree for the game of Craps.

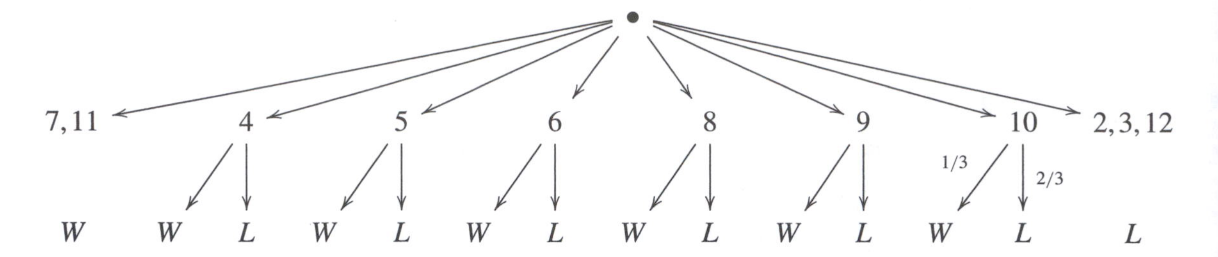

Compute the probability $p = P$("Shooter wins Craps"). You should get $p < 0.5$.

14.2 Side bettors in craps may bet against the shooter, putting money on the "Don't pass" line. The "Don't pass" bet wins if, on the first roll, the shooter throws a sum of 2 or 3 and loses if, on the first roll, a sum of 7 or 11 is thrown. If a sum of 12 is thrown on the first roll, the bet is a "push" – the "Don't pass" bettor breaks even. If neither a $2, 3, 7, 11$, or 12 is rolled on the first roll, then the first roll is, as usual, the point P. The "Don't pass" bet wins if the shooter throws a sum of 7 before P and loses if the shooter throws P before a 7. Compute the probability of winning a "Don't pass" bet.

14.3 Use Bayes' Law and Exercise 14.1 to compute the following conditional probabilities:

(a) P("Point = 4" | "Shooter wins Craps") =
(b) P("Point = 8" | "Shooter wins Craps") =
(c) P("Point = 9" | "Shooter loses Craps") =
(d) P("First-roll sum = 2, 3, or 12" | "Shooter loses Craps") =
(e) P("First-roll sum = 7" | "Shooter wins Craps") =

14.4 Complete the proof of Theorem 14.3 by computing the following probabilities:

(a) P("Four of a Kind") =
(b) P("Full House") =
(c) P("Straight") =
(d) P("Two Pairs") =
(e) P("Pair") =
(e) P("Nothing") =

For parts (a), (b), (d), and (e), use the values-first approach. For the Straight, notice that the values are determined by the low card. Be sure to subtract off the Straight Flushes. Finally, for part (e), you can use the complement rule. Can you calculate P("Nothing") directly?

14.5 Compute the probability of drawing 6 cards from a deck of 52 and getting Three of a Kind and no higher hand.

14.6 Compute the probability of drawing 7 cards from a deck of 52 and getting one suit of 5 cards and another suit with the other 2 cards.

14.7 Compute the probability of drawing 7 cards from a deck of 52 and getting 3 pairs and no Three-of-a-Kind hand.

14.8 (Bridge) The card game **Bridge** is played with four players split into two teams. The cards are dealt out completely so that each player has 13 cards. We will not describe the play of the game but simply mention that the number of cards a player has in each suit plays a critical role. Long suits are assets in bridge. The list of the number of cards in each suit is called the **distribution** of the hand. For example, a distribution of $[5,4,3,1]$ means that the player has one 5-card suit, one 4-card suit, one 3 card suit, and one suit with just one card. Notice that the four numbers describing a distribution always add to 13. Compute the probabilities of the following distributions. In each case, you should use the suits-first method. That is, assign suits to the distribution numbers (counting the number of ways to do this) and then choose the card values for each suit.

(a) $[5,4,3,1]$
(b) $[7,4,2,0]$
(c) $[6,6,1,0]$
(d) Two five-card suits
(e) One seven-card suit

14.9 Consider the following hands of Five-Card Stud. In each example, the pot is currently \$100,000. Your opponent has just bet \$50,000 more. Compute your expected value for calling your opponent's bet. Use the given value b for the bluffing probability.

(a)

	Up cards			Down cards	
You:	A♠	A♢	A♣	4♠	Q♠
Opponent:	2♠	5♠	7♠	?	?

Your opponent bluffs with probability $b = 0.01$.

(b)

	Up cards			Down cards	
You:	K♠	K♢	K♣	J♢	7♡
Opponent:	A♣	A♠	8♣	?	?

Your opponent bluffs with probability $b = 0.10$.

(c)

	Up cards			Down cards	
You:	2♡	2◇	2♣	4♡	J♣
Opponent:	5♠	5♡	4♣	?	?

Your opponent bluffs with probability $b = 0.05$. *Hint:* To compute the probability of getting at least one five, compute the probability of getting no fives and use the Complement Rule.

(d)

	Up cards			Down cards	
You:	7♠	7♣	7◇	6♡	J♣
Opponent:	A♠	2♡	3♡	?	?

Your opponent bluffs with probability $b = 0.10$.

(e) Determine the bluffing percentage b so that the expected value $E = 0$ in part (a).

14.10 When testing for a disease requiring immediate medical attention, the worst-case scenario is that the test is negative and yet the tested patient does have the disease. Bayes' Law can be used to determine the probability of this unfortunate outcome. Suppose that we have Test X, which returns either "Positive" or "Negative" when testing a patient for a certain disease. We know that Test X returns "Positive" 99 percent of the time when the patient is sick and "Negative" 85 percent of the time when the patient is not sick. Suppose that 95 percent of a population is not sick. Compute the probability $P(\text{"Sick"} \mid \text{"Negative"})$.

14.11 A population is split 60–40 between supporting Candidate Y and opposing his reelection. Of the supporters of Candidate Y, 90 percent support his tax plan, whereas of those who oppose him, only 45 percent support the tax plan. Suppose that a voter is chosen at random and supports the tax plan. Find the probability that the voter is a supporter of Candidate Y.

14.12 The voting population in a state is 55 percent registered Democrats and 45 percent Republicans. Among registered Democratic voters, 90 percent vote for the Democratic candidate, whereas 85 percent of the registered Republican voters vote for the Republican candidate. What is the probability that a voter who votes for the Republican candidate is a registered Republican?

14.13 Suppose that events $B_1, \ldots, B_n$ form a partition of the sample space for a probability experiment. Let W be another event for this experiment. Use the Principle of Induction (10.1) to prove that

$$P(W) = P(W \text{ and } B_1) + P(W \text{ and } B_2) + \cdots + P(W \text{ and } B_{n-1}) + P(W \text{ and } B_n)$$

15 Zero-Sum Games

Zero-sum games are encounters of total conflict. Whether it be opposing sides in a legal case, a negotiation between a seller and a buyer, or a fight for political turf, the outcome of a zero-sum game reveals a clear winner and a loser with corresponding gains and losses. Without the nuances that arise from the possibility of cooperation, zero-sum games are well suited for mathematical analysis. In this chapter, we discuss the Fundamental Theorem of Game Theory, also known as the *Minimax Theorem*. The theorem was proved in full generality by John von Neumann in 1928, an achievement often taken as the birth of modern game theory.

The Minimax Theorem assigns a *value* to every zero-sum game representing the expected value E of the game. In Theorem 9.9, we assigned the value $E = -\$.33$ to the Poker game Betting on Diamonds. In this case, the value signifies that Player 1 stands to lose 33 cents a game on average. When the two players in a zero-sum game each have only two strategy choices, finding the value of the zero-sum game is an interesting exercise. We discuss this procedure and use the ideas to prove the 2×2 *Minimax Theorem*. We then consider the case where one player can choose from more than two strategies. We compute the value of the Poker game Bluffing, Calling, and Raising (Example 8.8) to illustrate the technique in this case. We begin with a motivating example from the world of sports.

Example 15.1 Coaching Conflict

In the game of football, the battle for yardage represents a zero-sum game. We imagine that the coaches of the Offense and Defense on opposite teams in a football game engage in a simultaneous-move game of strategy. The Offense chooses between the strategies Run (R) and Pass (P). The Defense chooses between Blitz (B) and Zone (Z). The payoff is the yardage the Offense gains on the play. The payoff matrix is as follows:

Payoff Matrix for Coaching Conflict

Defense

Offense

Strategy	B	Z
R	8	2
P	−3	5

The Coaching Conflict might be replayed continually through the course of the football game, creating a repeated game. The repeated game creates a mixed-strategy game as in Definition 8.7. Recall that this means that we set

$$p = \text{probability the Offense chooses to Run}$$
$$q = \text{probability the Defense chooses to Blitz}$$

The mixed-strategy game is then the zero-sum game in which the Offense chooses a probability p and the Defense chooses a probability q. The payoff is the expected value

$$E(p,q) = \text{expected number of yards for the Offense per play}$$

As we did for the Poker game Betting on Diamonds in Chapter 9, we could compute $E(p,q)$ directly and try to solve for probabilities p^* and q^*, giving a Nash equilibrium for the mixed-strategy game. We give an alternate approach here. We imagine that the coaches play the mixed-strategy game as a sequential-move game with the Defense revealing q first and then the Offense revealing p.

Mixed-Strategy Version of Coaching Conflict: Defense First

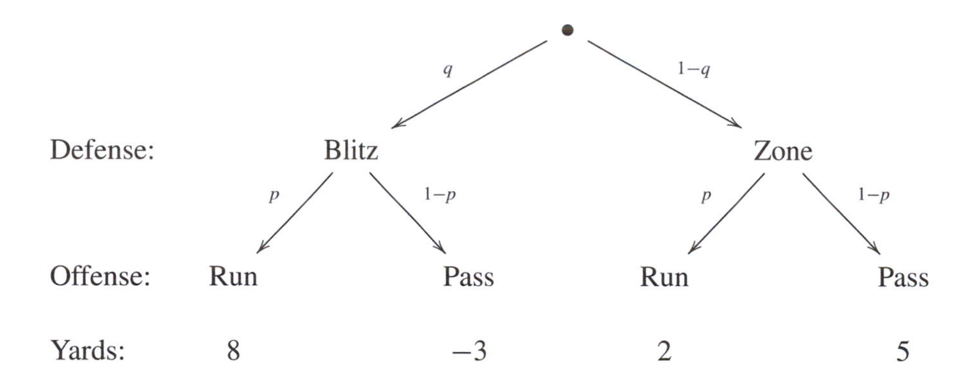

The expected value $E(p,q)$ can be computed directly from this tree. We compute $E(p,q)$ using the Linearity of Expectation (Theorem 3.13) instead. Let

$$E_B = \text{expected value when Defense chooses Blitz}$$
$$E_Z = \text{expected value when Defense chooses Zone}$$

Then $E = E(p,q)$ can be computed as follows:

Expected Value of Coaching Conflict: Defense First

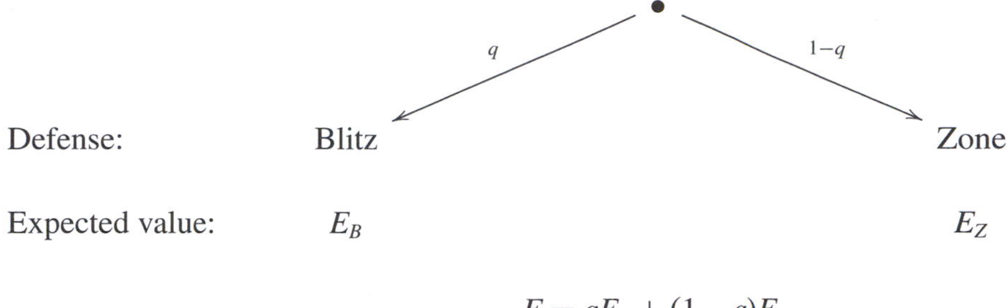

$$E = qE_B + (1-q)E_Z$$

The Offense can neutralize the Defense by ensuring that the subgames corresponding to the Blitz and Zone produce the same expected values, that is, $E_B = E_Z$. Observe that when $E_B = E_Z$, we have

$$\begin{aligned} E &= qE_B + (1-q)E_Z \\ &= qE_B + (1-q)E_B \\ &= E_B \end{aligned}$$

The variable q has been removed from the equation! Substituting, $E_Z = E_B$ similarly gives $E = E_Z$ as well. The mixed-strategy choice q by the Defense does not matter when $E_B = E_Z$.

Now the subgame expected values E_B and E_Z depend on the Offense's choice of p. Using the second stage of the game tree, we compute

$$\begin{aligned} E_B(p) &= 8p - 3(1-p) \qquad & E_Z(p) &= 2p + 5(1-p) \\ &= 11p - 3 & &= -3p + 5 \end{aligned}$$

To neutralize the Defense, the Offense sets $E_B(p) = E_Z(p)$ and solves for $p = p^*$. That is,

$$E_B(p) = E_Z(p): \qquad 11p - 3 = -3p + 5 \implies 14p = 8 \implies p^* = \frac{4}{7}$$

When the Offense chooses the probability $p^* = 4/7$, the Defense's choice of q is irrelevant. The expected value E when the Offense chooses p^* is the common value

$$E = E_B(4/7) = E_Z(4/7) = 3.3 \text{ yards per play}$$

We can repeat the same process to compute the probability q^* that the Defense can choose to neutralize the Offense. Simply reverse roles and write the following:

Mixed-Strategy Version of Coaching Conflict: Offense First

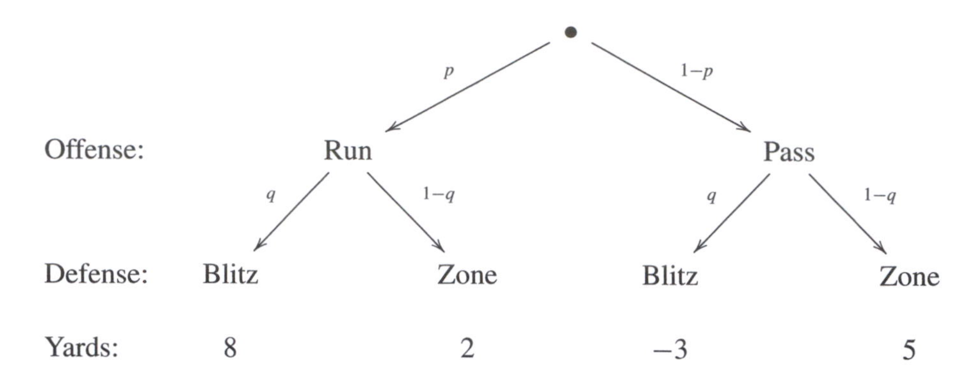

Again, the expected value can be computed as follows from the subgame diagram using the Linearity of Expectation:

Expected Value of Coaching Conflict: Offense First

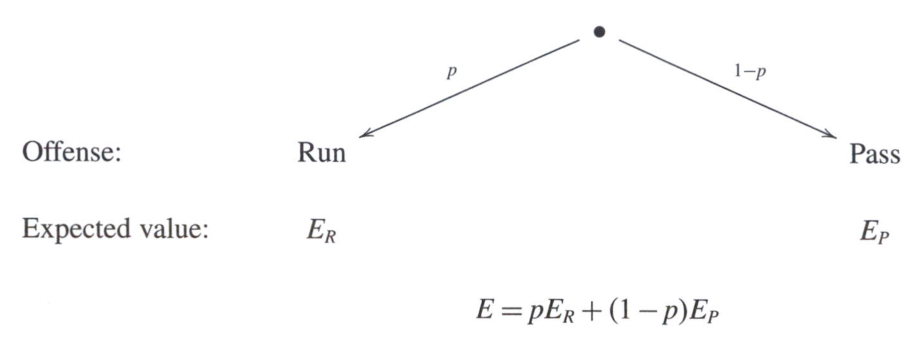

$$E = pE_R + (1-p)E_P$$

In this case, the expected values E_R and E_P depend on the probability q. That is,

$$\begin{aligned} E_R(q) &= 8q + 2(1-q) \qquad & E_Z(q) &= -3q + 5(1-q) \\ &= 6q + 2 & &= -8q + 5 \end{aligned}$$

The Defense can neutralize the Offense by setting $E_R(q) = E_P(q)$ and solving for q^*. That is,

$$E_R(q) = E_P(q): \qquad 6q + 2 = -8q + 5 \implies 14q = 3 \implies q^* = \frac{3}{14}.$$

When the Defense mixes strategies by playing Blitz with probability $q^* = 3/14$ and Zone with probability $1 - q^* = 11/14$, the expected value E of the game does not depend on p and is given by the common value

$$E = E_R(3/14) = E_P(3/14) = 3.3 \text{ yards}$$

It is reassuring but not surprising that we obtain the same expected value $E = 3.3$ yards from either computation. Suppose that the Defense chooses $q^* = 3/14$. Then the expected value E is the same for whatever value of p the Offense chooses. In particular, the Offense can choose $p = p^* = 4/7$. We have already seen that when the Offense chooses $p = p^*$, the expected value is $E = 3.3$ yards. The common value of all four subgame expected values

$$E = E_B(p^*) = E_Z(p^*) = E_R(q^*) = E_P(q^*) = \text{ 3.3 yards}$$

is called the *value* of the zero-sum game Coaching Conflict.

The preceding analysis gives an effective method for determining the value of a zero-sum game provided that each player has only two strategies. We summarize this method now.

Solving Zero-Sum Games

The subgame expected values $E_B(p), E_Z(p), E_R(q)$, and $E_P(q)$ used in the analysis of Coaching Conflict correspond to one player or the other choosing a pure strategy. Recall from Definition 15.6 that a player chooses a pure strategy in a repeated game simply by selecting the one strategy on every round.

By a **2** $\times$ **2 zero-sum game** we mean a zero-sum game in which the two players each choose between two strategies. We write the general 2×2 payoff matrix for a zero sum game as follows:

Payoff Matrix for General 2 x 2 Zero-Sum Game

Player 2

Player 1

Strategy	X	Y
A	a	b
B	c	d

We have four pure strategies to consider, Player 1 choosing A or B and Player 2 choosing X or Y. The expected values are computed directly from the payoff matrix:

Player 1 Mixes Strategies, Player 2 Chooses ***X***

Player 2

Player 1

Probability	Strategy	X	Y
p	A	a	b
$1-p$	B	c	d

$$E_X(p) = pa + (1-p)c$$

Player 1 Mixes Strategies, Player 2 Chooses Y

Player 2

Player 1

Probability	Strategy	X	Y
p	A	a	b
$1-p$	B	c	d

$$E_Y(p) = pb + (1-p)d$$

Player 1 Chooses A, Player 2 Mixes Strategies

Player 2

Player 1

Probability	q	$1-q$
Strategy	X	Y
A	a	b
B	c	d

$$E_A(q) = qa + (1-q)b$$

Player 1 Chooses B, Player 2 Mixes Strategies

Player 2

Player 1

Probability	q	$1-q$
Strategy	X	Y
A	a	b
B	c	d

$$E_B(q) = qc + (1-q)d$$

By Linearity of Expectations, we have two expressions for $E(p,q)$, the expected value of the mixed-strategy game. That is,

$$E(p,q) = E(p,q) = qE_X(p) + (1-q)E_Y(p) \quad \text{and} \quad E(p,q) = pE_A(q) + (1-p)E_B(q)$$

If we can find the probabilities p^* and q^* solving the linear equations

$$E_X(p^*) = E_Y(p^*) \quad \text{and} \quad E_A(q^*) = E_B(q^*) \tag{15.1}$$

then all four values expected values will be equal. Their common value $E = E(p^*,q^*)$ is the payoff for a Nash equilibrium (p^*,q^*) for the mixed-strategy game. We may formulate our progress as follows:

Theorem 15.2 *The mixed-strategy version of the general* 2×2 *zero-sum game has a Nash equilibrium* (p^*, q^*) *provided that both the equations*

$$E_X(p^*) = E_Y(p^*) \quad \text{and} \quad E_A(q^*) = E_B(q^*)$$

can be solved for probabilities p^* *and* q^*.

Proof When Player 1 chooses the mixed strategy $p = p^*$, we have

$$\begin{aligned} E(p^*, q) &= qE_X(p^*) + (1-q)E_Y(p^*) \\ &= qE_X(p^*) + (1-q)E_X(p^*) \\ &= E_X(p^*) \end{aligned}$$

The payoff of the mixed-strategy game is fixed at the value $E_X(p^*)$. Player 2 cannot unilaterally improve from $q = q^*$. In fact, Player 2 cannot change the value of E.

Similarly, when Player 2 plays $q = q^*$, then

$$\begin{aligned} E(p, q^*) &= pE_A(q^*) + (1-p)E_B(q^*) \\ &= pE_A(q^*) + (1-p)E_A(q^*) \\ &= E_A(q^*) \end{aligned}$$

Player 1's choice of p is now irrelevant to E. Since neither player can improve by a unilateral change of strategy, the pair (p^*, q^*) is a Nash equilibrium for the mixed-strategy game by definition. The second assertion follows, as before. ■

Theorem 15.2 is a special case of the famous Minimax Theorem. We state the general result here: define a **finite zero-sum game** to be a zero-sum game in which the players each choose from a finite list of probabilities. For instance, Player 1 may choose from m strategy choices and Player 2 from n strategy choices. The game in this case is called an ***m* x *n* game**.

Theorem 15.3 The Minimax Theorem *The mixed-strategy game arising from any finite zero-sum game has a Nash equilibrium.*

The name "Minimax" refers to the fact that, at the Nash equilibrium of a mixed-strategy game, the players have simultaneously optimized their expected payoff versus the other's mixed-strategy choice. The Minimax Theorem leads to the following important definition:

efinition 15.4 The **value** of a finite zero-sum game is the expected value E at the Nash equilibrium of the mixed-strategy version of the game. The probabilities giving this Nash equilibrium are called the **mixed-strategy equilibrium** of the zero-sum game.

For the Coaching Conflict, we have

$$p^* = \frac{4}{7}, \quad q^* = \frac{3}{14}, \quad \text{Value} = 3.3 \text{ yards}$$

Let us return to the 2×2 case. With Theorem 15.8 in hand, we are left to ask whether the pure-strategy expected-value equations in (15.1) can always be solved. Here is an example addressing this question:

Example 15.5 Coaching Conflict II

The Offense in Coaching Conflict has just signed the league's best running back. With this new weapon, the payoff matrix has become

Payoff Matrix for Coaching Conflict II

Defense

Offense

Strategy	B	Z
R	10	6
P	−3	5

We apply the method of pure strategies outlined earlier. First, compute p^* by solving

$$E_B(p) = E_Z(p): \quad 10p - 3(1-p) = 6p + 5(1-p) \implies p^* = \tfrac{2}{3}$$

All is well. Next, compute q^* by considering pure strategies for the Offense. That is,

$$E_R(q) = E_P(q): \quad 10q + 6(1-q) = -3q + 5(1-q) \implies q^* = -\tfrac{1}{12}$$

We now have a problem – a negative probability!

You may have already noticed the reason for this problem. With the new running back, the Offense has no reason to pass: Run dominates Pass for the Offense. The original game has a Nash equilibrium, a saddle point, at Run and Zone. For the mixed-strategy game, the pure strategies "Always run" and "Always zone" are then Nash equilibria. Now these pure strategies correspond to the probabilities $p^* = 1$ and $q^* = 0$, and these are then the mixed-strategy equilibrium for Coaching Conflict II. The value of this game is simply the saddle-point value

$$p^* = 1, \quad q^* = 0, \quad \text{Value} = 6 \text{ yards}$$

We summarize this case in the following definition:

Definition 15.6 A zero-sum game with a saddle point is said to have a **pure-strategy equilibrium**. In the 2×2 case, this corresponds to a mixed-strategy equilibrium (p^*, q^*) when $p^* = 0$ or 1 and $q^* = 0$ or 1. The **value** of a zero-sum game with a saddle point is just the value of the saddle point.

The connection here between the failure to find probability q^* and the dominated strategy in the 2×2 game is not a coincidence. Notice that we failed to find a probability for the Defense's mixed strategy q^*; and yet it is the Offense with a dominated strategy. In fact, this is how it works generally.

Theorem 15.7 *Suppose that we are given a 2×2 zero-sum game in which Player 1 chooses between A and B and Player 2 chooses between X and Y.*

1. *Suppose that $E_X(p) = E_Y(p)$ does not have a solution for any probability p. Then Player 2 has a weakly dominant strategy.*
2. *Suppose that $E_A(q) = E_B(q)$ does not have a solution for any probability q. Then Player 1 has a weakly dominant strategy.*

Proof Suppose that (1) is true so that $E_X(p) = E_Y(p)$ does not have a solution for $0 \leq p \leq 1$. This means that, for all p in this range, either $E_x(p) > E_Y(p)$ or $E_x(p) < E_Y(p)$.

Let us assume that $E_X(p) > E_Y(p)$ for $0 \leq p \leq 1$. This means that Y always produces a smaller expected value than X no matter how Player 1 mixes strategies with p. Player 2 has a dominated strategy to play Y. We can picture this graphically. The expected value functions $E_X(p) = pa + (1-p)c$ and $E_Y(p) = pb + (1-p)d$ for $0 \leq p \leq 1$ represent line segments. Notice that $E_X(0) = c$ and $E_X(1) = a$. Similarly, $E_Y(0) = d$ and $E_Y(1) = b$. We mark these outputs on the vertical lines corresponding to $p = 0$ and $p = 1$. The picture must be as follows.

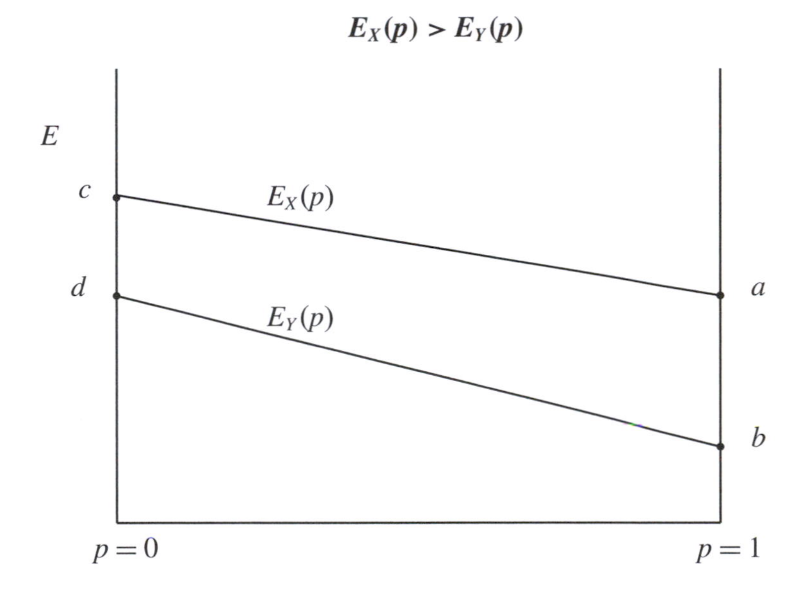

At the level of the payoff matrix, notice that the inequalities $c > d$ and $a > b$ already directly imply that strategy Y dominates strategy X for Player 2. This completes the proof in one case. We leave the remaining, similar cases for you in Exercise 15.4 ∎

We can now prove the Minimax Theorem for 2×2 games.

Theorem 15.8 2×2 **Minimax Theorem** *The mixed-strategy game for the general 2×2 zero-sum game has a Nash equilibrium.*

Proof If the equations $E_X(p^*) = E_Y(p^*)$ and $E_A(q^*) = E_B(q^*)$ can both be solved for probabilities p^* and q^*, then this pair is a Nash equilibrium for the mixed-strategy game by Theorem 15.2. Otherwise, if either equation cannot be solved, the original game has a Nash equilibrium, giving a pure-strategy equilibrium for the mixed-strategy game. ■

Graphical Representation

As we saw in the proof of Theorem 15.7, it is helpful to picture the pure-strategy method graphically. We consider the General 2×2 Zero-Sum Game, as earlier. When Player 1 mixes strategies according to the probability $p = P(\text{"Player 1 plays } A\text{"})$, the pure-strategy expected values are

$$E_X(p) = pa + (1-p)c \quad \text{and} \quad E_Y(p) = pb + (1-p)d$$

Again, $E_X(p)$ has value c when $p = 0$ and value a when $p = 1$. Similarly, $E_Y(0) = d$ and $E_Y(1) = b$. We mark these outputs on vertical lines corresponding to $p = 0$ and $p = 1$. Here is one possible graph:

Graph of General 2 x 2 Zero-Sum Game

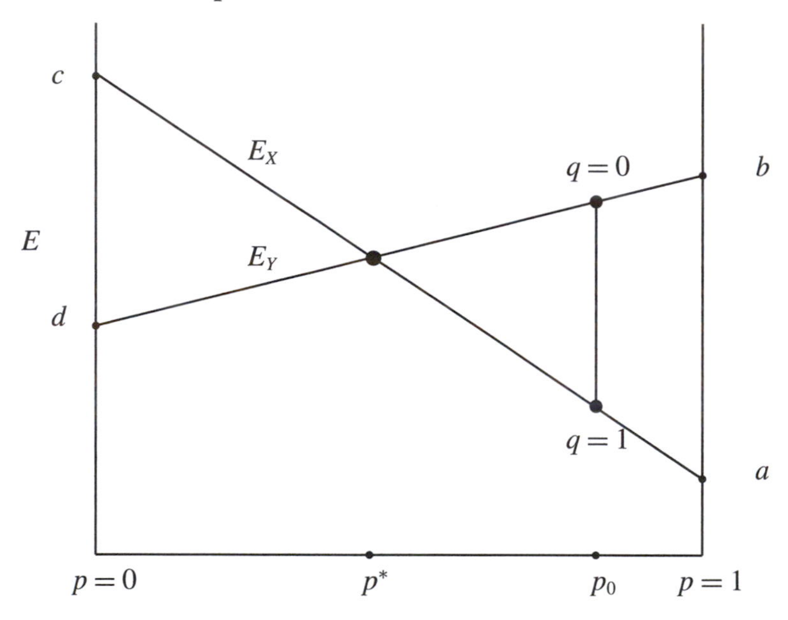

Suppose now that Player 1 chooses $p = p_0$. Then the expected value of the mixed-strategy game is determined by Player 2's choice of q according to the formula

$$E(p_0, q) = qE_X(p_0) + (1-q)E_Y(p_0)$$

Player 2's choice of q gives expected values ranging from $E_Y(p_0)$ when $q = 0$ to $E_X(p_0)$ when $q = 1$. That is, Player 2 controls the position of the expected value on the vertical line segment shown. Now observe that when Player 1 chooses the probability $p = p^*$, Player 2 cannot change the expected value. This observation is the first part of the proof of Theorem 15.8. The second part, regarding q^*, is pictured similarly.

The proof of the Minimax Theorem for the general $m \times n$ zero-sum games requires advanced analysis; finding the mixed-strategy equilibrium for such a game requires much more effort than in the 2×2 case. We conclude this chapter by showing how the graphical representation leads to a method for solving $2 \times n$ games. We focus on the gambling game Bluffing, Calling, and Raising introduced in Example 8.8. Recall that the game leads to the following 2×3 zero-sum game:

Pure-Strategy Zero-Sum Game for Bluffing, Calling, and Raising

Player 2

Player 1

Strategy	Raise	Call	Fold
Bluff	−9	−5	2
Fold	3	1	−1

We again imagine that Player 1 mixes strategies according the probability $p = P$ ("Player 1 bluffs"). We graph the expected values $E_R(p), E_C(p)$, and $E_F(p)$ corresponding to the pure strategies of Player 2. The picture is as follows:

Mixed-Strategy Game for Bluffing, Calling, and Raising

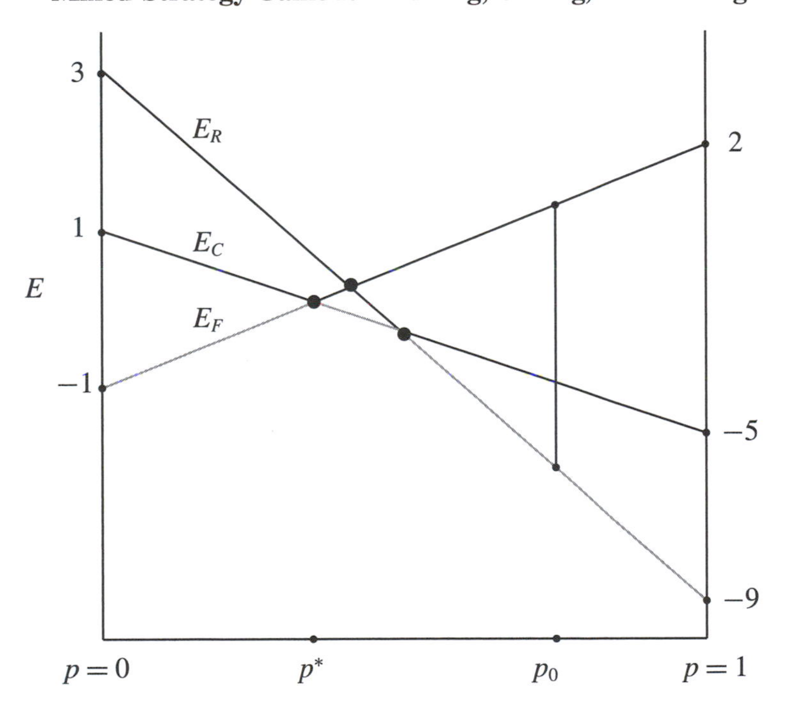

Suppose that Player 1 fixes on the strategy $p = p_0$. Player 2 can then mix strategies to obtain any expected value on the vertical line segment shown. Recall that Player 2 wants the *smallest* value of E possible. If p_0 is fixed, Player 2 will choose the lowest possible point on each such vertical line segment. For Player 1, the lower boundary of the region between the lines is the worst-case scenario. Player 1 maximizes the value of E on this lower boundary when he or she chooses the intersection point between E_C and E_F. The intersection between E_R and E_F has a higher value but is not on the lower boundary.

The value p^* is the found as usual. That is,

$$E_C(p) = E_F(p): \quad -5p + (1-p) = 2p - (1-p) \quad \Longrightarrow \quad p^* = \tfrac{2}{9}$$

You may recognize this as the same value for p^* we calculated for Betting on Diamonds.

Now turn to Player 2. For this player, a mixed strategy has two probabilities:

$$q_1 = \text{probability Player 2 raises} \quad \text{and} \quad q_2 = \text{probability Player 2 calls}$$

with

$$q_3 = \text{probability Player 2 folds} = 1 - q_1 - q_2$$

Suppose that Player 1 chooses $p = p^*$ as earlier. Then Player 2 minimizes E by not raising. Observe that $E_R(p^*)$ is greater than the common value of E_C and E_F at their intersection. We conclude that $q_1 = 0$.

With the possibility of Player 2 raising removed, the calculation of a mixed-strategy equilibrium for Player 2 reduces back to the familiar case of a 2×2 game.

Bluffing, Calling, and Raising

Player 2

Player 1

Strategy	C	F
B	−5	2
F	1	−1

In fact, this game is just Betting on Diamonds in another guise. We have computed the mixed-strategy equilibrium in Theorem 9.9. We conclude that

Solution to Bluffing, Calling, and Raising

$$p^* = \frac{2}{9}, \quad q_1^* = 0, \quad q_2^* = \frac{1}{3}, \quad q_3^* = \frac{2}{3}, \quad \text{Value} = -\$0.33$$

The preceding analysis extends to a general $2 \times n$ zero-sum games.

Theorem 15.9 *The mixed-strategy equilibrium for a $2 \times n$ zero-sum game is the Nash equilibrium of the 2×2 zero-sum game obtained by eliminating all but two of Player 2's strategies.* ■

A careful proof of this result requires quite a few steps. We will only summarize the procedure for finding the mixed-strategy equilibrium for a given $2 \times n$ zero-sum game:

Player 2

Player 1

Strategy	C_1	C_2	...	...	C_n
A	a_1	a_2	...	...	a_n
B	b_1	b_2	...	...	b_n

First, check to see if the game has a saddle point. If so, this outcome is the value of the game. If there is no saddle point, we search for the mixed-strategy equilibrium. Draw vertical lines for $p = 0$ and $p = 1$. Plot the second-row values b_i on the line $p = 0$ and the values a_i on the line $p = 1$. Connect corresponding row values with line segments, and find the lower boundary of the region enclosed by the lines you have drawn. Next, locate the point of intersection of the two lines, which represents the highest point on this lower boundary. This pair of lines arises from two strategies C_i and C_j for Player 2. These two strategies together with A and B for Player 1 are the needed 2×2 zero-sum game. The mixed-strategy equilibrium for this 2×2 game is that of the original.

Exercises

15.1 Find the pure- or mixed-strategy equilibrium and value for the following 2×2 zero-sum games:

(a)

Player 2

Player 1

Strategy	X	Y
A	−5	3
B	4	−1

(b)

Player 2

Player 1

Strategy	X	Y
A	−2	10
B	4	6

(c)

Player 2

Player 1

Strategy	X	Y
A	3	−3
B	2	1000

(d)

Player 2

Player 1

Strategy	X	Y
A	12	−2
B	4	5

(e)

Player 2

Player 1

Strategy	X	Y
A	−4	−3
B	2	−1

(f)

Player 2

Player 1

Strategy	X	Y
A	1	−3
B	4	−7

15.2 View the battle between a Batter and Pitcher in a baseball game as 2×2 zero-sum game. The Batter can guess $f =$ Fastball or $c =$ Curveball and swing accordingly. The Pitcher can throw $F =$ Fastball or $C =$ Curveball. When the batter guesses Fastball and the pitcher throws a Fastball, the batter is hitting .400. When the batter guesses Fastball and the pitcher throws a Curveball, the batter is hitting .150. When the batter guesses Curveball and the pitcher throws a Curveball, the batter is hitting .360. When the batter guesses Curveball and the pitcher throws a Fastball, the batter is hitting .220. Write down the payoff matrix for this game, and find the mixed-strategy equilibrium and solution.

15.3 In a tennis match, the Server can either rush the net (R) or stay on the baseline (S). The Returner can either lob (L) the return or hit the return normally (N). The probabilities that the Server wins the point based on the various combinations are summarized in the following table:

Returner

Server

Strategy	L	N
R	0.25	0.60
S	0.80	0.50

Find the mixed-strategy equilibrium strategies for both players.

15.4 Prove the remaining cases for Theorem 15.7 regarding the the General 2×2 Zero-Sum Game.

(a) $E_Y(p) > E_X(p)$ for $0 \leq p \leq 1$ implies that X dominates Y for Player 2.

(b) $E_A(q) > E_B(q)$ for $0 \leq q \leq 1$ implies that A dominates B for Player 1.
(c) $E_B(q) > E_A(q)$ for $0 \leq q \leq 1$ implies that B dominates A for Player 1.

15.5 Prove that a 2×2 zero-sum game has a saddle point if and only if a player has a weakly dominated strategy.

15.6 Find the pure- or mixed-strategy equilibrium and value for the following $2 \times n$ zero-sum games:

(a)

Player 2

Player 1

Strategy	X	Y	Z
A	6	3	2
B	1	4	7

(b)

Player 2

Player 1

Strategy	X	Y	Z
A	−3	4	5
B	2	−1	0

(c)

Player 2

Player 1

Strategy	W	X	Y	Z
A	8	2	−1	5
B	0	5	6	3

(d)

Player 2

Player 1

Strategy	V	W	X	Y	Z
A	8	6	0	4	10
B	−4	0	4	2	−6

15.7 Consider the following variation on Bluffing, Calling, and Raising (Example 8.8). The betting is the same. Both players ante \$2. Here Player 1 rolls a die and does not reveal its value to Player 2. Player 1 wins if the die is a ⚅. As before, Player 1 can bet \$8 or fold. If Player 1 bets, Player 2 has three options: (1) he or she can fold and lose the ante, (2) he or she can call by placing \$8 in the pot, or (3) he or she can raise and place \$18 in the pot. If Player 1 bets and Player 2 raises, Player 1 must match the extra \$10. Compute the expected values for the pure-strategy choices in this game to obtain a 2×3 zero-sum game. Find the mixed-strategy equilibrium and value of this game.

15.8 Consider another variation on Bluffing, Calling, and Raising. Both players ante \$2. Player 1 draws a card and does not reveal it to Player 2. As usual, Player 1 wins if the card is a Diamond. Player 1 can bet \$8 or fold. If Player 1 bets, Player 2 has three options: (1) he or she can fold and lose the ante, (2) he or she can call by placing \$8 in the pot, or (3) he or she can raise and place \$10 in the pot. If Player 2 raises, Player 1 must match the extra \$2. Compute the expected values for the pure-strategy choices in this game to obtain a 2×3 zero-sum game. Find the mixed-strategy equilibrium and value of this game.

15.9 Solving a 3×3 zero-sum game often requires newer techniques than those we have seen in this chapter. However, if one player has a strategy that is weakly dominated, we can eliminate that strategy and solve the corresponding 2×3 or 3×2 game as in the preceding exercise. Use this approach to determine the value of the following 3×3 zero-sum games:

(a)

Player 2

Player 1

Strategy	*X*	*Y*	*Z*
A	6	3	2
B	1	4	7
C	−1	3	5

(b)

Player 2

Player 1

Strategy	*X*	*Y*	*Z*
A	−3	2	2
B	4	7	8
C	−5	10	−6

(c)

Player 2

Player 1

Strategy	*X*	*Y*	*Z*
A	1	3	6
B	6	−1	2
C	−7	5	7

16 Partial-Conflict Games

The political and military conflicts of the twentieth century presented stark and often tragic examples of strategic encounters played out on the world stage. Mathematically, these conflicts correspond to partial-conflict games. Well-documented examples such as the Cuban missile crisis, the nuclear arms race, and the Iran hostage crisis served as excellent test cases for game theory. Partial-conflict games involve a psychological dimension not present in zero-sum games. There is no overarching theorem such as the Minimax Theorem characterizing all partial-conflict games. Rather, the theory of these games presents a rich variety of dilemmas, paradoxes, and competing solution concepts. We explore this theory here.

The application of game theory to the study of real-world strategic conflict follows a three-step process. The first step is to *model* the conflict mathematically. The rules of the game, strategy choices, and payoffs for outcomes all must be defined based on the motivations of the various players in the conflict. Generally, the representation of a real-world situation in mathematical terms is called a **mathematical model.** Once the mathematical model (in our case, the game) has been defined, the game-theory analysis can begin. The last step is the interpretation of the conclusions of the analysis within the original framework. Here is a schematic for the process:

Social conflict ⟶ Game ⟶ Analysis ⟶ Interpretation

We illustrate the first step in this process beginning with two simplified examples from the field of international relations. We emphasize that the modeling step for a real-world situation is generally a difficult problem requiring a careful accounting of the circumstances surrounding the conflict.

Example 16.1 Free-Rider Problem

Two countries share a large, clear-water lake. The economies of the two countries depend on their fishing industries, and the countries compete for market share. The lake is highly susceptible to overfishing. The countries have agreed to a "catch limit." In practice, it is easy to exceed this limit without discovery.

Example 16.2 Arms Race

Two countries share a common border. A signed treaty establishes a "no-arms zone" near this border. An unopposed army would provide either country with a strong political advantage, allowing the armed country to take advantage of the country that has no arms. If both countries arm the border, war is certain.

In both examples, the two countries each choose between two strategies, making these 2×2 games. In both cases, the countries in conflict are interchangeable. The corresponding notion in game theory is defined as follows:

Definition 16.3 A two-player simultaneous-move strategic game is said to be **symmetric** if both players have the same strategy choices and there is no change in the game if the players switch roles.

Finally, the countries in these examples are not in total conflict. Distilled to their basic structure, the conflicts are examples of an important class of strategic games, defined as follows:

Definition 16.4 A 2×2 **symmetric game** is a simultaneous-move, symmetric, partial-conflict game of strategy in which each player has two strategy choices.

We refer to the two strategy choices, when appropriate, as **cooperate** (C) and **defect** (D). The payoff matrix now takes the following form.

2 x 2 Symmetric Game Payoff Matrix

Strategy	C	D
C	(a,a)	(c,d)
D	(d,c)	(b,b)

Payoffs for 2 x 2 Symmetric Games

The game is then determined by the assignment of four payoffs.

$$a = \text{mutual-cooperation payoff}$$
$$b = \text{mutual-defection payoff}$$
$$c = \text{lone-cooperator payoff}$$
$$d = \text{lone-defector payoff}$$

It is often easier to rank the outcomes in terms of desirability for the respective players in a social conflict than it is to assign actual payoffs. A payoff matrix for a 2×2 symmetric game is said to be in **ordinal form** if the payoffs are just the ordinals 1, 2, 3, and 4 in some order, with 4 understood as the payoff for the most desirable outcome on down to 1, the payoff for the least desirable outcome. We carry out this assignment process for our two examples.

For the Free-Rider problem example, we define

C = "Obey catch limit and D = "Overfish"

We suggest that the worst-case scenario for either country is to obey the fishing limit while the other country overfishes – that is, to be the lone cooperator – because the cooperating country loses market share while their defecting neighbor exhausts future resources. We set $c = 1$. Of course, we could also argue that mutual defection is the worst outcome because the lake is then dangerously overfished. We make mutual defection the second worst outcome, and set $b = 2$. Notice that this is a judgment call.

Next, compare the remaining two outcomes: mutual cooperation or lone defection. In the first case, the countries are eco-friendly but not getting rich. In the second case, the lone defector enjoys a market advantage, and the neighbor is not overfishing. We will designate taking the "free ride" as the most desirable outcome, and set $d = 4$. This leaves mutual cooperation as the second-best outcome: $a = 3$. The result is the following game:

Free-Rider problem in Ordinal Form

Country 2

Country 1

Strategy	C	D
C	(3,3)	(1,4)
D	(4,1)	(2,2)

For the Arms Race, we define

C = "Don't arm" and D = "Arm"

The worst-case scenario for both countries is mutual defection. When both countries arm the border, the result is war. We assign $b = 1$. The best-case scenario for either country is to be the lone defector because the country that arms the border unopposed is rewarded with political advantage. We set $d = 4$. After lone defection, the next-best outcome occurs for both countries is mutual cooperation. We set $a = 3$ and conclude that $c = 2$. We have the following game:

Arms Race in Ordinal Form

Country 2

Country 1

Strategy	C	D
C	(3,3)	(2,4)
D	(4,2)	(1,1)

The ordinal forms for the Free-Rider Problem and the Arms Race should be familiar. As you have probably noticed, the Free-Rider Problem corresponds to the Prisoner's Dilemma (Example 4.1) and the Arms Race to the game Chicken (Example 9.2):

Prisoner's Dilemma

Strategy	*C*	*D*
C	(3,3)	(1,4)
D	(4,1)	(2,2)

Chicken

Strategy	*C*	*D*
C	(3,3)	(2,4)
D	(4,2)	(1,1)

The next step in the process is the mathematical analysis. Here we may summarize our analysis from Chapter 9. The Prisoner's Dilemma has a dominated-strategy solution at mutual defection. The game Chicken has two Nash equilibria at *CD* and *DC* and no dominated strategies. Although these observations are certainly on point, they are not completely satisfying. As was the case for our analysis of zero-sum games, it is helpful to imaging playing these games repeatedly. In this context, we call the repeated game an *iterated game* to emphasize that the game is played a fixed number of times. Generally, an **iterated game** is the repeated game obtained by playing a fixed number of rounds of a partial-conflict game. The most famous example is the *Iterated Prisoner's Dilemma*, which has been the subject of important psychological studies and deep theoretical research. To make the game more interesting, we introduce dollar payoffs in the following example:

Example 16.5 Iterated Prisoner's Dilemma

Two players face off in 100 consecutive games of the following version of the Prisoner's Dilemma:

Payoff Matrix for Iterated Prisoner's Dilemma

		Player 2	
	Strategy	Cooperate	Defect
Player 1	Cooperate	(\$30, \$30)	(\$0, \$50)
	Defect	(\$50, \$0)	(\$10, \$10)

The goal for each player is to maximize their total payoff after 100 rounds.

The game Iterated Prisoner's Dilemma models the tension between greed and cooperation in social or political interaction. Two players who faithfully cooperate earn \$3,000 each, an excellent payoff for this game. The temptation to defect is always lurking, however. Once the choice to defect is made, the players quickly find themselves in the trap of mutual defection.

The search for cooperative strategies for this game is an active subfield of computer science. A simple but surprisingly effective strategy is called **Tit-for-Tat**: a player cooperates until his or her opponent defects and then defects until his or her opponent cooperates. Tit-for-Tat has two key features: on the one hand, Tit-for-Tat encourages cooperation; on the other hand, Tit-for-Tat punishes defection by defecting on the next round. We refer to the latter as an act of **retaliation.** We encourage you to compare some strategies for the Iterated Prisoner's Dilemma and the corresponding game of Iterated Chicken in the exercises.

In Chapter 4, we introduced the notion of a backward induction solution for sequential move games. We may extend this idea to iterated games as well. We motivate the idea using the Iterated Prisoner's Dilemma as in Example 16.5. Begin with the one-hundredth round. Since this is the last round, neither player needs to fear retaliation for defecting. We can remove this consideration, at least for the one-hundredth round. Following the logic of dominated strategies, both players defect on the one-hundredth round, just as in a single round of the game.

Next, consider the ninety-ninth round of the game. Each player knows that on the one-hundredth round, the rational strategy is to defect. Why cooperate on the ninety-ninth round, then? There is no point worrying about retaliation on the next round because mutual defection is inevitable. We conclude that both players defect on the ninety-ninth round in rational play. We can continue this argument to conclude that both players defect on the ninety-eighth round, then on the ninety-seventh round, and so on until eventually we arrive at the first round of the game. We reach the conclusion that both players should follow the strategy "Always defect". We summarize this approach in the following definition:

efinition 16.6

The **backward induction solution** to an iterated game of strategy occurs when players analyze rational play for the last round of the game and then, perhaps using this information, determine rational play for the next-to-last round and so on until the first round of the game is reached.

Of course, the term "rational play" must be interpreted for partial-conflict games. We used the dominated-strategy concept of a solution for the Prisoner's Dilemma. The backward induction solution of the Iterated Prisoner's Dilemma then simply replicates the original dilemma. In fact, in the iterated game, the cynical calculation to defect seems even more shortsighted or *myopic*. In game theory, **myopia** occurs when a player focuses on the gains of the present round at the expense of the overall game. We explore the issue of myopia further later but turn now to another class of partial-conflict games.

Sequential-Move Games

We originally encountered the notion of a backward induction solution in our analysis of the sequential-move game Splitting the Allowance (Example 4.8).

Recall that we reasoned that the younger son has no reason to reject the older son's offer, whether it be \$5 or \$10. Knowing this, the older son can offer the lesser amount and maximize his payoff.

The backward induction solution for Splitting the Allowance reveals a basic flaw for this solution concept: the younger son may be inclined to sacrifice his own allowance just to spite his brother. The analysis is correct mathematically but ignores the psychological aspects of the game. The following example has been used by Reinhard Selten to critique the backward induction solution concept:

Example 16.7 Chain-Store Game

This is a partial-conflict sequential-move game of pure strategy with 11 players. Player B is the owner of a chain of 10 big-box hardware stores in 10 towns. The other players, $A_1, \ldots, A_{10}$, are individuals in the 10 towns, each deciding whether to open a small specialty hardware store to compete with B's chain store. The competitions take place consecutively. First, A_1 and B play a sequential-move game. The prospective owner A_1 chooses between opening a store ("In") and not opening a store ("Out"). Then the chain-store owner B decides on a strategy. He has an advantage over the individual owners in that he can lower prices to cost in the one store. While the lack of profits is not desirable, this strategy will effectively drive the individual owner out of business. The chain-store owner chooses between "Cooperate," by which we mean that he will not lower prices, and "Defect," which means that he lowers prices and puts A_1 out of business. The next game is the same except now between A_2 and B and so on.

Here is one version of a single stage of the game represented as a tree. The payoffs are somewhat arbitrary but, hopefully, plausible. We assume that the individual store owner A_n loses a large franchise fee of \$250,000 if he or she goes in and B defects. We assume that the hardware market in a town is worth \$200,000 at normal prices.

A Stage of the Chain-Store Game

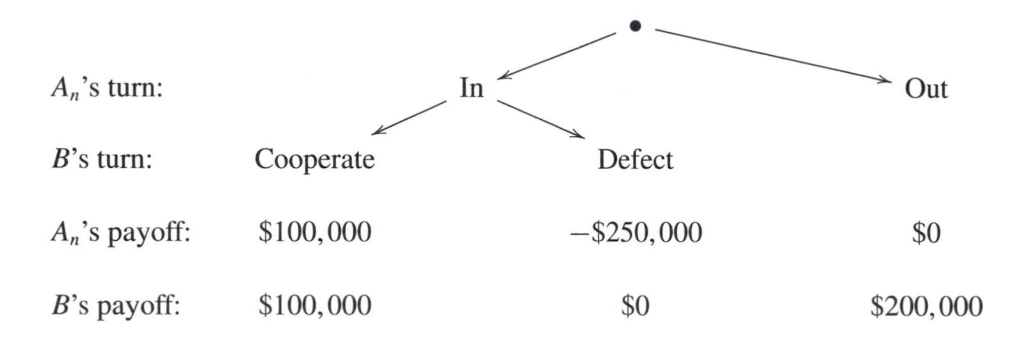

Each particular stage of this game can be analyzed itself using backward induction as in Definition 4.10.

Backward Induction for a Stage of the Chain-Store Game

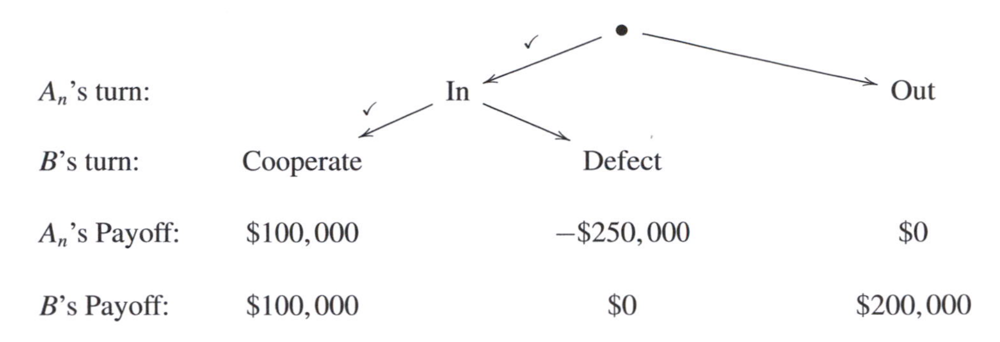

The backward induction solution for any given round of the game calls for A_n to go in and for B to cooperate.

For the chain-store owner, the game is not one round but rather 10. Obviously, B would prefer that each of the A_n owners opted out. Suppose that A_1 opts in. Then B can set an example for A_2, A_3, and so on by defecting and punishing A_1. While defection hurts B in the short term, the later profits obtained if the other stores respond to this warning would more than compensate for a zero payoff in round one. The idea of choosing a strategy to set an example for future rounds of a game is called **deterrence**.

The backward induction solution to the iterated game is reminiscent of that for the Iterated Prisoner's Dilemma. Begin with the last round of the game: since there are no further rounds to play, A_{10} knows that B has no incentive to defect. The only reason B defects is to deter future players from opting in, and there are no future players. Consequently, A_{10} opts in, and B cooperates. Next, consider the penultimate round. Since A_{10} will definitely opt in in the next round, A_9 reasons that there is no incentive for B to defect if A_9 opts in: A_{10} will not be deterred from opting in, as argued earlier. Thus A_9, knowing that B has no incentive to defect, opts in as well, and in this analysis, B cooperates in the ninth round. Continuing in this way, we see that the backward induction solution of the Chain-Store Game is for each of the players to opt in and for B to cooperate in each case.

As was the case with the Iterated Prisoner's Dilemma, the backward induction solution to the Chain-Store Game is unsatisfying. The advantage that B might gain by using deterrence in early rounds of the game is ignored by the analysis, just as the possibility of faithful cooperation is ignored for the Iterated Prisoner's Dilemma. This game and the flaw in the analysis are related to a logical paradox called the *Surprise Quiz Paradox* (Exercise 16.8).

We next consider a sequential-move gambling game related to the game Splitting the Allowance.

Example 16.8 Betting for the Pot

This is a two-player sequential-move game. A third player, the "house," does not take action but can profit. Each player antes $5 so that the pot contains $10 to start. On his or her turn, each player can place up to $10 in the pot or can pass. The first player to pass loses, the other player wins and takes the pot. There is one catch. As soon as the pot contains $30 or more, both players lose, and the house keeps the pot.

In Betting for the Pot, we do not know when the last stage will occur or even which player's turn it will be. We can, however, identify the worst possible situation that can face a player on his or her turn:

Worst-case scenario for Player X:

Pot = $29   Player X: Bet or Pass?

Here Player X could be either Player 1 or 2. This player faces two unappealing choices: pass and lose all the money bet so far or bet and lose this additional bet along with all the money already bet so far. Optimal play in terms of payoff is to pass and let Player Y take the pot.

Next, let us consider the situation with one fewer dollar in the pot:

Another scenario:

Pot = $28   Player Y: Bet or Pass?

Here Player Y seems to be in a more favorable position. Player Y can bet $1 and place Player X in the preceding scenario. If Player X plays to minimize his or her loss, Player Y wins a $29 pot, the maximal win possible. In fact, because Player Y can bet up to $10, he or she can increase the pot to $29 if he or she faces Pot = $28, $27, $26, . . ., $19.

This last observation, in turns, leads to the backward induction solution to the game: Player 1 should bet $8 on his or her first turn. Facing Pot = $18, Player 2 should pass on his or her turn. Otherwise, by betting, Player 2 will leave Player 1 facing a pot containing between $19 and $28. By betting $8, Player 1 *threatens* to bid the pot up to $29. Player 2 then faces the worst-case scenario. Seeing the writing on the wall, Player 2 should pass before losing more than the ante.

This analysis is the basis for the following definition:

Definition 16.9 An opening play in a sequential-move partial-conflict game that compels Player 2 to pass by a backward induction analysis will be called a **threatening opening play**.

We remark that the backward induction analysis for Betting for the Pot overlooks the possibility of spiteful action as in Splitting the Allowance. When faced with a pot of $29, a player may be inclined to pay one extra dollar just to punish his or her opponent. Of course, this type of behavior is irrational from the point of view of maximizing payoff. Still, Player 1's strategy for maximizing his or her payoff is based on the idea of threatening Player 2. Given the combative stance taken by his or her opponent, it is reasonable to imagine that Player 2 views his or her subsequent bid as a counterthreat, a promise to bid $1 more if ever faced with a $29 pot.

The psychological dynamic of threat and punishment is elegantly captured by our next example, a famous auction game introduced by Martin Shubik.

Example 16.10 The Dollar Auction

Two players alternately bid for a one-dollar bill provided by the house. On his or her turn, a player may either pass or bid a value higher than the previous bid. (Player 1 can pass or bid any positive amount for his or her first turn.) If a player passes, the other player gets the dollar and pays his or her last bid. Here is the catch: the player who passes must pay his or her last bid to the house as well.

For example, a possible round of bidding is as follows:

A Possible Round of the Dollar Auction

Player	Bid
1	$0.25
2	$0.30
1	$0.35
2	Pass

Here Player 1 wins the dollar, paying 35 cents – a net profit of 65 cents. Player 2 has a net loss of 30 cents.

Playing the Dollar Auction can create strange behavior. To illustrate, let us suppose that the bidding has reached 99 cents, say this is Player 1's bid, and it is Player 2's turn. If Player 2 passes, he or she loses his or her previous bid, say, a cents. By bidding a dollar instead, Player 2 stands to break even. Thus, if $a > 0$, bidding one dollar or more is a rational bid for Player 2 at his or her turn. The strangeness may then continue: since Player 1 is out 99 cents if he or she passes, he or she is still motivated to outbid Player 2 to gain the dollar, albeit at a steep price. Depending on the level of the bet, this option may be better than passing and losing 99 cents. This behavior of bidding beyond the value of the object is called **escalation**. The Dollar Auction is a model of escalatory behavior in social or political conflict.

We can modify the Dollar Auction by restricting bidding and introducing a maximum bid. This restriction allows for a backward induction analysis. We discover the idea by means of the following example:

Example 16.11 Dollar Auction with Maximum Bid

The game is the same as the Dollar Auction except that now the highest allowable bid is \$15. If one player bids \$15, that player receives the dollar and pays \$15. The other player pays the amount of his or her last bid.

As we did for the game Betting for the Pot, we look for a threatening opening bid by Player 1 that will compel Player 2 to pass. A first example is the opening bid of 99 cents. When Player 1 bids 99 cents, Player 2 only stands to lose by bidding. Generally, we make the following statement:

Rationality Assumption:

No player will bid \$1 or more than his or her previous bid.

Notice that the payoff to Player 1 from this bid is only 1 cent, and this is provided that Player 2 plays according to this logic. We look for a smaller threatening opening bid to increase Player 1's profit.

Consider the effect of the maximum bid of \$15. Notice that if the bidding were ever to reach above the \$14 dollar range, the smart move would be to just bid \$15. While "smart" may seem like a poor description of either player's actions in this hypothetical game, it is better to pay \$15 for a one dollar bill than to lose over \$14 because you lose the bidding. We start a backward induction with auctions that end up in this neighborhood. Specifically, suppose that in the course of the auction, Player 1 makes the bid \$14.01. We claim that Player 2 should pass regardless of his or her previous bid. The reason is simple: having bid \$14.01, Player 1 now prefers bidding \$15 and winning the auction to passing. Thus the only way that Player 2 can win the bidding is to bid \$15 first. This is not a rational bid for Player 2 because his or her last bid must have been \$14 or less. By our "Rationality Assumption" we conclude that any opening bid by Player 1 that eventually results in a bid of \$14.01 by Player 1 is a threatening opening bid.

We next argue that any opening bid that eventually results in Player 1 bidding

$$\$13.02 = \$14.01 - \$0.99$$

is a threatening opening bid. Having made the bid of \$13.02, Player 1 prefers to bid \$14.01 and win rather than pass. Now Player 2's previous bid is less than \$13.02. By the "Rationality Assumption," Player 2's possible response bids to Player 1's \$13.02 would be strictly less than \$14.01. Such a bid by Player 2 leaves Player 1 to bid \$14.01, at which point Player 2 will be compelled to pass, as argued earlier. We conclude that any opening bid by Player 1 that eventually results in Player 1 making the bid \$13.02 is a threatening opening bid.

The logic of induction continues this pattern: opening bids that result in Player 1 bidding

$$\$12.03, \$11.04, \$10.05, \ldots, \$2.13, \$1.14, \$0.15$$

are all threatening opening bids. We can argue that 15 cents is the smallest possible threatening opening bid. If Player 1 bids less than 15 cents, then Player 2 can bid

15 cents. The preceding argument now shows that Player 2 has made a threatening bid for Player 1, so Player 1 should pass. The opening bid of 15 cents is called the **minimal threatening opening bid.**

The general result extends this analysis to any value of the maximum bid M and the payoff P, with $M > P$. We will think now exclusively in cents, so the preceding case is $M = 1,500$ and $P = 100$. Let

$$b = \text{minimal threatening opening bid for Player 1}$$

To find b, we begin by taking $P - 1$ (99 cents, in our special case) and subtracting this from M. We continue doing this until further subtraction would yield a negative number. This last positive value is b. Arithmetically, observe that

$$1,500 = 15(99) + 15$$

This means that we subtract 99 cents 15 times from 1,000 cents (or \$15) and are left with 15 cents. Finding the minimal threatening bid is then just a long division problem:

$$\begin{array}{r|l} & 15 \\ \hline 99 & 1,500 \\ & -1,485 \\ \hline & 15 = \text{remainder} \end{array}$$

The minimal threatening opening bid b is the remainder of division of $P-1$ into M.

The following theorem was proved by Barry O'Neill.

Theorem 16.12 O'Neill's Theorem *Consider a two-player auction for a payoff of P cents in which, as in the Dollar Auction, the winning bidder gets the payoff, and both players pay their last bids. Let M be the maximum bid. Let b be the remainder on division of P − 1 into M. Then the backward induction solution of the game occurs when Player 1 bids b and Player 2 passes.*

Proof The proof is the same argument as the special case. We show that any opening bid that results in Player 1 eventually bidding $M - (P - 1)$ is threatening. We then argue the same for $M - 2(P - 1)$ and so on until we reach the stage, say, the nth, when $M - n(P - 1) = b$. Minimality is argued as before. ■

Theory of Moves

We apply backward induction to the study of social dilemmas such as the Prisoner's Dilemma and Chicken. Start with any 2×2 symmetric game:

Player 2

	Strategy	C	D
Player 1	C	(a,a)	(c,d)
	D	(d,c)	(b,b)

We create a sequential-move game by allowing players the opportunity to make unilateral changes in strategy. For example, starting at the node mutual cooperation (CC), if the players alternate changing strategies, they create the following tour of outcomes:

$$\begin{array}{ccc} CC & \xleftarrow{\text{Player 2}} & CD \\ \text{Player 1} \downarrow & & \uparrow \text{Player 1} \\ DC & \xrightarrow[\text{Player 2}]{} & DD \end{array}$$

We use this idea to create the following game:

Example 16.13 Moving Game

This is a two-player sequential-move game of strategy based on a given 2×2 symmetric game. The game begins at any one of the four outcomes CC, CD, DC, or DD. Each player, on his or her turn, chooses between two strategies, "Pass" and "Move." If a player chooses "Move," the game switches to the new outcome corresponding to a unilateral change in strategy (from C to D or vice versa) by this player. The game ends after either (1) both players pass on the first move, (2) either player passes subsequent to the first move, or (3) both players move twice. In case (3), the game has completed a cycle, and the players receive no payoff. In cases (1) and (2), the players receive the payoffs at the node where the game ended.

Here is a sample round of the Moving Game using the Prisoner's Dilemma for the payoff matrix. We take the starting outcome to be DD. We suppose that both players elect to move once before Player 1 passes on his or her second turn.

A Sample Round of the Moving Game

Action	Outcome	Payoffs
	DD	(2,2)
Player 1 moves	CD	(1,4)
Player 2 moves	CC	(3,3)

Let us examine the outcome of this game. Both players have earned 3 points, a respectable payoff. However, suppose that Player 1 had not passed and instead moved to DC with payoff (4, 1). Player 2 would have no choice but to pass because one more move results in no payoff to either player by rule (3). Player 1 gains the maximal payoff with this move, whereas Player 2 receives only 1. Knowing this, Player 2 should pass on his or her first turn. Working backward, we see that Player 1 should pass on his or her first turn as well. We have arrived at the backward induction solution to this game for Player 1, namely, to pass. By the symmetry of the payoff matrix, Player 2 has the same backward induction solution. We conclude that for the Moving Game starting at outcome *DD* in the Prisoner's Dilemma, the backward induction is for both players to pass. This analysis leads to a new notion of an equilibrium.

efinition 16.14 An outcome in a 2×2 symmetric game is a **nonmyopic equilibrium** if the backward induction solution for the Moving Game starting at this outcome is for both players to pass.

We have shown that mutual defection is a nonmyopic equilibrium for the Prisoner's Dilemma. Of course, we already knew that mutual defection was the Nash equilibrium for this game, so perhaps this should not come as much of a surprise. A more interesting result is the following:

Theorem 16.15 *Mutual cooperation is a nonmyopic equilibrium for the Prisoner's Dilemma.*

Proof We show that the backward induction solution for Player 1 in this game is to pass. The argument for Player 2 is the same. Here is what happens when Player 1 moves:

Player 1 Moves from *CC* in the Prisoner's Dilemma

Action	Outcome	Payoffs
	CC	(3,3)
Player 1 moves	*DC*	(4,1)
Player 2 moves	*DD*	(2,2)

If Player 1 moves, then Player 2 is placed in the least desirable position, the lone cooperator. Player 2 will move. Player 1 can then move again, but doing so places Player 2 in his or her best-case scenario *CD*, which pays Player 2 the payoff 4. We conclude that Player 1 should pass to get the payoff 3 as opposed to the payoff 2 that will result from moving. ■

Theorem 16.15 shows how mutual cooperation can be interpreted as rational play in the Iterated Prisoner's Dilemma. A player who moves from mutual cooperation to lone defection faces retaliation and will soon end up with the lower payoff from mutual defection. Therefore, it is reasonable to stay put and continue to cooperate.

We give one further example of a nonmyopic equilibrium arising in the game of Chicken.

Theorem 16.16 *Mutual cooperation is a nonmyopic equilibrium for Chicken.*

Proof Here are the possible moves and payoffs:

Players 1 and 2 Move from *CC* in Chicken

Action	Outcome	Payoffs
	CC	(3,3)
Player 1 moves	*DC*	(4,2)
Player 2 moves	*DD*	(1,1)
Player 1 moves	*CD*	(2,4)

Suppose that Player 1 moves. If Player 2 passes, he or she settles for payoff 2. However, consider what happens when Player 2 moves to *DD* as shown. Player 1 will then be compelled to move to *CD* because his or her payoff will increase from 1 to 2. Player 2 thus has obtained the highest payoff of 4. We conclude that if Player 1 moves, his or her payoff reduces from 3 (at the starting outcome *CC*) to 2 (at the final outcome *CD*). Player 1 thus should pass. Player 2 also should pass by the same argument. ■

The Moving Game is a simplified version of a framework called the *Theory of Moves* introduced by Steven Brams who applied the theory to study political conflict. We have avoided various subtleties in the analysis. In particular, we have introduced the rule (3) to end the game after two moves by each player to avoid the problem of repetition. The penalty of no payoff to both players who complete a full cycle is intended to enforce the position that there is no sense moving at all if the game just returns to the original outcome.

Exercises

16.1 In the 2008 Democratic Primary, the opponents Barak Obama and Hillary Clinton had to choose between two basic strategies:

$$C = \text{“Stay positive”} \quad \text{or} \quad D = \text{“Go negative”}$$

Model this game as a 2×2 symmetric game in ordinal form. Explain your reasoning for ranking the outcomes. Find the Nash equilibria.

16.2 Two friends argue a question of trivia. Both are completely convinced that they are right and that the other is wrong. Their options are

$$C = \text{“Concede the point”} \quad \text{or} \quad \text{D} = \text{“Continue arguing”}$$

Suppose that the friends would prefer to continue arguing rather than have both concede the point simultaneously. Model this game as a 2×2 symmetric game in ordinal form. Is this the same game you found in Exercise 16.1? Find the Nash equilibria.

16.3 The conflict known as Hunting Stag dates back to the philosopher Jean-Jacques Rosseau. Two hunters go out to the woods, where they have two possible quarry: stag or hare. In order to kill a stag, both hunters must work together. Either hunter can kill a hare by himself or herself, but a hare is not nearly as valuable as a stag. The hunters choose between two strategies:

$$C = \text{“Hunt stag”} \quad \text{and} \quad D = \text{“Hunt hare”}$$

If the hunters work together, they will kill a stag, the most desirable outcome. Assume that it is easier to shoot a hare if the other hunter is looking for a stag. Find the 2×2 symmetric game in ordinal form representing Hunting Stag. Find all Nash equilbria.

16.4 Consider the following five strategies for the Iterated Prisoner's Dilemma:

(a) "Always defect" = defect on every turn
(b) "Tit-for-Tat" = cooperate until the other player defects; then defect until he or she cooperates
(c) "Random defector" = defect every third turn
(d) "Hopeful cooperator" = cooperate until opponent defects twice in a row; then defect until opponent cooperates again
(e) "Always cooperate" = cooperate on every turn

Run the strategies off in round-robin pairs, 10 rounds each, and keep track of the total score for each strategy. Can you devise a strategy that will produce a higher total score than the winner?

16.5 Treat the Iterated Prisoner's Dilemma as a zero-sum game in which the player with the lower total score on 100 rounds pays the player with the higher total score the difference in their scores. In the event of a tie, there is no payoff to either player. Prove that playing the strategy "Always defect" is guaranteed to never result in a negative payoff. Further, prove that "Always defect" is the only strategy with this property. Argue that mutual defection is the Nash equilibrium for this zero-sum game.

16.6 Define the Iterated Chicken Game using the following payoff matrix.

Payoff Matrix for Iterated Chicken Game

Player 2

	Strategy	C	D
Player 1	C	(3,3)	(1,5)
	D	(5,1)	(0,0)

Consider the following five strategies for the iterated game:

(a) "Always defect" = defect on every turn
(b) "Tit-for-Tat" = cooperate until the other player defects; then defect until he or she cooperates

(c) "Cooperate until defect" = cooperate until the other player defects, then defect; every time
(d) "Hopeful cooperator" = cooperate until opponent defects twice in a row; then defect until opponent cooperates again
(e) "Always cooperate" = cooperate on every turn

Run the strategies off in round-robin pairs, 10 rounds each, and keep track of the total score for each strategy. See if you can devise a strategy that will produce a higher total score than the winner.

16.7 Is there a backward induction solution for the Iterated Chicken Game? Does the iterated game have any Nash equilibria? Explain your answers.

16.8 (The Surprise Quiz Paradox) The professor of a game-theory seminar announces that there will be a surprise quiz on one of the five days next week. The professor emphasizes that the day on which the quiz is given will definitely be a surprise to the students. A group of students meets to study over the weekend. Before the night is over, they arrive at an unshakable conviction: there can be no quiz next week. Explain why the students believe that there can be no surprise quiz next week. *Hint:* Can there be a surprise quiz on Friday? Suppose that the professor gives the quiz on Tuesday. Was the quiz a surprise? Explain the paradox.

16.9 Here is a famous old puzzle with connections to backward induction. Four students in a game-theory seminar are invited to play a game. The four will be placed in a room, and the professor will put either a red or a white hat on each student's head. No one can see his or her own hat, but each student can see the other three hats. The first one to guess the color of his or her own hat correctly wins an A for the course. An incorrect guess results in the student receiving an F, and the game is over. The professor promises that at least one hat of each color will be used. When the hats are placed on the students' heads, no one says anything for a moment. A second later, one student correctly names the color of her hat. How did this student know that she was getting the A? Can you generalize this idea to five players?

16.10 We can create variations on the game Betting for the Pot (Example 16.8) by changing the quantities *maximum bid and maximum pot* involved in defining the game. In the current version, maximum bid = \$10 and maximum pot = \$30. Find the backward induction solution, that is, the opening bid O for Player 1 for the variations corresponding to the following values for these maxima:

(a) Maximum bid = \$10; maximum pot = \$50
(b) Maximum bid = \$10; maximum pot = \$100
(c) Maximum bid = \$7; maximum pot = \$40
(d) maximum bid = \$12; maximum pot = \$65

16.11 Find the optimal strategy for Player 1 in the Dollar Auction with Maximal Bid in the following cases for P = payoff amount and M = maximum bid. This is just arithmetic using Theorem 16.12. Remember to convert to cents.

(a) $P = 25$ cents and $M = \$10$
(b) $P = 50$ cents and $M = \$20$

(c) $P = \$10$ and $M = \$20$
(d) $P = \$1$ and $M = \$20$

16.12 Consider the Dollar Auction with payoff $P = \$1$ and maximum bid M with bids in cents. Find two values of M so that Player 1's payoff in rational play is again 50 cents. Can you describe all such M?

16.13 Prove that the outcomes *CD* and *DC* are not nonmyopic equilibria for the game Prisoner's Dilemma.

16.14 Prove that the outcomes *CC*, *CD* and *DC* are all nonmyopic equilibria for the game Chicken.

16.15 Consider the following 2×2 symmetric game:

Player 2

Player 1	Strategy	*C*	*D*
	C	(2,2)	(3,4)
	D	(4,3)	(1,1)

Find any Nash equilibria and all nonmyopic equilibria for this game.

16.16 Prove that there are 24 different 2×2 symmetric games in ordinal form.

16.17 Define a 2×2 symmetric game to be a **social dilemma** if the following three requirements are satisfied

Requirement 1: The strategy *C* does not weakly dominate *D*.
Requirement 2: The payoff from lone defection is larger than that from mutual defection.
Requirement 3: The payoff from lone cooperation is less than that from mutual cooperation.

Prove that there are four distinct social dilemmas in ordinal form.

16.18 The notion of a Prisoner's Dilemma generalizes to partial-conflict games with n players. Here is a motivating example: a **Vickrey auction** is a sealed-bid auction in which the high bidder wins the object but pays the second-highest bid. Suppose that there are n bidders in a Vickrey auction. Assume that each bidder has a true valuation v for the object. As usual, the payoff for the winning bidder is that bidder's valuation minus the amount paid (the second-highest bid). The losing bidders get zero payoff. Consider the following two strategies for a given bidder:

D = "Bid your true valuation" or C = "Bid one-half your true valuation"

Treat the Vickrey auction as an n-player two-strategy partial-conflict game.

(a) Prove that D weakly dominates C for each individual bidder.
(b) Prove that all bidders are at least as well off and one bidder is strictly better off if all bidders play C.

A partial conflict game with multiple players each choosing between strategies C and D satisfying these conditions is called a **Generalized Prisoner's Dilemma.**

17 Takeaway Games

We have seen that gambling games and zero-sum games are described by a real number, the expected value E. For a gambling game, E is the average payoff for a round of the game. For a zero-sum game, E is the expected value at the mixed-strategy equilibrium (Definition 15.4).

In this chapter, we consider a class of strategic games described by an **ordinal number**, by which we mean a nonnegative integer $n = 0, 1, 2, \ldots$. Recall that a takeaway game is a finite two-player sequential-move strategic game with perfect information for which the last player to move wins. In Chapter 10, we proved Zermelo's Theorem for this class of games (Theorem 10.9), establishing that one player has a strategy that guarantees victory. Our proof of Theorem 10.9 depended on the Binary Labeling Rule (Defintion 4.16) and the Principle of Induction for Trees. In this chapter, we introduce a refinement of the Binary Labeling Rule we call the *Ordinal Labeling Rule*. This new rule assigns an ordinal number called the *Sprague–Grundy number* to a takeaway game. Like the Binary Labeling Rule, the Ordinal Labeling Rule determines the winner and winning strategy for a takeaway game.

We then revisit the prototypical takeaway game: Nim. With many piles and many chips, the game Nim is too large to solve directly by labeling rules. Charles Bouton produced an elegant solution to the general game in 1901 involving *binary numbers*. We introduce binary numbers here and use them to prove Bouton's Theorem. We apply Bouton's Theorem, in turn, to address Question 1.4 from Chapter 1.

Sprague–Grundy Numbers

We introduce a new labeling system for game trees of takeaway games. Given a finite set of ordinal numbers S, the number $\text{Mex}(S)$ is defined to be the smallest ordinal not in S. The term $\text{Mex}(S)$ stands for the **minimum excluded number** of S. Here are some examples:

Examples of Mex S

$$\text{Mex}(\{0,1,2,4,5,6\}) = 3$$
$$\text{Mex}(\{1,3,4,6,7,10\}) = 0$$
$$\text{Mex}(\{0,1,2,3,4,6,12,18\}) = 5$$

We use the function $\text{Mex}(S)$ for the following definition:

Definition 17.1 The **Ordinal Labeling Rule** assigns ordinal labels to the nodes of a game tree for a takeaway game as follows: label the end nodes 0. Move backward up the tree toward the root. Given a node A, let S denote the set of ordinal labels on the nodes to which A directly branches. The **ordinal label** for A is then $\text{Mex}(S)$.

Here are some examples of labeling node A with the Ordinal Labeling Rule:

Ordinal Labelings for Node A

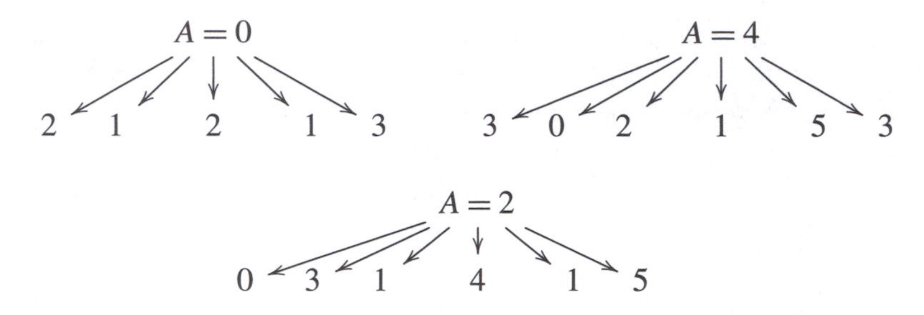

Here is an example of the Ordinal Labeling Rule applied to a game tree:

Ordinal Labeling of a Game Tree

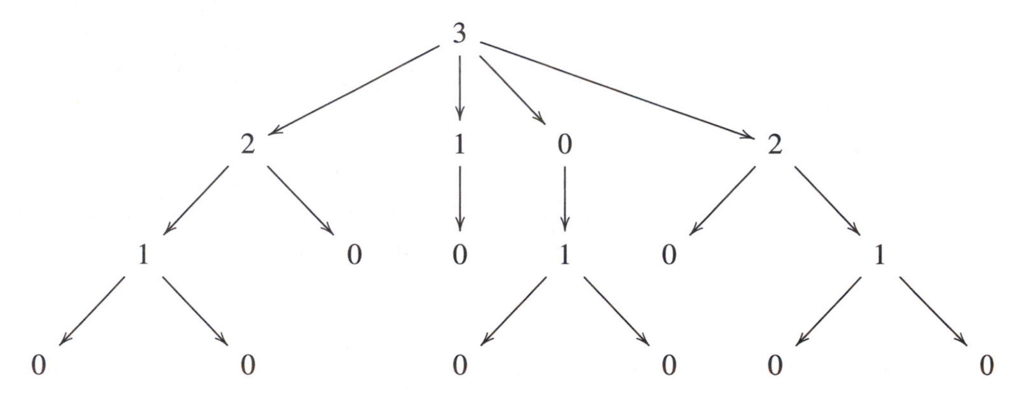

Notice that the Binary Labeling Rule is directly recovered from the Ordinal Labeling Rule. We just change the nodes with nonzero labels to the label 1 and leave the nodes labeled 0 alone. In particular, we have the following consequence of Theorem 10.9:

Theorem 17.2 *Suppose that we are given a takeaway game with a game tree labeled according to the Ordinal Labeling Rule. Then Player 1 can guarantee victory if and only if the root node is labeled with a nonzero number. The winning strategy is "Move to a 0 node."* ■

We introduce the following definition:

Definition 17.3 The **Sprague–Grundy number** for a takeaway game is the ordinal label assigned to the root of the game tree by the Ordinal Labeling Rule.

When positive, the Sprague–Grundy number gives some measure of the complexity of a takeaway game. For example, notice that an ordinal label of n on the root node of the game tree requires that there be at least n branches from the root and that the game tree be of depth at least n. Here are two examples:

Example 17.4 One-Pile Nim

Start with one pile of n chips. Each player, on his or her turn, can take as many chips (but at least one) from the pile. The last player to take chips wins.

Recall that we write this game as Nim(n) (Definition 4.18). The subgames of Nim(n) are One-Pile Nim games Nim(k) for $k < n$. An abridged game tree is as follows:

Abridged Game Tree for Nim(n)

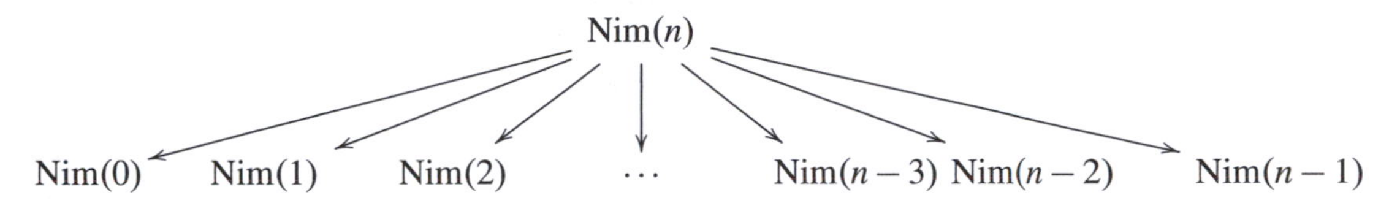

Now the subgame Nim(0) corresponds to an end node for Nim(n), so this node is labeled 0. It is easy to see that the subgame Nim(1) has an ordinal label 1 because there is exactly one branching in the game tree, to Nim(0). Continuing in this way, we see that the ordinal label for the subgame Nim(2) is 2, the ordinal label for Nim(3) is 3 and so on. This leads to the following theorem:

Theorem 17.5 *The Sprague–Grundy number of Nim(n) is n.*

Proof The proof is left for you in Exercise 17.8. ■

Here is another example of a takeaway game from Chapter 4:

Example 17.6 Taking Chips

We introduced this game in Example 4.14. Again, we start with a pile of n chips. Let k be a given integer. In this case, players, on their turns, must take at least one chip and no more than k chips from the pile. The last player to take chips wins.

Let us write $TC(n,k)$ for the game Taking Chips starting with n chips with a maximum k taken on a turn. Notice that the subgames of $TC(n,k)$ are games $TC(m,k)$ for $m < n$. With this notation, we can write down an abridged game tree for $TC(n,k)$ as we did above for Nim(n). We consider the special case where $n = 9$ and $k = 5$.

Abridged Game Tree for Taking Chips: $TC(9,5)$

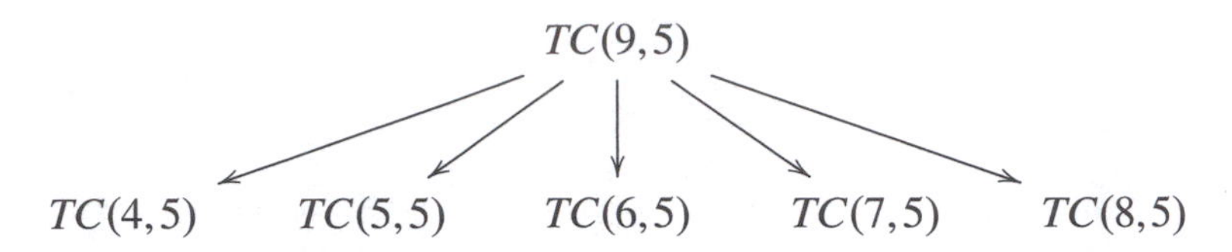

Observe that when $k \geq m$, the maximum constraint becomes meaningless. Generally, we have for $k \geq m$,

$$TC(\text{m, k}) = \text{Nim}(m)$$

By Theorem 17.5, the ordinal labels on $TC(4,\ 5)$ and $TC(5,\ 5)$ are 4 and 5, respectively.

Next, consider the subgame $TC(6,\ 5)$. The nodes directly adjacent to $TC(6,\ 5)$ are $TC(1,\ 5)$ through $TC(5,\ 5)$. Based on the preceding, these nodes have ordinal labels 1 through 5, respectively. Thus, the

$$\text{Sprague–Grundy number for } TC(6,\ 5) = \text{Mex}(\{1,2,3,4,5\}) = 0$$

Continuing, we see $TC(7,\ 5) = 1$, TC(8, 5) = 2, and finally, $TC(9,\ 5) = 3$.

The Ordinal Labeling Rule introduces an interesting new exercise for a given takeaway game. It is natural to wonder, however, what we gain by using ordinal labels because the winner and winning strategy are already determined by the Binary Labeling Rule. We will eventually see that the Sprague–Grundy number encodes key information for analyzing complicated games built from simpler ones. We return to the realm of 0's an 1's with a brief digression on binary numbers.

Binary Numbers

The ordinal numbers are arguably the most fundamental abstract idea we have. The meaning of "three" or "seventeen" is understood regardless of the word used to describe it. Mathematically, the notion of number makes sense without a commitment to the **representation** of the number, the specific notation used to mention it. Nonetheless, a commitment to some representation must be made, and diverse systems have been invented for the purpose. Our system of representation is, of course, the **decimal system**. We have symbols for 10 digits $0, 1, \ldots, 9$. We then represent a general number using these digits and powers of 10.

For example,

$$3{,}456 = 3 \cdot 10^3 + 4 \cdot 10^2 + 5 \cdot 10^1 + 6 \cdot 10^0$$

Standard algorithms allow us to perform the basic arithmetic operations of addition and multiplication in this representation.

The number 10 is called the **base** of the decimal system The fact that we use 10 is simply a historical artifact. Humans prefer counting in tens for obvious biologic reasons. We can just as well use any other positive ordinal (except 1) as the base. Arguably, the most useful alternative to the decimal system is the **binary system**, wherein the base is 2. In this case, an integer is written using only 0's and 1's.

The idea is simple enough. Take a number such as 19. Write

$$19 = 16 + 2 + 1 = 2^4 + 2^1 + 2^0.$$

Consequently, in *binary form*, we have

$$19 =_2 10011$$

We use the notation to "$=_2$" to indicate that the right-hand side is a binary number.

Definition 17.7 The **binary form** of an ordinal number n is the expression

$$n =_2 x_k x_{k-1} \cdots x_2 x_1 x_0$$

where each x_i is either 0 or 1, and

$$n = x_k \cdot 2^k + x_{k-1} \cdot 2^{k-1} + \cdots + x_2 \cdots 2^2 + \cdots + x_1 2^1 + x_0 \cdot 2^0$$

The x_i are called the **binary digits** of n.

There is an algorithm for converting a decimal number to a binary number. First, we continually divide the number by 2, keeping track of the remainders. Then we convert the column of remainders to binary digits. We illustrate the method for the numbers 43 and 123:

Conversion of 43 to Binary

Division by 2	Remainder
43 = 2(21) + 1	1
21 = 2(10) + 1	1
10 = 2(5) +0	0
5 = 2(2)+1	1
2 = 2(1) + 0	0
1 = 2(0) + 1	1

Power of 2	Binary digit
2^0	1
2^1	1
2^2	0
2^3	1
2^4	0
2^5	1

$$43 =_2 101011$$

Conversion of 123 to Binary

Division by 2	Remainder	Power of 2	Binary digit
123 = 2(61) + 1	1	2^0	1
61 = 2(30) + 1	1	2^1	1
30 = 2(15) +0	0	2^2	0
15 = 2(7)+1	1	2^3	1
7 = 2(3) + 0	1	2^4	1
3 = 2(1) + 1	1	2^5	1
1 = 2(0) + 1	1	2^6	1

$$123 =_2 1111011$$

We introduce the *binary column form* for an ordinal.

Definition 17.8 The **binary column form** for an ordinal m with binary form $m =_2 x_k x_{k-1} \cdots x_1 x_0$ is the column of binary digits:

Binary Column Form of *m*

Power of 2	Binary digit
2^k	x_k
2^{k-1}	x_{k-1}
$\vdots$	$\vdots$
2^1	x_1
2^0	x_0

Notice that we list the binary digits in this format from most significant digit 2^k down to least 2^0. This is the reverse order from which the digits are computed using our algorithm.

Bouton's Theorem

We now turn to the game of Nim in its most general form.

Definition 17.9 Given k positive integers $m_1, m_2, \ldots, m_k$, the game

$$\mathbf{Nim}(m_1, m_2, \ldots, m_k)$$

is the sequential-move game played with k piles of chips of sizes m_1 through m_k. On his or her turn, each player takes as many chips as desired but at least one from one of the remaining piles. The last player to take chips wins.

We remark that the game tree for a general game of Nim is quite large. In fact, it is easy to see that

$$\text{Depth (game tree for Nim}(m_1, m_2, \ldots, m_k)) = m_1 + m_2 + \cdots + m_k$$

Our labeling rules will be of little use for such a large game. Recall that Theorem 4.20 gives a complete solution to the game Two-Pile Nim. Player 1 is the winner in rational play precisely when the two piles are unequal. The winning strategy is to always balance the piles. Bouton's Theorem extends the idea of "balance" to give a solution to the general game. Consider our first version the game Nim(6,5) (Example 4.19). We know that the first rational move for Player 1 is to remove one chip from the pile of six. Here is this first move represented in binary:

First Move in Binary Column Form for Nim(6,5)

Power of 2	6	5
2^2	1	1
2^1	1	0
2^0	0	1

$\Longrightarrow$

Power of 2	5	5
2^2	1	1
2^1	0	0
2^0	1	1

We see that this play makes the binary digits agree in each row. This is the idea of "balance" that we would like to generalize to the case of multiple piles (columns).

Let us examine this process in another case, this time with three piles. Suppose that we play the game Nim(1,2,5). It is not difficult to check that Player 1 wins this game in rational play by first moving to the node (1,2,3), which has binary label 0. That is, Player 1 takes two chips from the pile of five:

First Move in Binary Column Form for Nim(1,2,5)

Power of 2	1	2	5
2^2	0	0	1
2^1	0	1	0
2^0	1	0	1

$\Longrightarrow$

Power of 2	1	2	3
2^2	0	0	0
2^1	0	1	1
2^0	1	0	1

In this case, we can see that the first move in the winning strategy yields a configuration (1,2,3), in which all the rows have an even number of 1s. The top row has no 1s, and the second and third rows each have two 1s. This observation is the key to the general solution for Nim.

Let us define some terminology. We refer to the sequence $(m_1, m_2, \ldots, m_k)$ of pile sizes as a **configuration** for a game of Nim. We define the **binary representation of Nim$(m_1, \ldots, m_k)$** to be the table whose ith column is the binary column form for m_i. Here is an example:

Binary Representation of Nim(2,4,5,6)

2	4	5	6
0	1	1	1
1	0	0	1
0	0	1	0

For convenience, we omit the column corresponding to the powers of 2.

Given a configuration for a game of Nim, we focus on the rows in the binary representation. There are three rows for $(2,4,5,6)$. We define a row to be **balanced** if the number of ones that occur in the row is even. The middle row for $(2,4,5,6)$ is balanced because it has two ones. We say that the row is **unbalanced** if the number of ones is odd. The top and bottom rows are unbalanced (with three ones and one one, respectively) for $(2,4,5,6)$.

Definition 17.10 We say that a configuration $(m_1, m_2, \ldots, m_k)$ of Nim is **balanced** if all the rows in the binary representation are balanced. If one or more rows have unbalanced status, we say that the configuration is **unbalanced**.

For the preceding example, we have

(2,4,5,6) Is an Unbalanced Configuration

2	4	5	6	Row status
0	1	1	1	Unbalanced
1	0	0	1	Balanced
0	0	1	0	Unbalanced

Now we note that, by removing chips from the pile of four chips, Player 1 can render all three rows balanced in one move. On the binary level, this amounts to removing the 1 from the first row and inserting a 1 in the last row:

4		1
0	$\Longrightarrow$	0
1		0
0		1

With this move, the new configuration is then

(2,1,5,6) Is a Balanced Configuration

2	1	5	6	Row status
0	0	1	1	Balanced
1	0	0	1	Balanced
0	1	1	0	Balanced

Now it is Player 2's turn. He or she must remove chips from some pile. This move will, of necessity, leave one of the rows unbalanced. Can you see why?

Player 1 will then be in the position to rebalance all the rows. By following this strategy of "balancing the configuration," Player 1 wins the game Nim(2,4,5,6).

Here is another example:

Balancing Nim (28, 13, 22, 10)

Unbalanced Configuration $\Longrightarrow$ **Balanced Configuration**

28	13	22	10	Row status
1	0	1	0	Balanced
1*	1	0	1	Unbalanced
1*	1	1	0	Unbalanced
0	0	1	1	Balanced
0*	1	0	0	Unbalanced

17	13	22	10	Row status
1	0	1	0	Balanced
0*	1	0	1	Balanced
0*	1	1	0	Balanced
0	0	1	1	Balanced
1*	1	0	0	Balanced

In both cases, we have reduced a pile with a 1 in the highest unbalanced row (shaded) to balance the configuration. Specifically, we have changed this binary digit 1 to a 0 and then changed the starred digits in this column necessary to balance the rows below. The general solution of Nim follows this algorithm.

Theorem 17.11 Bouton's Theorem *In the game of Nim, Player 1 can guarantee victory if and only if the binary representation of the pile amounts is an unbalanced configuration. The winning strategy for Player 1, in this case, is to balance the configuration.*

Proof We prove the result by induction on the depth of the game tree. Recall that the depth of the game tree of Nim $(m_1, m_2, \ldots, m_k)$ is the sum $m_1 + m_2 + \ldots + m_k$.

Base Case: The game Nim(1) with one pile containing one chip has an unbalanced configuration. Of course, Player 1 wins this game by balancing the configuration.

Induction Hypothesis: Let $n > 1$ be given, and suppose that the statement of the theorem is true for all games of Nim with a game tree of depth $\leq n$.

Induction Step: Suppose that we are given a configuration for Nim with a game tree of depth $= n + 1$. We have two cases:

Case 1: Suppose that the binary representation of the game is unbalanced. We argue that we may then identify one of the piles and, removing chips only from that pile, create a new configuration that is balanced. The result in this case then follows from the induction hypothesis.

Locate the highest row in the binary representation that is an unbalanced row. We take our pile to be one of the piles of chips whose binary representation has a 1 in this highest unbalanced row (there is at least one such pile). Let n be the number of chips in this chosen pile. We take chips from this pile. First, we change the 1 that occurs in the highest unbalanced row to a 0. This balances the highest unbalanced row. Next, we focus on the row below this one. If this row is unbalanced, we change the corresponding digit in n's column from a 0 to a 1 or vice versa. This change will balance the next unbalanced row. We continue in this way working down the rows, balancing them, if necessary, by switching the binary digit for n in this row. When we are done, we have a new value m for our pile. This value m balances the configuration. We leave the proof that $m<n$ for Exercise 17.7. Player 1 reduces the pile with n chips to a pile with m chips. The result is a balanced configuration. Player 1 wins in rational play by the induction hypothesis.

Case 2: Suppose that the configuration is balanced so that all the rows in the binary column form are balanced. Taking chips from a pile means switching at least one of the binary digits in the column corresponding to this pile. Each switch of digits in a single column changes the status of the corresponding row from balanced to unbalanced. Thus Player 1 must move to an unbalanced configuration. By our induction hypothesis, Player 2 wins this game. ■

We can now address Question 1.4. Recall the game introduced in Example 1.3: Player 1 splits 11 chips into piles. Player 2 then goes first in the resulting game of Nim.

Question 1.4 *In A Game of Nim, which player has the advantage, Player 1 or Player 2?*

We prove the following result.

Theorem 17.12 *Suppose that a collection of odd chips is split into piles. Then the resulting Nim configuration is unbalanced.*

Proof The proof follows from two easy facts left for you to justify in Exercise 17.10: First, the 2^0 (ones) digit x_0 of the binary form of an ordinal m is 1 if and only if m is odd. Second, given a collection of numbers $m_1, \ldots, m_k$, the ones digit in binary form of the sum $m_1 + \cdots + m_k$ is equal to 1 if and only if the sum of the ones digits from the binary form of the m_i is odd.

Combining these facts, we see that if the odd number n is split according to $n = m_1 + \ldots + m_k$, then the row corresponding to the 2^0 = ones digits for the configuration $(m_1, \ldots, m_k)$ is unbalanced because the sum of these entries is an odd number. ■

By Theorem 17.12 and Bouton's Theorem (Theorem 17.11), we see that no matter how Player 1 splits the 11 chips into piles, Player 2 will win the resulting Nim game in rational play.

We conclude by offering one further question motivated by our work in this chapter:

What is the Sprague–Grundy number of $\text{Nim}(m_1, \ldots, m_k)$?

The beautiful answer to this question is the subject of the Sprague–Grundy Theorem (Theorem 22.4). We prove this result in Chapter 22.

Exercises

17.1 Compute the Sprague–Grundy numbers for the following versions of Two-Pile Nim:

(a) Nim(3, 1)
(b) Nim(4, 2)
(c) Nim(4 ,3)
(d) Nim(5, 4)
(e) Nim(5, 1)
(f) Nim(10, 4)

17.2 Compute the Sprague–Grundy numbers for the following versions of Taking Chips. Recall that $TC(n,k)$ denotes the game starting with a pile of n chips with maximum k taken on a turn.

(a) $TC(7, 3)$
(b) $TC(10, 5)$
(c) $TC(10 ,3)$
(d) $TC(12, 4)$
(e) $TC(20, 6)$
(f) $TC(41, 4)$

17.3 Compute the Sprague–Grundy numbers for the following versions of Three-Pile Nim:

(a) Nim(1, 1, 2)
(b) Nim(1, 2, 2)
(c) Nim(1, 2 ,3)
(d) Nim(1, 2, 4)
(e) Nim(2, 3, 3)
(f) Nim(2, 3, 4)

17.4 Convert the following integers n to binary:

(a) $n = 37$
(b) $n = 42$
(c) $n = 120$
(d) $n = 233$

17.5 Use Bouton's Theorem (Theorem 17.11) to decide which player wins the game of Nim with the following initial configurations. When Player 1 is the winner in rational play, indicate his or her first move.

(a) Nim $(1,3,10)$
(b) Nim $(2,4,15,7)$
(c) Nim $(13,25,26,17)$
(d) Nim $(12,13,15,16,19)$

17.6 Consider the following unbalanced configurations of Nim. Write down the sequence of moves taken in rational play by Player 1, assuming that on his or her turn, Player 2 either takes three chips from the largest pile or, if the largest pile contains fewer than three chips, Player 2 takes the largest pile.

(a) Nim $(4,5,8)$
(b) Nim $(5,6,9,14)$

(c) Nim $(4,8,15,16)$
(d) Nim $(3,3,6,9,13)$

17.7 Prove the following result about binary numbers used in the proof of Theorem 17.11. Start with an integer $n =_2 x_j x_{j-1} \ldots x_{k+1} x_k x_{k-1} \ldots x_1 x_0$ in binary form. Suppose that the binary digit $x_k = 1$. Prove that if m has binary digits agreeing with n in all digits above k, has the kth binary digit = 0, and has arbitrary binary digits below k, that is, $m =_2 x_j x_{j-1} \ldots x_k y_k y_{k-1} \ldots y_1 y_0$, where $y_k = 0$ and $y_{k-1}, \ldots, y_0$ are either 0 or 1, then $m < n$.

17.8 Prove Theorem 17.5 using the Principle of Induction for Trees (10.2).

17.9 Let n and k be positive integers with $k < n$. Let R denote the remainder upon division of n by $k+1$. Prove that the R equals the Sprague–Grundy number of $TC(n,k)$, the game Taking Chips starting with n chips and with maximum k taken per round.

17.10 Prove the following facts used in the proof of Theorem 17.12:

(a) An integer n is odd if and only if the 2^0 (ones) binary digit x_0 for n is a 1.
(b) Let $m_1, \ldots, m_k$ be a list of integers. Then the ones digit in the binary form of the sum $m_1 + \cdots + m_k$ is 1 if and only if the sum of the ones digits of the binary forms of the m_i is odd.

17.11 Let m be a positive, even integer. Prove that $(m, m+1, m+2)$ is an unbalanced configuration.

17.12 Characterize integers m and n so that the Nim configuration $(1,m,n)$ is balanced.

17.13 Consider the following two-player strategic game. Player 1 is given 4 chips and Player 2 is given 10 chips. The first stage of the game is simultaneous move. Player 1 builds one pile using some (at least one) of his or her chips. Player 2 makes two piles using all of his or her chips. The players then play Nim with the three piles, with Player 1 going first. Can either player guarantee victory in this game? Explain your answer.

18 Fairness and Impossibility

The basic principle of fairness for a game is easy to articulate: Each player should have an equal opportunity to win. For a social choice, the meaning of fairness is a far richer question. The criteria introduced in Chapter 12 focus on the suitability or unsuitability of certain candidates to be the social choice. In this chapter, we introduce various new fairness criteria that focus on the effect of changes in the preference table. We begin with a notion called *Monotonicity* concerning the invariance of election outcomes when voters increase their preferences for the winning candidate. We then introduce two criteria, *Anonymity* and *Neutrality*, that capture two fundamental democratic principles. These criteria focus on the existence of election biases, either for particular voters or for candidates. We prove *May's Theorem* (Theorem 18.8) to the effect that the only voting method for two-candidate elections satisfying these three fairness criteria is the Simple Majority Method.

We then turn to the notion of an *impossibility theorem*, an idea made famous by Arrow's thesis. We introduce Arrow's criterion of *Independence* and prove a simplified version of his impossibility theorem for social choice methods (Theorem 18.12) proved by Alfred Mackay.

Arrow's original theorem focuses on social welfare methods as opposed to social choice methods. We introduce Arrow's Independence Criterion for social welfare methods as well as the *Pareto Criterion* in this setting. We then state *Arrow's Impossibility Theorem* (Theorem 18.19). We conclude by proving an impossibility theorem formulated by Amartya Sen that reveals a basic incompatibility between individual liberties and collective rights.

Before beginning, we recall that we do not allow ties in either social choice or social welfare outcomes. We follow Convention 5.9 and use alphabetical order to break all ties arising in the final outcome of an election. The problem of ties will be a more substantive issue in this chapter. Indeed, we will see that our convention leads directly to certain violations of fairness criteria.

Monotonicity

In Chapter 6, we proved that various "democratic" social choice methods are susceptible to strategic manipulation in the sense of Definition 6.3. Individual voters (or voting blocs) can change their true preferences to improve the outcome from their perspective. Indeed, this is true for all nondictatorial voting methods by the Gibbard-Satterthwaite Theorem (Theorem 6.4).

If we inspect our examples of manipulability, however, we notice one common feature: the voting bloc chosen to vote insincerely was the bloc for whom the true social choice, say, X, was the least preferred candidate. This bloc then altered its preferences for the other candidates to tip the favor away from X. Since any outcome is better for this bloc than X, the insincere votes count as manipulation.

It is perhaps not surprising that a voting bloc intent on not electing candidate X can, in some cases, vote so as to ensure that X loses. We ask now whether it is ever possible for a winning candidate X to lose by virtue of a move *up* in the rankings. Can an increase in support ever hurt a candidate in an election?

Definition 18.1 A voting method satisfies **Monotonicity** if the following condition holds for all preference tables: if X is the social choice, then X will remain the social choice even if voters who have a candidate A ranked directly above X in their individual preferences alter their preferences so that X is ranked directly above A.

Monotonicity, simply put, requires that an increase in support never hurts a candidate. A violation of Monotonicity, then, is a scenario in which increasing the support for a winning candidate X somehow manages to make X a loser. We easily obtain the following positive result:

Theorem 18.2 *The Plurality Method and Borda Count satisfy Monotonicity.*

Proof Assume that X is the Plurality social choice so X has the most first-place votes. Suppose that a bloc of voters P ranks A over X. When P votes insincerely, ranking X over A instead, X either gains A's first-place votes or there is no overall change to the first-place vote counts. Either way, X still has a plurality in the new preference table and so is again the Plurality social choice.

Similarly, if X is the Borda Count social choice, any switch from A above X to X above A has the effect of increasing X's Borda points at the direct expense of A's Borda points. Other candidates' Borda points are unaffected. Thus, if X originally has the most Borda points, then X will again have the most in the new election. ■

We next show that the Hare and Coombs Methods can violate Monotonicity because of the effect of indirect causes arising from multiple rounds. We explain a scenario: suppose that we have a preference table, and candidate X is the winner of, say, the Hare Method. In the final round, X beats B. Suppose that the win for X was fortunate in the following sense: if X had faced A in the final round, A would have beaten X head to head and won the election. Now suppose further that increasing X's support at the expense of B actually ends up helping A to make it to the final round. If this scenario is achieved, then A will beat X in the final round, and we will have our violation. We use this idea to prove the following theorem:

Theorem 18.3 *The Hare Method violates Monotonicity.*

Proof Consider the following preference table, arranged to create the scenario just described:

Original Preference Table

Number of voters	7	5	6	2
1st choice	X	B	A	B
2nd choice	A	A	X	X
3rd choice	B	X	B	A

The Hare social choice is X. However, if the bloc of two voters moves X up at the expense of B, then we obtain

New Preference Table

Number of voters	7	5	6	2
1st choice	X	B	A	X
2nd choice	A	A	X	B
3rd choice	B	X	B	A

In this case, A survives the first-round elimination and beats X head to head in the final round. ■

We can use the same idea to prove the following theorem:

Theorem 18.4 *The Coombs Method violates Monotonicity.*

Proof We begin with

Original Preference Table

Number of voters	10	5	4	7
1st choice	X	B	A	B
2nd choice	A	A	B	X
3rd choice	B	X	X	A

The Coombs social choice is X.

Now consider the bloc consisting of the five voters who have A ranked directly above X in their true preference order. If we switch this ranking, we obtain

New Preference Table

Number of Voters	10	5	4	7
1st choice	X	B	A	B
2nd choice	A	X	B	X
3rd choice	B	A	X	A

This change has the effect of making A the first candidate eliminated in the Coombs Method. When we run X versus B head to head, the winner is B. ■

May's Theorem

We have seen a number of fairness criteria capturing various desirable properties of a voting method. We introduce here two natural criteria focusing on the method by which the votes are counted. The first requires that the identity of the voters be irrelevant to the outcome.

Definition 18.5 A voting method satisfies **Anonymity** if the outcome of an election does not change if any two voters swap their rankings in the sense that each voter submits the ranking previously submitted by the other voter.

All of our "democratic" voting methods (Plurality, Borda Count, Hare, Coombs, and Pairwise Comparisons with Agenda) satisfy Anonymity, as can be seen directly. In fact, it is this feature of these voting methods that permits us to keep track only of the number of voters in a given voting bloc without worrying about the identities of the individual voters. A Dictatorship clearly violates Anonymity: the outcome of the election certainly can change if the dictator swaps ballots with another voter.

Anonymity requires that a voting method not show bias among voters. We can also expect a democratic voting method to be unbiased as regards the candidates. This ideal is the basis of the following definition:

Definition 18.6 A voting method satisfies **Neutrality** if whenever a candidate X wins an election and all voters change their votes only by swapping candidates X and Y in their individual rankings, then candidate Y wins the second election.

Neutrality is the requirement that a candidate's identity does not affect the outcome of the election. If candidate X wins in some scenario, then Y also wins if only the roles of X and Y are swapped. While Neutrality seems like a reasonable requirement for a voting method, in fact, each of our "democratic" voting methods violates this criterion for technical reasons. Recall in Definition 2.7 that we require a voting method to produce a unique social choice with no ties allowed. To break ties, our convention is to use the alphabetical order of candidate names: A beats B, who beats C, and so on. This device represents a direct violation of Neutrality.

Other than this last example, all violations of fairness criteria given in this text have involved elections with three or more candidates. When only two candidates are present, the Simple Majority Method (Definition 5.12) provides an easy and equitable voting method. When there are an odd number of voters, the issue of ties does not arise, and we can observe the following:

Theorem 18.7 *For an election with two candidates and an odd number of voters, the Simple Majority Method satisfies Anonymity, Neutrality, and Monotonicity.*

Proof Anonymity is obvious from the definition of the Simple Majority Method. Neutrality is also clear because there is no possibility of a tie and thus no need to introduce a nonneutral tie-breaking method. Finally, Monotonicity is here a special case of the corresponding result for the Plurality Method (Theorem 18.2). ■

We now consider the question of whether there is another voting method for two-candidate elections other than the Simple Majority Method that satisfies these three criteria. For definiteness, let us call the candidates X and Y and suppose that there are, say, nine voters. If the imagined alternate voting method satisfies Anonymity, then we can simply count the votes for each candidate; we need not keep track of the voters by name. Suppose that five voters prefer X to Y, with the other four preferring Y to X. We would expect X to win this election. Suppose that, in fact, Y wins the election. If the voting method satisfies Monotonicity, then we can conclude that Y wins any election with the support of four or more voters. Now suppose that the voting method also satisfies Neutrality. Then X also should win with four or more votes of support because swapping candidate names results in the reverse outcome. We conclude that X and Y must both win the original election. This conclusion contradicts our overriding prohibition on ties. Returning to the original election, we conclude that, in fact, X wins with the simple majority. We convert this logic to a proof of the following result proved by Kenneth May:

Theorem 18.8 May's Theorem *Suppose that we are given a voting method for a two-candidate election with an odd number of voters. As usual, we assume that the voting method does not produce ties. If the voting method satisfies Anonymity, Neutrality, and Monotonicity, then the voting method is the Simple Majority Method.*

Proof Let $n = 2m + 1$ be the (odd) number of voters, and let X and Y be the candidates. We show that X is elected if and only if $m+1$ or more voters prefer X to Y. The corresponding statement for Y follows directly by Neutrality.

First, by Anonymity, the outcome of the election depends only on the number of voters q who prefer X to Y. Second, by Monotonicity, if X wins for some value of q, then X also wins with the support of more than q voters. It makes sense, then, to set

$$q_0 = \text{minimum number of voters preferring } X \text{ that results in } X \text{ winning the election}$$

Suppose that $q_0 < m + 1$. By Neutrality, both X and Y win with the support of q_0 or more voters. Consider the situation where m voters prefer X and the other $m + 1$ prefer Y. Then, since $m, m + 1 \geq q_0$, we conclude that both X and Y must win this election. Since a voting method cannot produce a tie, this is a contradiction.

If $q_0 > m+1$, then Y wins with the support of at least $r_0 = n - q_0 - 1 = 2m - q_0$ voters. By Neutrality, X is also elected with the support of at least r_0 voters. But

$r_0 < q_0$ because $q_0 > m+1$. This contradicts the definition of q_0 as the minimum number of votes needed for X to win. Since the alternatives lead to contradictions, we conclude that $q_0 = m+1$. It follows that the voting method is the Simple Majority Method. ■

We remark that with an even number of voters and two candidates, there is no available voting method that satisfies Anonymity, Neutrality, and Monotonicity without allowing ties.

Independence and Impossibility

Kenneth Arrow created a new paradigm in social choice theory with his proof of his famous impossibility theorem sketching the limitations of democratic voting methods. A novelty of Arrow's theorem is that it establishes a universal statement about voting methods. Instead of fixing a voting method and considering all possible preference tables, Arrow focused on specific fairness criteria and considered all possible voting methods. We make this into the following general definition:

Definition 18.9 An **impossibility theorem** in social choice theory is a theorem of the form: all voting methods of a certain type violate at least one of a list of specified fairness criteria.

Arrow introduced a principle of fairness that seems innocent enough at first glance. Suppose that a voting method produces a social choice X for a given preference table. Let Y be a losing candidate. Suppose that some voters alter their preferences, but all voters retain their original preferences for X versus Y. Arrow's criterion requires that such changes should not result in Y winning the election. The changes in the preference table involve only preferences for **irrelevant alternatives**, that is candidates other than the winner X and the chosen loser Y.

Definition 18.10 A social choice method satisfies the **Independence of Irrelevant Alternatives** if the following holds for all preference tables: if X is the social choice and Y is a losing candidate, then when voters alter their preferences but no voter changes his or her preferences for X versus Y, then Y should not be the new social choice. For brevity, we refer to this as the **Independence Criterion.**

Notice that we do not require X to win the new election, only that Y continues to be a loser. We use an example of a Condorcet Paradox to obtain the following theorem:

Theorem 18.11 *The Plurality, Hare, and Coombs Methods all violate the Independence Criterion.*

Proof The following version of the paradox will suffice in all cases:

Original Preference Table

Number of voters	5	4	3
1st choice	A	B	C
2nd choice	B	C	A
3rd choice	C	A	B

Notice that A is the Plurality, Hare, and Coombs social choice. We take C to be the chosen loser. Suppose that the voting bloc of four voters retains their preference for C over A but moves the irrelevant alternative B from the first to second. We obtain the following:

New Preference Table

Number of voters	5	4	3
1st choice	A	C	C
2nd choice	B	B	A
3rd choice	C	A	B

The new social choice is C in all three cases. ■

We observe that the Borda Count also violates the Independence Criterion. We leave the proof for you in Exercise 18.4.

The preceding argument can be applied to prove the following simplified version of Arrow's Theorem:

Theorem 18.12 Mackay's Theorem *All social choice methods violate either the Majority Winner Criterion or the Independence Criterion.*

Proof We prove this theorem by contradiction. Specifically, we assume that we are given a voting method that satisfies both fairness criteria. We then prove that this voting method cannot possibly produce a social choice when applied to the preference table from the preceding proof:

Original Preference Table

Number of voters	5	4	3
1st choice	A	B	C
2nd choice	B	C	A
3rd choice	C	A	B

We now prove that each possible winning candidate A, B, and C leads to a contradiction. Actually, we treat the case that the social choice is C and leave similar arguments for the other two cases for you in Exercise 18.11.

Suppose that C wins the election. Consider the losing candidate B. Since the voting method satisfies the Independence Criterion, B should not be the social choice when the bloc of five voters switches their preferences for B and the irrelevant alternative A, leaving their preference for B over C unchanged. We obtain the following preference table:

New Preference Table

Number of voters	5	4	3
1st choice	B	B	C
2nd choice	A	C	A
3rd choice	C	A	B

However, since the voting method satisfies the Majority Winner Criterion, B must win this new election. We conclude that C cannot be the social choice. ■

Fairness and Social Welfare

In Chapter 5, we saw that our voting methods extend to give a full ranking of the candidates called a *social welfare* (Definition 5.17). Here is an example:

Example 18.13

Consider the following preference table:

Number	7	5	8	2
1st choice	A	C	D	B
2nd choice	B	A	C	D
3rd choice	C	B	A	C
4th choice	D	D	B	A

The Plurality Method ranks the candidates according to the number of first-place votes: Thus we see that

Plurality social welfare: $D \quad A \quad C \quad B$

The Borda Count ranks the candidates according to the number of Borda points. Here we obtain

Borda Count social welfare: $C \quad A \quad D \quad B$

For the Hare and Coombs Methods, the candidates are ranked according to the round in which they are eliminated. Thus we have

Hare social welfare: $A \quad D \quad C \quad B$
Coombs social welfare: $C \quad A \quad B \quad D$

Arrow's Independence Criterion extends to social welfare methods as follows:

inition 18.14 A social welfare method satisfies the **Independence Criterion** if the following holds: in any election, if candidate A is ranked above candidate B in the social welfare, then A remains ranked above B if voters change there preferences for other candidates, but no changes are made by voters in their rankings of A versus B.

Essentially the same argument given in Theorem 18.11 proves the following negative result:

Theorem 18.15 *The Plurality, Borda Count, Coombs, and Hare Methods violate the social welfare version of the Independence Criterion.*

Proof The proof is left for you in Exercise 18.6. ■

We next introduce a famous and eminently reasonable notion of fairness named for economist and philosopher Vilfredo Pareto. The criterion is based on the principle that if society expresses a unanimous preference for one candidate over another, then that ordering of the candidates should occur in the social welfare.

inition 18.16 A social welfare method satisfies the **Pareto Criterion** if, whenever all voters rank candidate A over candidate B in a preference table, candidate A is ranked above candidate B in the social welfare.

The Pareto Criterion is a natural measure of the democratic tendencies of a voting method. It is perhaps surprising then that, of our democratic social welfare methods, only the Borda Count satisfies this fairness criterion. The problem, again, is the issue of ties. We prove:

Theorem 18.17 *The Borda Count satisfies the Pareto Criterion. The Plurality, Hare, and Coombs Social Welfare Methods violate the Pareto Criterion.*

Proof We leave the proofs that the Borda Count satisfies the Pareto Criterion to you (Exercise 18.12). For the violations, consider an election with one voting bloc P and three candidates A, B, and C. Assume that P ranks A first, C second, and B third. Since only A receives first-place votes, we must resort to the tie breaker, alphabetical order, to determine the rankings of B and C. The Plurality, Hare, and Coombs social welfare each has A first, B second, and C last. This social welfare violates the Pareto condition because P ranks C above B. ■

It is natural to ask for a social welfare method that satisfies both the Independence and Pareto Criteria. A less than satisfying example is given by the following theorem:

Theorem 18.18 *A Dictatorship, as a social welfare method, satisfies both the Independence and Pareto Criteria.*

Proof See Exercise 18.3. ■

We can now state Arrow's Impossibility Theorem, which ends the search for more examples.

Theorem 18.19 Arrow's Impossibility Theorem *The only social welfare method with three or more candidates that satisfies both the Independence and Pareto Criteria is a Dictatorship.* ■

We prove Theorem 18.19 in Chapter 23. We conclude this chapter with an impossibility theorem proved by Amayrta Sen that appeared after Arrow's Theorem.

Sen's Impossibility Theorem

Sen's theorem concerns a question that lies at the heart of many hot-button political issues. Take, for example, the question of gun control. Most people may agree that it is in the best interests of a society to restrict access to guns. In our terminology, the social preference is in favor of restrictive gun laws. At the same time, individual citizens feel that it is well within their rights to own a gun. We are faced with the question

> How do we balance the twin goals of preserving collective rights while also ensuring individual liberties?

Sen proved that striking this balance is ultimately an impossibility. In fact, the goal of obeying the collective will and the goal of guaranteeing individual voter choice are fundamentally incompatible. Sen illustrated this dilemma with an example along the following lines:

Example 18.20 Big Sister's Diary

Two brothers find their big sister's diary and try to decide whether to read the diary or not. The plan for reading the diary is for one brother to remain on the lookout while the other reads. This creates three alternatives: Older brother reads the diary (O), younger brother reads the diary (Y), or nobody reads the diary (N). The older brother respects his sister's privacy, so he favors N. However, if anyone is going to read the diary, he wants to read it. Older brother's second preferred outcome is O. Younger brother is a prankster and would like the diary to be read. He is afraid of his big sister, though, so he would prefer O and then Y. In summary, the boys preferences are as follows:

Preference Table for Big Sister's Diary

Brother	Older	Younger
1st choice	N	O
2nd choice	O	Y
3rd choice	Y	N

The brothers try to rank the alternatives N, O, and Y. They agree that each should be allowed to control the ranking of the outcome of reading the diary themselves or nobody reading the diary. Thus older brother ranks N over O, and younger brother ranks Y over N. They find that the only possible social ranking that preserves these two individual preferences is

Social welfare
Y
O
N

The brothers now confront Sen's dilemma. Both brothers prefer O to Y. Their desire to preserve individual preferences, however, leads to Y ranked above O in the social welfare.

Motivated by this example, we formulate and prove Sen's Impossibility Theorem. First, we define the notion of "obeying the collective will" as satisfying the Pareto Criterion. We next formalize the notion of individual liberties. Fix a social choice method, and suppose that we are given a collection of voters and a list of at least three candidates A, B, and C. We say that a particular voter p has **individual liberties** if there are two candidates, say, A and B, such that if voter p ranks A over B, then A is ranked over B in the social welfare (regardless of how the other voters rank the candidates), whereas, if voter p ranks B over A, then B is ranked over A in the social welfare (again, regardless of the rankings of the other voters). In other words, voter p controls the social ranking of A and B. In this situation, we say that voter p has **rights over** candidates A and B.

Of course, in a Dictatorship, the dictator has individual liberties, but everyone else is powerless. At the other extreme is the ideal of **libertarianism**, a political system in which all voters have individual liberties. We introduce a notion that lies somewhere in between.

Definition 18.21 A social welfare method satisfies **Minimal Liberalism** if, in any collection of voters choosing from among three or more candidates, we can always identify two voters who have individual liberties.

Notice that Minimal Liberalism makes no requirements when there are just two candidates. In this case, two voters having individual liberties leads directly to a contradiction. In other words, a violation of Minimal Liberalism requires an example with at least three candidates and two voters.

With these definitions in place, we are prepared to prove the following theorem:

Theorem 18.22 Sen's Impossibility Theorem *No social welfare method satisfies both the Pareto Criterion and Minimal Liberalism.*

Proof Suppose, for a contradiction, that there is a social welfare system that satisfies both the Pareto Criterion and Minimal Liberalism. Suppose that we are given a collection of at least two voters voting on three candidates A, B, and C. By Minimal Liberalism, there are two voters, say, p and q, with individual liberties.

There are two cases to consider. First, suppose that p has rights over A and B and q has rights over B and C. Consider the following preference table:

Voter	p	q	**All other voters**
1st choice	C	B	C
2nd choice	A	C	B
3rd choice	B	A	A

Since p ranks A over B and q ranks B over C, the social welfare must preserve these rankings. The only possibility then is

Social welfare
A
B
C

This ranking violates the Pareto Criterion, because all voters rank C above A. We conclude there can be no such social welfare method.

It remains to consider the case where p has rights over A and B and q has rights over A and C. We leave the proof in this case to Exercise 18.16. ■

The main theorems in this chapter (Theorems 18.8, 18.12, and 18.22) illustrate the art of formulating theorems in social choice theory. While the proofs of these results are not difficult, each theorem reveals how certain combinations of fairness criteria can place severe restrictions on voting methods.

Exercises

18.1 Use the following preference table to give another violation of Monotonicity by the Hare Method:

Number of voters	9	7	11	3	6
1st choice	*A*	*B*	*C*	*D*	*A*
2nd choice	*D*	*D*	*D*	*B*	*C*
3rd choice	*B*	*C*	*B*	*C*	*D*
4th choice	*C*	*A*	*A*	*A*	*B*

18.2 Use the following preference table to give another violation of Monotonicity by the Coombs Method:

Number of voters	4	6	7	15	6
1st choice	*A*	*C*	*B*	*D*	*C*
2nd choice	*C*	*A*	*A*	*B*	*A*
3rd choice	*D*	*B*	*C*	*C*	*D*
4th choice	*B*	*D*	*D*	*A*	*B*

18.3 Prove that a Dictatorship, as social choice method, satisfies Monotonicity and the Independence Criterion.

18.4 Give an example to prove that the Borda Count violates the Independence Criterion for social choice methods.

18.5 Prove that the Borda Count satisfies the Pareto Criterion for social welfare methods.

18.6 Prove that the social welfare methods defined by the Plurality, Borda Count, Hare, and Coombs Methods all violate the Independence Criterion for social welfare methods.

18.7 The Pareto Criterion for a social welfare method can be formulated for social choice methods as follows: a social choice method satisfies the **Pareto fairness criterion** if whenever all voters rank candidate *A* over candidate *B*, then *B* is never the social choice. Prove that the Plurality, Borda Court, Hare, and Coombs Methods all satisfy the Pareto fairness criterion for social choice methods.

18.8 Recall that the Pairwise Comparisons with a Fixed Agenda is the social choice method in which the candidates are first ranked according to the agenda. The last-ranked candidate then runs head to head against the second-to-last-ranked candidate, with the winner moving on to face the third-to-last-ranked candidate and so on until the final round, when the first-ranked candidate faces the survivor of the earlier rounds to determine the social choice. Use the following preference table to prove that this social choice method violates the Pareto Criterion for social choice methods (Exercise 18.7):

Number of voters	10	9	8
1st choice	D	C	B
2nd choice	A	D	C
3rd choice	B	A	D
4th choice	C	B	A

Use the reverse alphabetical agenda D, C, B, A. That is, run D versus C first, with the winner facing B and the winner of that head-to-head contest facing A.

18.9 Prove that the Pairwise Comparisons with a Fixed Agenda satisfies Monotonicity.

18.10 We define a new fairness criterion for social choice methods called **Nonperversity** as follows: suppose that we give a voting method and a preference table with candidate X as the social choice of an election. Suppose that some voting bloc has X ranked directly behind another candidate A. Suppose that this voting bloc instead submits its preferences ranking X above A in the new preference order. Then the social choice satisfies Nonperversity if candidate A is not the new social choice. Prove that the Plurality, Borda Counts, Hare, and Coombs Methods all satisfy Nonperversity.

18.11 Complete the proof of Theorem 18.12 by arguing that neither A nor B can be the social choice for the given election.

18.12 The Double Plurality voting method was introduced in Exercise 12.8. Decide whether the Double Plurality Method satisfies or violates Monotonicity, Independence, and the Pareto fairness criteria for social choice methods. Prove your answer in each case.

18.13 Prove that a Dictatorship satisfies the Independence and Pareto fairness criteria for social welfare methods.

18.14 Prove that there is no social choice method for two candidates and an even number of voters that satisfies Monotonicity, Anonymity, and Neutrality and also avoids ties.

18.15 Monotonicity extends to a fairness criterion for social welfare methods as follows: a social welfare method satisfies **Monotonicity** if whenever a given preference table produces a social welfare with A ranked above B, then A remains ranked above B even if voters who have B ranked above A reverse their rankings for these candidates. Decide whether the social welfare methods defined by the Plurality, Borda Count, Hare, and Coombs Methods satisfy or violate the Monotonicity Criterion for social welfare methods. Prove your answer.

18.16 Complete the proof of Theorem 18.22 by arguing that there can be no winner of the given election when p has rights over A and B and q has rights over B and C.

Suggestions for Further Reading

Numbered references refer to the Bibliography.

Chapter 12

Amayarta Sen's published lecture [46] gives an account of the debate between Borda and Condorcet. Hannu Nurmi's article [30] provides a useful overview of voting methods. Dan Felsenthal's article [14] gives a comprehensive list of violations of fairness criteria.

Chapter 13

Banzhaf's paper [3] summarizes his use of the power index in legal cases. The paper includes an analysis of the power of states in the U.S. Electoral College. The text by Alan Taylor and Allison Pacelli [49] contains much more material on yes-no voting systems, trade robustness, and power indices, including an analysis of several real-world voting systems. Philip Straffin's proof that, after adjusting denominators, power indices correspond to probabilities is in his 1988 paper [47].

Chapter 14

For more on gambling games pitched at the level of this text, see Edward Packel's book [35].

Chapter 15

Philip Straffin's text [48] has a wealth of accessible material on zero-sum games. The famous treatise by J. D. Williams [52] is highly readable. The proof of the general Minimax Theorem (Theorem 15.3) is advanced. See Appendix 2 of the classic game theory text by Duncan Luce and Howard Raiffa [23]. For a careful statement and examples of Nash's Equilibrium, see Chapter 3 of the advanced text by Martin Osborne and Ariel Rubinstein [34].

Chapter 16

An overview of the vast literature on the Prisoner's Dilemma is available online [22]. A major breakthrough on the Iterated Prisoner's Dilemma by William Press and Freeman Dyson is described in their 2012 paper [37]. Steven Brams has written extensively on partial-conflict games, applying his Theory of Moves to political and literary conflicts. [8, 9]. Reinhard Selten's famous analysis of the

Chain-Store Game is found in his 1978 paper [42]. Also see Barry O'Neill's paper on the Dollar Auction [32].

Chapter 17

Charles Bouton's original proof of Theorem 17.11 is found in his 1901 paper [4]. For more on Nim and combinatorial games, see Chapters 20 and 22 in this text.

Chapter 18

Kenneth May's Theorem (Theorem 18.8) originally appeared in his 1952 paper [25]. Alfred Mackay's Theorem (Theorem 18.12) can be found in his 1973 paper [24]. Sen's Impossibility Theorem (Theorem 18.22) appeared in his 1970 paper [44]. Many authors have given further analyses of the problem of individual rights and Sen's Impossibility Theorem. See, for example, the papers of Julian Blau [7] and Donald Saari in [40].

Part III

Special Topics

19 Paradoxes and Puzzles in Probability

A gambler may not know, off the top of his or her head, the exact probability of the winning cards in a hand of Poker or a doctor the exact likelihood of a particular diagnosis. They both may have a gut feeling for how likely or unlikely such an event is to happen. Calculating a probability provides, in this sense, a precise quantification for our intuition about the likelihood of an uncertain event.

In this chapter, we pursue a class of examples designed to stretch intuition and test beliefs about the idea of probability. Many of the examples presented here and in the exercises are commonly referred to as *paradoxes*. A **paradox** is a statement or problem leading to contradictory conclusions or solutions. True paradoxes in mathematics are rare and significant phenomena. For example, a simple paradox about "truth" known as the *Liar Paradox* (see Exercise 10.6) ultimately led to Gödel's remarkable Incompleteness Theorem in mathematical logic. In our own study, the Condorcet Paradox is the basis for Arrow's Impossibility Theorem (Theorem 18.19) and the central example in social choice theory. While we retain the term "paradox" in the names when this is common use, the question of which examples to follow are truly paradoxical is an interesting one and left to your interpretation. We begin with what is perhaps the most popular example of a probability puzzle.

Example 19.1 The Monte Hall Problem

Monte Hall was the flamboyant host of a television game show created in the 1960s called *Let's Make a Deal*. Contestants on the show were led by Hall through a series of guessing games and deals for cash and prizes. The scenario for our problem is the following: the prize is a new car, which is hidden behind one of three identical doors. Behind the other two doors are cows. The contestant must guess the correct door to win the car. Once the contestant has chosen a door, Monte Hall opens one of the remaining doors, and a cow walks through onto the stage. Notice that this is always possible because there are two doors concealing cows. The contestant is now given the opportunity to stay with his or her original choice or to switch to the remaining closed door.

Should the contestant stay or switch?

A first analysis appeals to simple intuition. Suppose that the contestant chose, say, Door 2. The car is either behind Door 2 or behind the remaining unopened door. The odds are even. Since there is no reason to switch doors, the contestant may as well stay with Door 2. We call this the *argument for staying*.

A second approach considers more carefully the actions that led up to this offer to switch doors. When the door was first chosen, the probability that it was the door concealing the car was $1/3$. The odds were against the contestant. Now the contestant is offered the chance to take the other side of the original bet. Thus the contestant should switch and improve the probability of winning the car to $2/3$. One way to view this is to picture the choice between staying and switching as strategies in a game with Door 2 as the chosen door and the car randomly placed behind one of the three doors. The situation is summarized as follows:

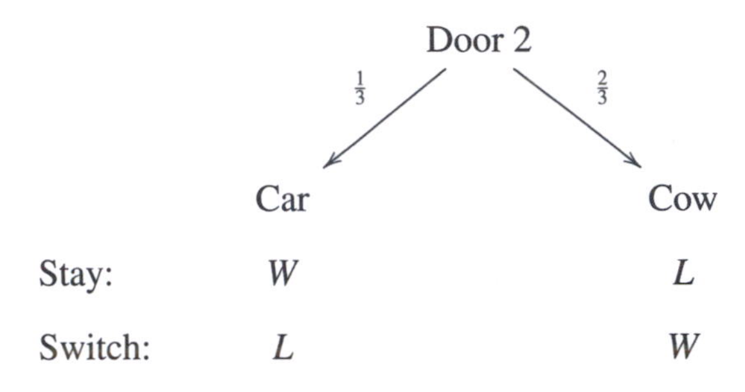

The strategy of switching by the contestant doubles her chance of winning the car.

The Monte Hall Problem illustrates the importance of considering underlying conditions when computing probabilities. The first point of view ignores the information present in the original choice of a door and focuses narrowly on the remaining choice, to stay or switch. This temporary amnesia leads us to an illusion of a 50–50 proposition, or $p = 1/2$. The tension between the competing probabilities of $p = 1/2$ and $p = 1/3$ will appear in several of our examples here and in the exercises.

The Monte Hall Problem is not a true paradox. The correct strategy can be confirmed by running multiple trials and computing expected values. The puzzle arises if one forgets the information present. Indeed, if a second contestant is kept in a soundproof booth off stage during the first stages of the Monte Hall Problem and is only brought in when there are two doors, one with a cow and one with a car, then this contestant does have reason to believe that the doors are equally likely.

The next example examines these ideas of forgetting and belief by creating a version of the sound proof booth.

Example 19.2 The Sleeping Beauty Problem

Sleeping Beauty sleeps until she is woken up and has no memory. On Sunday, a coin is tossed. If it is "Heads," Sleeping Beauty is woken up on Monday and again on Tuesday. If the coin is "Tails," she is only woken up on Tuesday. When Sleeping Beauty is woken up, she is informed of the experiment. She is then asked the probability that the coin came up "Heads". What probability p should Sleeping Beauty assign to "Heads"?

Notice that Sleeping Beauty has full information on the experiment but no information about what has happened in the past. She believes that the coin has been tossed and that she is awake. Since the coin is fair, Sleeping Beauty can believe that the probability it came up "Heads" is $p = 1/2$.

Adam Elga argued that Sleeping Beauty should instead compute the probability of "Heads" to be $p = 1/3$ by the following logic: since Sleeping Beauty does not know the day of the week, she may treat this as a second act of chance and create the following tree:

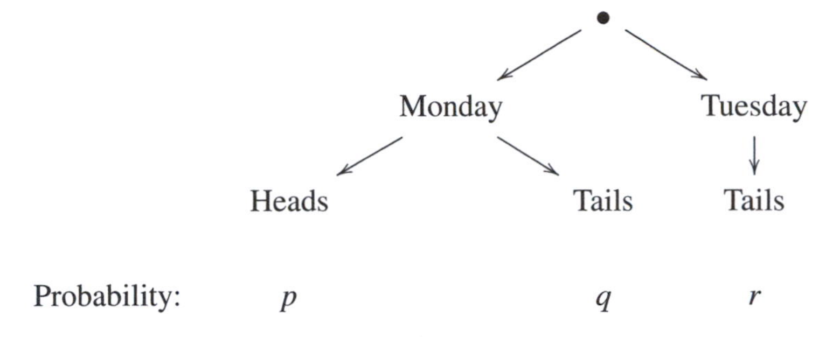

First, consider the probabilities q and r. These are the probabilities of "Tails and Monday" and "Tails and Tuesday," respectively. In other words, the coin came up "Tails," and Sleeping Beauty is assigning probabilities to the days of the week. Since Sleeping Beauty has no reason to believe that one day is more likely than the other, she can conclude $q = r$.

Next, we compare p and q. That is, suppose that it is Monday. Sleeping Beauty knows that she will be woken on Monday with either "Heads" or "Tails," as determined by the experiment. That fact that she is awake on Monday provides no new information and she concludes that $p = q$. But now $p = q = r$ and $p + q + r = 1$. We conclude that $p = 1/3$.

The Sleeping Beauty Problem has generated extensive debate and discussion. Proponents of opposing values for p are known as "Halfers" and "Thirders." What side are you on?

We turn to a mathematical puzzle in the form of the following question:

What probabilities can be produced by flipping a coin?

At first glance, the answer may seem obvious. Flipping a fair coin surely can only produce the probability $p = 1/2$. However, if we allow multiple flips, we can create many more probabilities. For example, the probability p of getting exactly two "Heads" with three flips is $p = 3/8$. Using $n = 4$ flips, we can label outcomes carefully and produce probabilities such as $5/16, 7/16$, and $3/8$ again. Generally, n flips produces probabilities of the form

$$p = \frac{k}{2^n} \qquad \text{for } k = 0, 1, 2, \ldots, 2^n$$

The following simple game produces a surprising probability p.

Example 19.3 Flipping for Heads

Flip a coin until "Heads" appears. Let n denote the number of flips. The player wins if n is even and loses if n is odd.

Flipping for Heads represents a creative interpretation of the original question. While rounds of the game will usually be short, the length of a round is indefinite. In other words, the game is not finite. A portion of the (infinite) probability tree is as follows:

Partial Probability Tree for Flipping for Heads

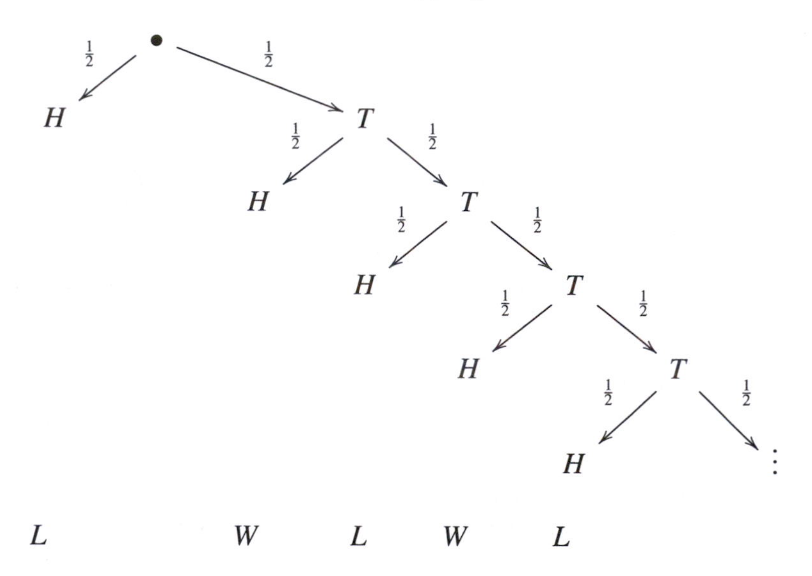

As in our solution of the game Craps (Example 14.1), we introduce a conditional to compute

$$p = \text{probability of winning Flipping for Heads}$$

Suppose that the first two flips are "Tails." We denote this event by TT. When TT occurs, the game Flipping for Heads starts over. That is, the player loses if the next flip is "Heads" and wins if the next two flips come up "Tails" and then "Heads," and so on. Precisely, $P(W \mid TT) = p = P(W)$, where W = "Win Flipping for Heads." Using this observation gives a simpler probability tree for the game:

Abridged Probability Tree for Flipping for Heads

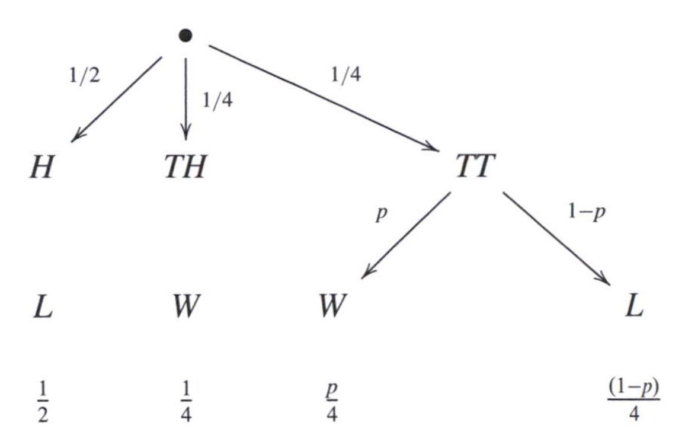

As usual, we can add the probabilities for end nodes labeled W to compute p. This gives the equation

$$p = P(W) = \frac{1}{4} + \frac{p}{4}$$

We solve this equation for p to find

$$4p = 1 + p \quad \text{which implies that} \quad p = \frac{1}{3}$$

Where the Monte Hall Problem creates the illusion of a 50-50 proposition in a game with three options, Flipping for Heads takes the prototype of an even-chance experiment, a coin flip, and produces the probability $p = 1/3$.

The game Flipping for Heads admits endless variations. We consider those in which the basic play is the same – the coin is flipped until "Heads" appears – but the labeling of the end nodes changes. We call the sequence of W's and L's that occur at the bottom of the game tree the **winner sequence**. The winner sequence for Flipping for Heads is then

Original winner sequence: $\quad LWLWLWLW \cdots$

Here is one variation:

Example 19.4 Variation on Flipping for Heads

We specify a new winner sequence for Flipping for Heads:

New winner sequence: $\quad WWLLLWWLLL \cdots$

In this version of the game, the player wins if "Heads" appears on the first or second flip, loses if on the third, fourth, or fifth flip, and then the pattern repeats. We again set $p = P(W)$ and use the repetition in the winner sequence to solve for p. Since the winner sequence repeats after five terms, we summarize these with the following abridged tree. For convenience, we list only the winning end nodes here.

Abridged Probability Tree for Variation 1

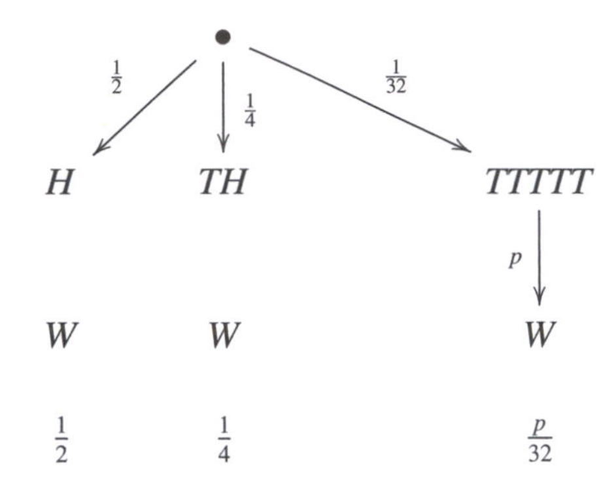

Adding these probabilities, we obtain

$$p = \frac{1}{2} + \frac{1}{4} + \frac{p}{32}$$

Multiplying both sides by 32 gives

$$32p = 16 + 8 + p \qquad \text{which implies that} \qquad p = \frac{24}{31}$$

Here is another version of the game

Example 19.5 Another Variation on Flipping for Heads

We introduce a more complicated winner sequence:

Winner sequence for another variation: $WLLLW\,\overline{LWL}\cdots$

Here we have used the notation $\overline{LWL}$ to indicate the repeating pattern.

We here introduce two probabilities. As usual, we set $p = P(W)$. We also define a second, conditional probability

$$q = P(W \mid TTTTT)$$

We then obtain the following abridged probability tree:

Abridged Probability Tree for Another Variation

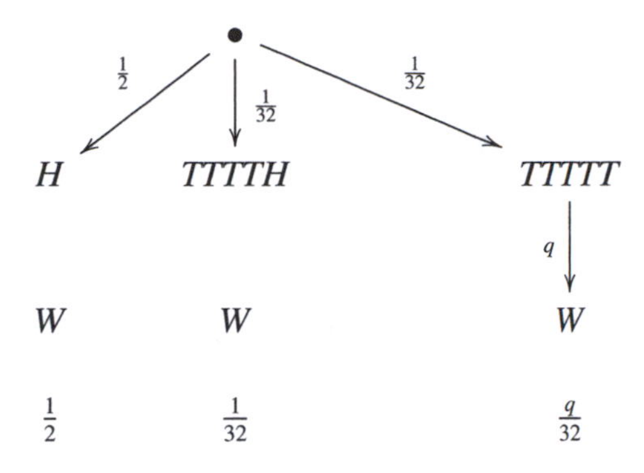

We conclude that

$$p = \frac{1}{2} + \frac{1}{32} + \frac{q}{32}.$$

It remains to compute q. Notice that q is just the probability of winning the game Flipping for Heads with winner sequence $\overline{LWL}$. We can compute this as in the previous two cases:

Abridged Probability Tree for $\overline{WLW}$

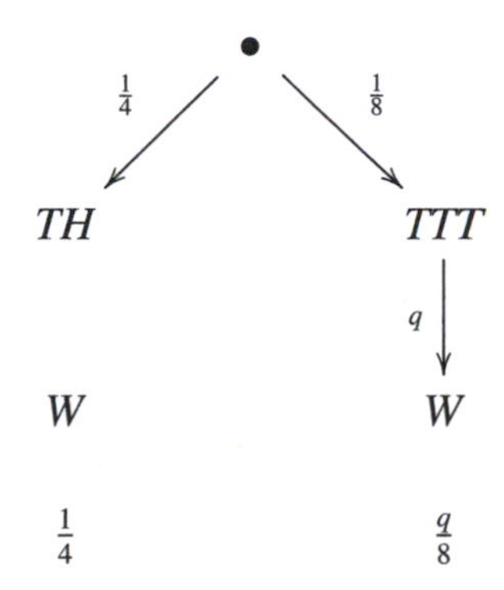

We compute $q = 1/4 + q/8$, so $q = 2/7$. Returning to our formula for p, we obtain

$$p \quad = \tfrac{1}{2} + \tfrac{1}{32} + \tfrac{q}{32} \quad = \tfrac{1}{2} + \tfrac{1}{32} + \tfrac{2}{7 \cdot 32} \quad = \tfrac{121}{224}$$

Further variations on the game of Flipping for Heads clearly abound. All we need to do is specify a winner's sequence, and we have a new version. We now refine the question asked earlier to the following:

What probabilities p can we create using variations on Flipping for Heads?

To answer this question, we introduce the notion of **binary expansion**. By a **probability**, we mean any number p with $0 \leq p \leq 1$. We have the following definition:

Definition 19.6 A **binary expansion** is an expression

$$p =_2 .x_1x_2x_3\cdots x_k\cdots$$

where the x_i are either 0 or 1 and where

$$p = x_1\left(\frac{1}{2}\right) + x_2\left(\frac{1}{2}\right)^2 + x_3\left(\frac{1}{2}\right)^3 + \cdots + x_k\left(\frac{1}{2}\right)^k + \cdots$$

For example,

$$\frac{1}{2} =_2 .1\overline{0} \quad \text{and} \quad \frac{5}{8} =_2 .101\overline{0}$$

As is the case with decimal expansion, binary expansion of a number is not unique. For instance, we can also write $1/2 =_2 .0\overline{1}$. In general, any binary expansion that terminates after finitely many digits can be represented in two different ways.

Just as every conceivable number can be expressed as a decimal, so too can every number be expressed in binary. Here is a sketch of the proof:

Theorem 19.7 *Every number p with $0 \leq p \leq 1$ has a binary expansion*

$$p =_2 .x_1x_2x_3\cdots$$

in which each $x_i = 0$ or 1.

Proof Find the first power n of $1/2$ such that $p > 1/2^n$. The first nonzero binary digit for p occurs in position n. That is, we set $x_1 = x_2 = \cdots x_{n-1} = 0$ and $x_n = 1$. To find the second nonzero digit, let $p_1 = p - 1/2^n$. We then find the first power $m > n$ such that $p_1 > 2^m$. Set $x_{n+1} = \cdots x_{m-1} = 0$ and $x_m = 1$. Next, take $p_2 = p_1 - 1/2^m$ and complete the same steps to compute the third nonzero binary digit. The process continues *ad infinitum* to yield the binary expansion of p. ■

For example, take the case $p = 1/3$. Since $1/3 > 1/2$, we set $x_1 = 0$. Next, since $1/3 > 1/4$, we set $x_2 = 1$. Now we subtract $1/3 - 1/4 = 1/12$ and continue. Since $1/12 < 1/16$ and $1/12 > 1/32$, $x_4 = 0$ and $x_5 = 1$. We must then subtract again to continue. As you may suspect, there is pattern. We have

$$\frac{1}{3} =_2 .010101\cdots = .\overline{01}$$

Notice the connection to the original version of Flipping for Heads. In fact, there is an easier method for computing binary expansions of fractions such as $p = 1/3$. We encourage you to explore this in Exercise 19.5. Here we apply Theorem 19.7 to obtain the following surprising answer to our questions:

Theorem 19.8 *Every probability p occurs as the probability of winning a variation on the game Flipping for Heads.*

Proof Suppose that p has binary expansion

$$p =_2 .x_1x_2x_3x_4\cdots$$

Define a winner sequence for a variation of Flipping for Heads as follows: If $x_1 = 1$ then the winner sequence starts with W. If $x_1 = 0$, then it starts with L. If $x_2 = 1$, the second term of our winner sequence is W; otherwise, if $x_2 = 0$, the second term is L. Continue in this fashion using the nth binary digit to decide whether to insert a W or an L as the nth term in the winner sequence. The resulting version of Flipping for Heads has probability $p = P(W)$. ■

It is worth comparing the calculations of p for our sample variations on Flipping for Heads with the proof of Theorem 19.7. For the calculations, we relied on the patterns in the winner sequence to compute p. For the general p, there may not be such a pattern and hence the unsatisfying nature of this proof. In fact, the probabilities p that exhibit a pattern in their binary expansions exactly correspond to fractions $p = a/b$, where a and b are integers $0 < a \leq b$. We refer to such p values as **rational probabilities**. We leave the question of the meaningfulness of an infinite game with no pattern to the labeling of the end nodes, that is, one with probability p of winning an irrational probability, for you to ponder.

We next give two famous examples of surprising phenomena with probabilities.

Example 19.9 Birthday Problem

The problem is to determine how many people n are needed so that the probability p_n of at least two having the same birthday (i.e., same day and month but not necessarily year) satisfies $p_n > 1/2$.

Of course, if n is larger than 365, then $p_n = 1$. (We ignore the issue of leap years here, for convenience). We compute p_n using the complement rule. That is, we compute the probability

$$q_n = \text{no two among the } n \text{ people have the same birthday}$$

We then obtain $p_n = 1 - q_n$.

Begin with the case $n = 2$. What is the probability that two people have different birthdays? We imagine choosing two dates from the 365 available and keeping track of the order. There are 365^2 possible ordered-pairs of birthdays. To find an ordered-pair of distinct birthdays, we choose the first one (365 choices) and then choose the second birthday from those remaining (364 choices). Thus

$$q_2 = \frac{365 \cdot 364}{365 \cdot 365} = 0.997$$

We next show how to compute q_n from q_{n-1}. Suppose that $n-1$ people are in a room and that no two of them have the same birthday. We would like to add one more person and still have no common birthdays. Since the original $n-1$ people have distinct birthdays, there are $365-(n-1)$ days left to choose for the nth person. Thus

$$q_n = \frac{365-(n-1)}{365} \cdot q_{n-1}$$

We can now work our way up to larger values of n:

$$q_3 = \frac{363}{365} \cdot q_2 = 0.992$$
$$q_4 = \frac{362}{365} \cdot q_3 = 0.983$$
$$q_5 = \frac{361}{365} \cdot q_4 = 0.973$$

The progression decreases from $q_1 = 1$ as n increases. Again, we know that $q_{365} = 0$. The question is when q_n first slips below $1/2$. That this happens rather quickly is something of a surprise.

Theorem 19.10 *Given $n = 23$ people in a room, the probability that two have the same birthday is greater than $1/2$.*

Proof The proof consists of computing q_{23}. We leave this for you as an exercise. We remark that, incidentally, $n = 23$ is the smallest n for which the theorem is true. ■

Next, we give an example of a strange wrinkle in the law of averages.

Example 19.11 Simpson's Paradox

Four softball players meet in a bar. Players A and B are on the Tigers, and Players X and Y are on their cross-town rival, the Lions. Players A and X are the stars of their respective teams, although Player X has been hurt and has missed part of the season. Player Y is an everyday player, and Player B rides the bench. The Tigers have just beaten the Lions. Player X reminds Player A that he or she (X) has a higher batting average than A. Player Y then brags that he or she (Y) has a higher batting average than B. The Tigers are quiet for a moment before Player B, known more for smarts than skill, states that while the Lions' claims are true, the two Tigers players are more likely to get a hit on average than the two Lions players. How is this possible?

Here is a scenario for which all the assertions are true:

Individuals

Player	Hits	At bats	Average
A	105	300	.350
B	10	50	.200
X	40	100	.400
Y	60	250	.240

Teams

Team	Hits	At Bats	Average
Tigers (*A* and *B*)	115	350	.329
Lions (*X* and *Y*)	100	350	.286

We can see that, when all the hits and at bats are combined, the Tigers hit a higher percentage than the Lions. Generally, **Simpson's paradox** is the phenomenon wherein a relationship between averages based on groupings reverses when the groupings are ignored. The paradox famously arose in a discrimination lawsuit against the University of California, Berkeley, Graduate School. See Exercise 19.10.

Bertrand's Paradox

We conclude with a beautiful example of a probability paradox arising in elementary geometry discovered by Joseph Betrand. We begin with a circle. For convenience, we assume the radius $r = 1$, although this is not necessary. A **chord** of the circle is a line segment between two points on the perimeter of the circle. Chords can be quite short or rather long, as follows:

Chords of a Circle

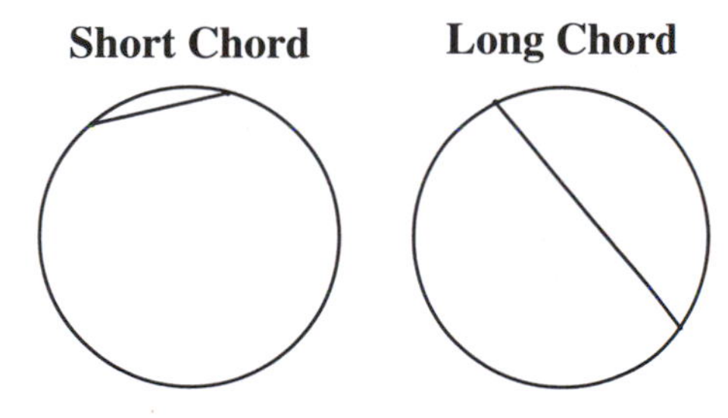

Bertrand's Paradox concerns the probability that a random chord is "long" in the following precise sense: inscribe an equilateral triangle in the circle, creating three chords of equal length a as follows:

Inscribed Equilateral Triangle

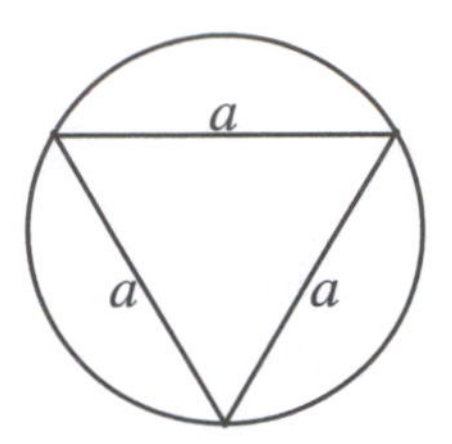

We may think of the length a as the middle value for chords of the circle. When $r = 1$, it is not hard to prove that $a = \sqrt{3}$, although we will not need this here (see Exercise 19.12). Define

$$p = \text{probability that a random chord of the circle has length} \geq a$$

Bertrand gave three separate calculations for p that we sketch briefly here.

The Argument for $p = 1/3$

We observe that it suffices to compare all chords for which one of the two points is a fixed point x on the perimeter. Note that the equilateral triangle divides the perimeter of the circle evenly into three sectors labeled A, B, and C.

Chord with One Point $= x$

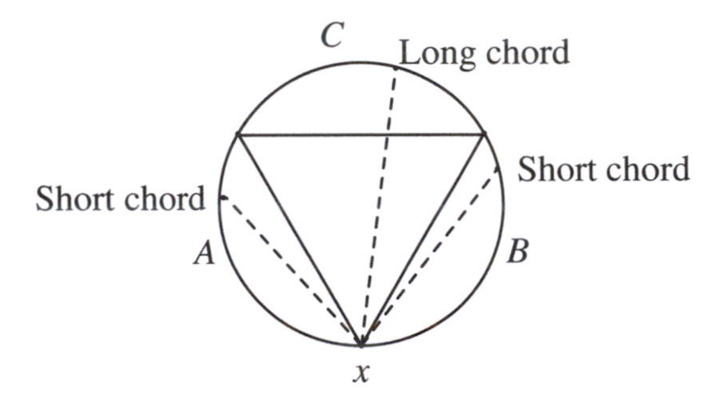

A chord with one endpoint x is longer than the triangle side length a if and only if the other endpoint of the chord lies in sector C. Since the second endpoint of a random chord starting at x can land in any of the three sectors A, B, or C, we conclude that the probability that such a chord has length $\geq a$ is given by $p = 1/3$.

The Argument for $p = 1/2$

In this case, we observe that every chord is perpendicular to some radius line of the circle. We thus compare the lengths of chords perpendicular to a fixed radius line.

Chords Perpendicular to a Radius

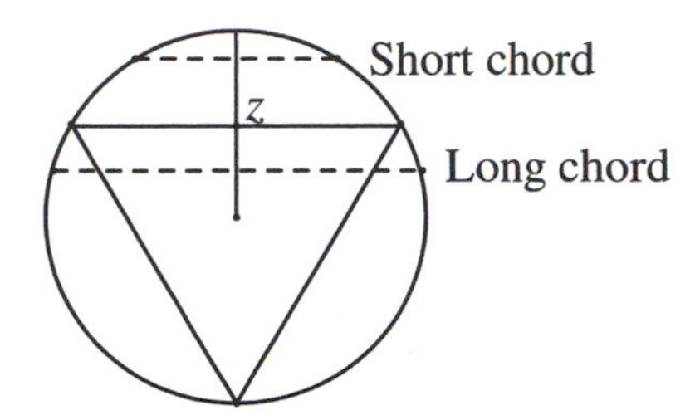

Here we notice that the long chords are precisely those that intersect the radius line below the point z. It is an easy exercise to prove that z is the midpoint of the radius line (see Exercise 19.12). Thus $p = 1/2$.

The Argument for $p = 1/4$

This time we view a chord of the circle as determined by its midpoint. We then see that a chord with midpoint inside the smaller circle of radius $r = 1/2$ has length $\geq a$, whereas a chord whose midpoint lies outside this circle have length $< a$.

Chords and Midpoints

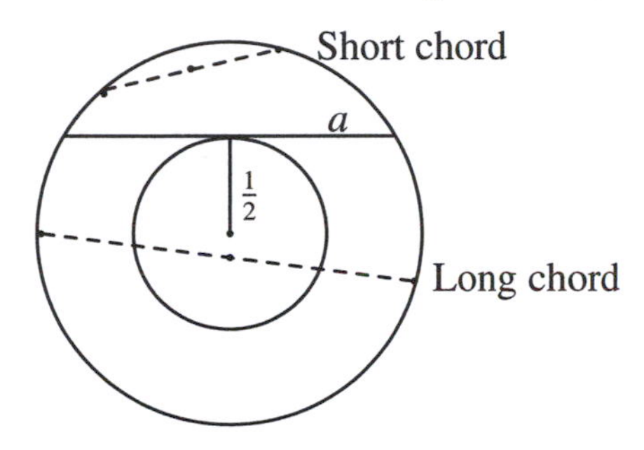

Thus the probability p of a long chord is the ratio of the area of the smaller circle (radius = 1/2) to the area of the full circe (radius = 1). That is,

$$p = \frac{\pi\left(\frac{1}{2}\right)^2}{\pi(1)^2} = \frac{1}{4}$$

Bertrand's Paradox gives a striking illustration of how the outcome of a probability calculation can depend on the perspective taken on the problem.

Exercises

19.1 (The Prisoner's Paradox) Three prisoners (X, Y, and Z) are serving life sentences. The judge has declared that two of the three will receive the death penalty and has revealed the identities of the doomed prisoners to the warden. Prisoner X has become friends with the warden and asks him a favor. Prisoner X knows that the warden will not reveal whether or not he (X) has been sentenced to death, so he asks the warden to reveal the identity of one of the two prisoners who has been so sentenced. The warden agrees and tells him that Prisoner Y has been sentenced to death. Prisoner X then responds happily, "The remaining death sentence is either for me or for Prisoner Z. My chances of living have improved to 50–50!" Do you agree with Prisoner X? How does this paradox compare with the Monte Hall Problem?

19.2 (Bertrand's Box Paradox) There are three closed boxes each containing two coins. One box contains two gold coins, the second box contains two silver coins, and the third box contains one gold and one silver coin. We choose a box at random and select a coin from this box. The coin is gold. The problem is to compute $p =$ probability that the second coin in this box is also gold.

The argument for $p = 1/2$ is similar to the flawed argument for the Monte Hall Problem. We received a gold coin. Therefore, the box with two silver coins has been eliminated. There are only two boxes to consider. We conclude that the probability that we find a second gold coin in our chosen box is $p = 1/2$.

Use Bayes' Law (Theorem 14.5) to compute the correct conditional probability

$$p = P(\text{“Chosen box has two gold coins”} \mid \text{“Chosen coin is gold”})$$

19.3 (Two-Envelopes Problem) Two sealed envelopes each contain a check. One envelope contains a check for the amount x, and the other a check for twice as much, or $2x$. The amounts x and $2x$ are not disclosed to the contestant. The contestant chooses an envelope and opens it. She is then given the option to either keep this check or trade in her envelope for the other envelope. What should the contestant do, stay or switch? Suppose that the amount on the check in the chosen envelope is y dollars. The question is whether $y = x$ or $y = 2x$.

(a) Argue that the choice between staying and switching is indifferent, and the contestant may as well stay.
(b) Compute the expected value when the contestant switches, and prove that this expected value is larger than y. Conclude that the contestant should always switch envelopes.

Do you feel that one of the arguments (a) or (b) is correct? Explain your reasoning.

19.4 Compute $p =$ probability that Player 1 wins the following variations on Flipping for Heads specified using a winner sequence. State the corresponding binary expansion $p =_2 .\_\_\_\_\_$ that you have computed.

(a) $\overline{LLWW}$
(b) $\overline{WLLWW}$
(c) $LWL\overline{LLLW}$
(d) $LWWLW\overline{WWLW}$
(e) $WLWWL\overline{WLLLL}$

19.5 Here is an algorithm for finding the binary expansion $p =_2 .x_1x_2x_3x_4\cdots$ for a fraction $p = a/b$. Set $n_1 = a$, the numerator of the fraction. Multiply n_1 by 2. If $2n_1 < b$, the first binary digit $x_1 = 0$. In this case, let $n_2 = 2n_1$. If $2n_1 \geq b$, then set $x_1 = 1$ and $n_2 = 2n_1 - b$. Now repeat the process with n_2 to get x_2, the second binary digit. Here is a schematic for the repeated step in the algorithm:

Computing the *k*th Binary Digit x_k and n_{k+1} for a/b

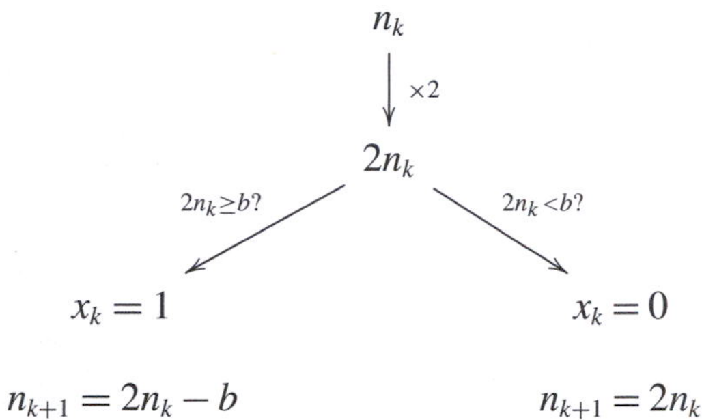

Use this algorithm, find the binary expansion for the following probabilities $p = a/b$:

(a) $p = 5/7$
(b) $p = 3/11$
(c) $p = 8/35$
(d) $p = 11/43$

19.6 (St. Petersburg Paradox) This paradox is related to the game Flipping for Heads. Flip a coin until "Heads" appears, and keep track of $n =$ number of flips until "Heads." The payoff for the game is 2^n dollars. What is the expected value for this game? How much would you be willing to pay up front to play? Can you formulate the paradox?

19.7 Roll a single die until you get a ⚀. Let n denote the number of rolls until a ⚀ appears. Given a winner sequence, we obtain the game Rolling for a ⚀. Determine the probability $p =$ "Player 1 wins the game" for the following winning sequences:

(a) $\overline{WL}$
(b) $\overline{LWW}$
(c) $WL\overline{LWL}$
(d) Find a winner sequence for Rolling for a ⚀ with probability of winning $p = 11/15$.

19.8 For the Birthday Problem (Example 19.9), prove that the probability q_n of n people all having distinct birthdays is given by

$$q_n = \frac{365 \cdot 364 \cdots (365-n+2) \cdots (365-n+1)}{365^n}$$

Use a calculator to confirm that $q_{22} > 1/2$ and that $q_{23} < 1/2$.

19.9 Assume that we are given M possible labels and n people. Let p_n be the probability that at least two people have the same label. Show that $p_n = 1 - q_n$, where $q_n =$ probability that no two people have the same label, is given by

$$q_n = \frac{M \cdot (M-1) \cdots (M-n+2) \cdots (M-n+1)}{M^n}$$

Apply the result to address the following problems:

(a) Find the probability p that at least two people of four were born on the same day of the week.
(b) Find the probability that in a group of 10, at least 2 were born in the same month.
(c) Find the minimum number of people so that the probability p that two were born in the same month satisfies $p > 1/2$.

19.10 This is a simplified version of the U.C. Berkeley discrimination case leading to Simpson's paradox. The data for men and women applicants to the Departments of Sciences and Humanities are as follows:

Sciences

	Men	Women
Applied	120	30
Accepted	90	25

Humanities

	Men	Women
Applied	40	100
Accepted	15	55

Evaluate the claims that (a) women are accepted to the university at a lower percentage than men and (b) women are accepted at a higher percentage in each division. Do you think that the data support a bias against women? Explain.

19.11 Show that integers a, b, c, d and w, x, y, z satisfying

$$\frac{w}{x} < \frac{a}{b}, \quad \frac{y}{z} < \frac{c}{d}, \quad \text{and} \quad \frac{w+y}{x+z} > \frac{a+c}{b+d}$$

give rise to a Simpson's Paradox. Devise another explicit example with meaningful categories.

19.12 Consider the following circle of radius 1 with inscribed equilateral:

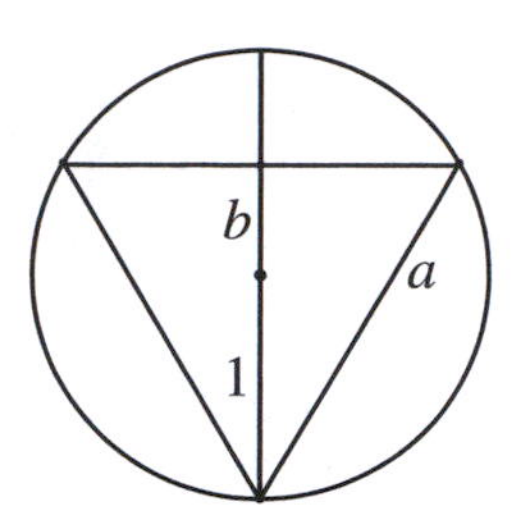

Prove that $a = \sqrt{3}$ and $b = 1/2$, as used in the arguments for Bertrand's Paradox.

20 Combinatorial Games

Combinatorial games offer an excellent example of the divide between theory and practice. On the one hand, we have Zermelo's Theorem (Theorem 4.17) asserting that one player can guarantee victory in any such game by following a particular series of moves. On the other hand, this theorem offers no practical help in executing the move-by-move play of a long and complicated game. While the theory of combinatorial games might be said to begin with Zermelo's Theorem, it certainly does not end with this result.

Recall that we have defined a combinatorial game to be a sequential-move finite win-lose game of perfect information. We begin this chapter by extending our proof of Zermelo's Theorem for takeaway games (Theorem 10.9) to the general case of combinatorial games. We then introduce an example of a Nim-like game called *Northcott's Game* and apply Bouton's Theorem (Theorem 17.11) to determine the winner and winning strategy. Each takeaway game gives rise to a new combinatorial game called the *Misère* version. We analyze the Misère version of Two-Pile Nim using both the Binary and Ordinal Labeling Rules. Finally, we introduce the idea of *strategy stealing* for takeaway games and apply this argument to give *weak solutions* for the games *Chomp* and *Hex*.

Zermelo's Theorem

Mathematically, a combinatorial game is modeled by a game tree. We recall a **game tree** is a rooted tree with end nodes labeled with the winning player (Definition 2.3). For takeaway games, the end-node labels are unnecessary: the player who moves to an end node on his or her turn is the winner. In Chapter 10, we used the Binary Labeling Rule (Definition 4.16) and the Principle of Induction for Trees (10.2) to prove Theorem 10.9. The idea was simply to label the game tree and then read off the winner and the winning strategy from the labels.

Here we extend this argument to prove Zermelo's Theorem for combinatorial games (Theorem 4.17). We illustrate the key idea with the following example:

Game Tree for a Combinatorial Game

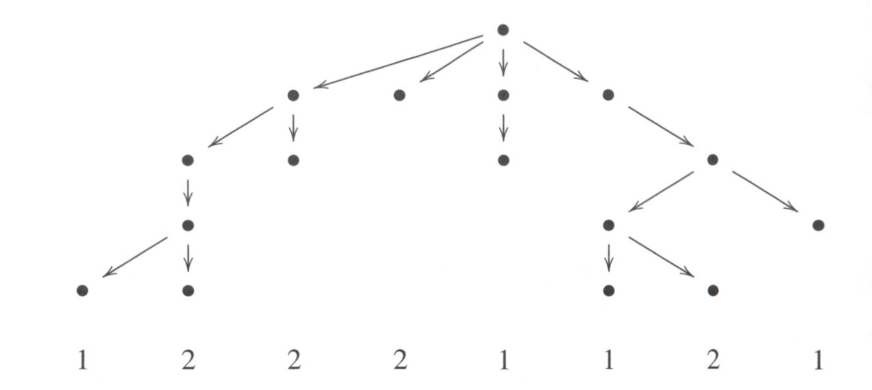

This game tree is not the game tree for a takeaway game for the simple reason that the end nodes are not labeled so that the last player to move always wins. We can identify the *problem end nodes* by considering the stages of the game.

Problem End Nodes in the Game Tree

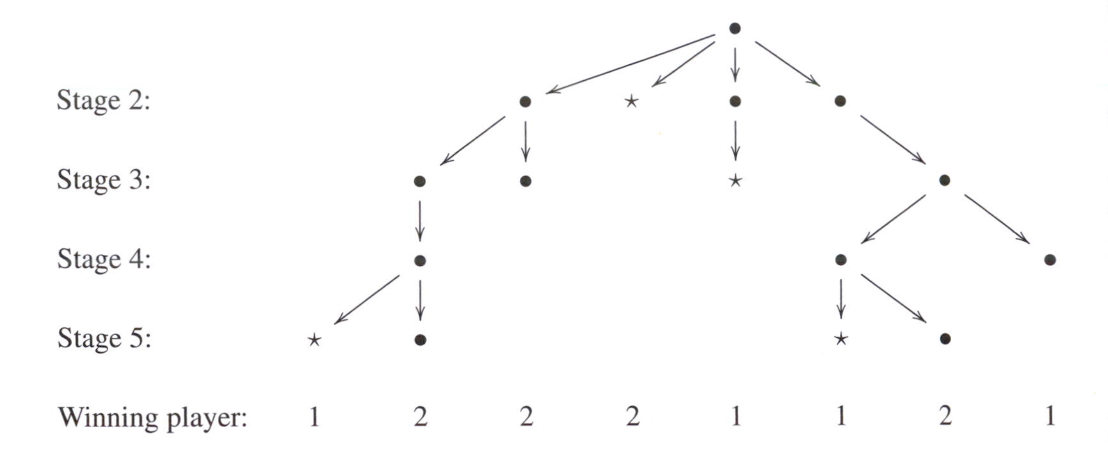

Here the starred end nodes have the "wrong" winner. More precisely, observe that for a takeaway game, Player 1 wins at end nodes occurring at an even stage of the game, and Player 2 wins at end nodes occurring at an odd stage of the game. Any end node that does not have the label determined by the stage of the game will be called a **problem end node**.

A simple trick now serves to convert the original game tree into a game tree for a takeaway game. We simply add an edge (indicated as a dotted edge here) to each of the starred end nodes, leading to a new end node at the next stage, as follows:

Conversion of a Game Tree to a Take away Game

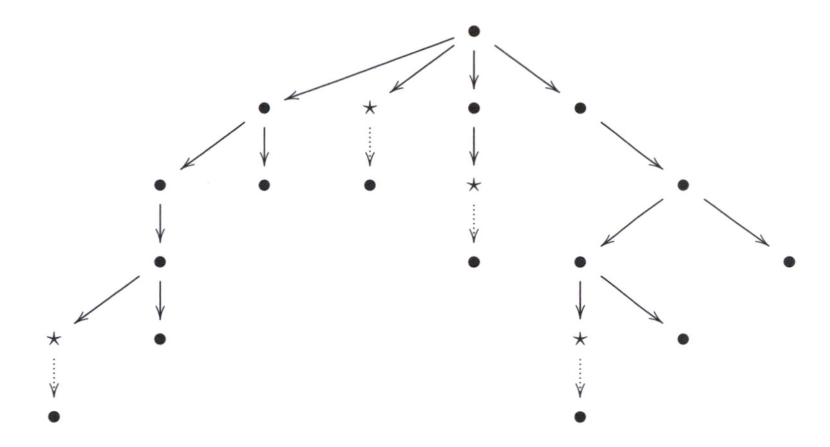

Recall from Definition 10.10 that the Tree Game corresponding to a finite, rooted tree is the takeaway game in which players take turns moving down the tree, with the player who moves to an end node as the winner. We note that the Tree Game played with the preceding tree is the same game as the original combinatorial game – the extra moves in the Tree Game do not represent a material change to the game.

We assert that this conversion can be performed on any game tree for a combinatorial game and so provide the following definition:

Definition 20.1 Given a game tree for a combinatorial game, define the **equivalent Tree Game** to be the Tree Game corresponding to the given tree in which new edges are added, leading to new end nodes for each of the problem end nodes in the original game tree, as earlier.

With this trick in hand, we deduce the following theorem:

Theorem 20.2 Zermelo's Theorem *In every combinatorial game, one player has a strategy that will guarantee victory.*

Proof A combinatorial game is represented by a game tree. Replace this game tree with the equivalent Tree Game. The result now follows from Theorem 10.9 applied to this Tree Game. ■

While Zermelo's Theorem assures us that every combinatorial game can be solved, finding the solution (the winning player and winning strategy) can be a challenging problem. Bouton's Theorem (Theorem 17.11) is the prototype of an elegant solution to a combinatorial game. We next consider an application of this result to a related game.

Northcott's Game

Bouton's ingenious use of binary numbers and his extension of the idea of balance to general configurations allow us to "scale the game tree" for the game of Nim and identify the winning player and winning strategy. The method applies to a larger class of games that we may call "Nim-like" games. We will not attempt to define this term but rather illustrate the idea with one example. Further examples of Nim-like games are given in the exercises.

Example 20.3 Northcott's Game

The game is played on an 8 × 8 checkerboard. Eight black (⊗) and eight red (⊙) checkers are arranged on the board, with one checker of each color in each column, as follows:

Northcott's Game

Player 1 (⊙)

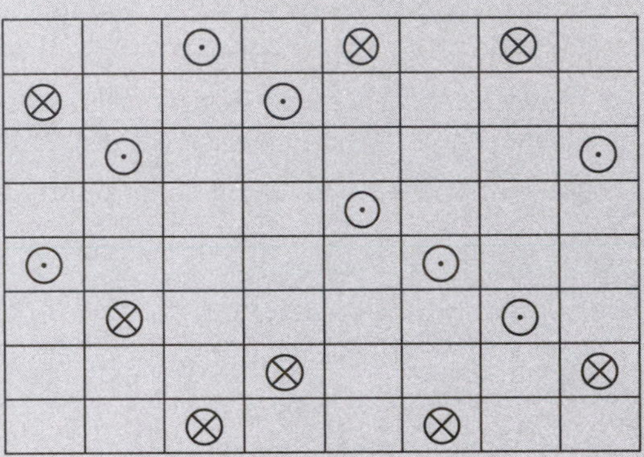

Player 2 (⊗)

A turn consists of moving one's checker in the given column. A player may move his or her checker up or down in the given column as many squares as desired, but he or she may not jump over the opponent's checker in this column. The game ends when one player cannot move a checker. This player is the loser of the game.

Northcott's game is not, strictly speaking, a combinatorial game because play can continue indefinitely. Notice, however, that either player can force the game to be finite by always moving his or her checkers toward the opponent's checkers. Northcott's Game is, in fact, a clever variation on the game Nim. Treat the spaces between the checkers as pile numbers. The preceding version of Northcott's game has Nim configuration

$$(2, 2, 6, 4, 2, 2, 4, 3)$$

As in Theorem 17.11, we represent these pile numbers in binary and check the balance of the rows:

Northcott's Game: Unbalanced Configuration

2	2	6	4	2	2	4	3	Row status
0	0	1	1	0	0	1	0	Unbalanced
1	1	1	0	1	1	0	1	Balanced
0	0	0	0	0	0	0	1	Unbalanced

Since the configuration is unbalanced, Player 1 can guarantee victory in this version of Northcott's game. The strategy for Player 1 is that prescribed by Bouton's Theorem, namely, to balance the rows in the configuration. At the binary level, a first move is to remove three from the pile of six, yielding the following:

Player 1 Balances the Configuration

2	2	3	4	2	2	4	3	Row status
0	0	0	1	0	0	1	0	Balanced
1	1	1	0	1	1	0	1	Balanced
0	0	1	0	0	0	0	1	Balanced

On the checkerboard, this amounts to Player 1 moving his or her checker in the third column down three spaces until it is two spaces from Player 2's checker. Now we can see that any move that Player 2 makes, whether he or she moves his or her checker up or down, will result in an unbalanced configuration. Player 1's winning strategy is to continue to balance the configuration on his or her turn, always moving the checkers closer to Player 2's checkers. Once the gaps are all closed (i.e., the piles are all empty), the end game consists in Player 1 forcing Player 2 down to the bottom row.

We next turn to a particularly confounding class of combinatorial games obtained by a simple change of the rules to a takeaway game.

Misère Games

Given a takeaway game, we obtain a new game by declaring the last player to move the loser.

Definition 20.4 The **Misère game** for a given takeaway game has the same rules of play as the original except that the player who moves last now loses.

It is instructive to consider the effect of converting the Misère version of a Tree Game as in Definition 10.10 back to the form of a Tree Game. Recall that a Tree

Game is the takeaway game determined by a rooted tree in which players alternate advancing from the current node to an adjacent node. The winner is the last player to move. In our terminology, every end node is now a "problem end node." The conversion is thus accomplished by adding an edge (shown here as a dotted edge) to each end node, as follows:

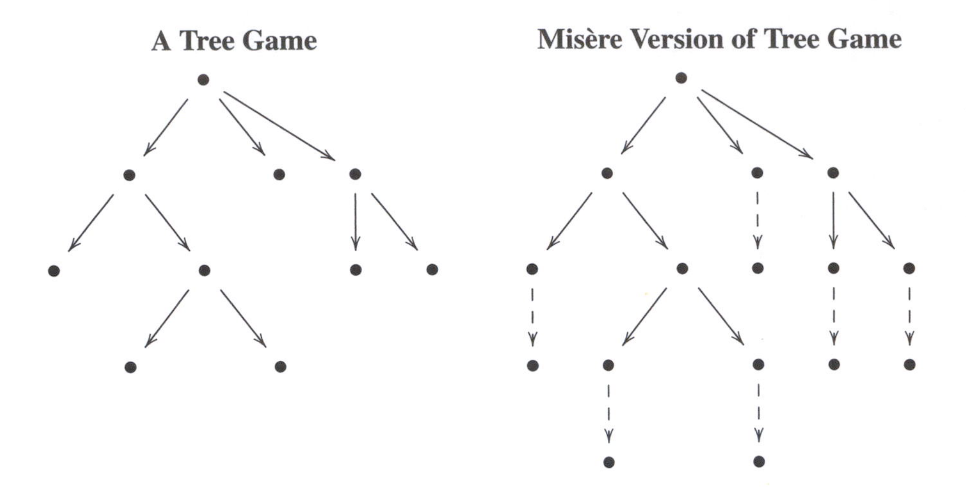

A similar adjustment may be made to the Binary and Ordinal Labeling Rules. For a Misère game, the end nodes all receive the label 1 (as opposed to 0). The rules for labeling the nodes higher in the tree remain the same. Here is the Ordinal Labeling the preceding tree, we obtain the following:

Misère Tree Game Ordinal Labels

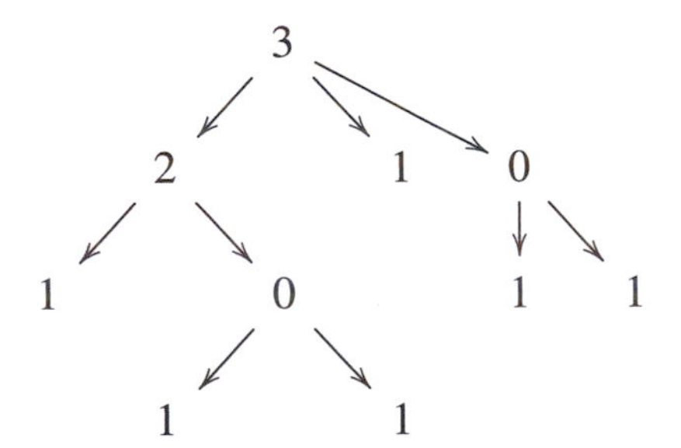

It is natural to ask whether our knowledge of a solution of a takeaway game is of any help in solving the Misère version. We address this question with the example of the game of Two-Pile Nim. By Theorem 4.20, Player 1 wins this game in rational play when the piles are unequal and loses when they are equal. Does the situation simply reverse for Two-Pile Nim Misère? It is easy enough to check some smaller cases. Here is what we find:

Comparing Labels for Nim and Nim Misère

	Binary labels		Ordinal labels	
Node	**Nim**	**Nim Misère**	**Nim**	**Nim Misère**
(0, 0)	0	1	0	1
(1, 0)	1	0	1	0
(1, 1)	0	1	0	1
(2, 0)	1	1	2	2
(2, 1)	1	1	3	3
(2, 2)	0	0	0	0
(3, 0)	1	1	3	3
(3, 1)	1	1	2	2
(3, 2)	1	1	1	1
(3, 3)	0	0	0	0

Notice that the labels are reversed for the three nodes above the horizontal line (nodes (0, 0), (1, 0), and (1, 1)). Below the horizontal line, the labels agree, at least for those that we have reached. We focus on the binary labels and prove that this pattern is, in fact, a theorem.

Theorem 20.5 *Player 1 wins the Misère version of Nim(m,n) in rational play if and only if (1) $m \neq n$ and either $m > 1$ or $n > 1$ or (2) $(m,n) = (1,1)$ or* $(m,n) = (0,0)$.

Proof We proceed by induction on the depth of the (original) game tree for Nim(m,n). Recall that the game tree has depth $= m+n$.

Base Case: We extend the usual base case here to include the nodes $(0,0),(1,0)$, and $(1,1)$. We have already observed that the theorem is true for these cases. That is, the winner is Player 1 when balanced and Player 2 when unbalanced: the winner in the Misère version is the opposite from the original version.

Induction Hypothesis: We assume that the statement is true for all games of Misère Nim(m,n) with $m+n \leq k$ for some fixed $k \geq 2$.

Induction Step: Suppose that $m+n = k+1$. We show that Misèrer Nim(m,n) has binary label 1 if and only if $m \neq n$. First, suppose that $m \neq n$, say, $m < n$. If $m > 1$, then, by the induction hypothesis, (m,m) has binary label 0. In this case, Player 1 should balance $(m,n) \rightarrow (m,m)$ and by the induction hypothesis will leave Player 2 with no winning move. If $m = 1$, then Player 1 simply moves $(1,n) \to (1,0)$, and Player 2 loses on his or her next turn.

Now suppose that $m = n$. We may assume that both $m,n \geq 2$ by the base case. The only possible moves for Player 1 are of the form $(m,m) \to (l,m)$, where $l < m$. Since $m \geq 2$, our induction hypothesis indicates that these nodes all have binary label 1. We conclude that (m,m) has binary label 0. Player 1 loses the game Misère Nim(m,m) in rational play for $m > 1$. ■

Although the solution of ordinary Two-Pile Nim certainly informed our discovery of Theorem 20.5, the Misère version presented a new challenge. In fact, Misère versions of takeaway games are often harder to solve and, in many cases, represent challenging open problems in game theory.

Chomp and Hex

We conclude by introducing two combinatorial games with visual appeal and considerable complexity. First, we have the game popularized by David Gale.

Example 20.6 Chomp

The game is played starting with a rectangular grid of squares (thought of as a sectioned chocolate bar) in which the lower-left entry is a poison square. Here is an example:

A 4 x 6 Bar for Chomp

Each player, on his or her turn, chooses a square, and "eats" this square plus all squares above and to the right of this square. The player forced to eat the poison square loses. Chomp is thus a Misère game. The non-Misère version in which the poison square is edible is an easy win for Player 1.

Here is a sample round of the game. Player 1 chooses squares with an X, and Player chooses squares with a Y.

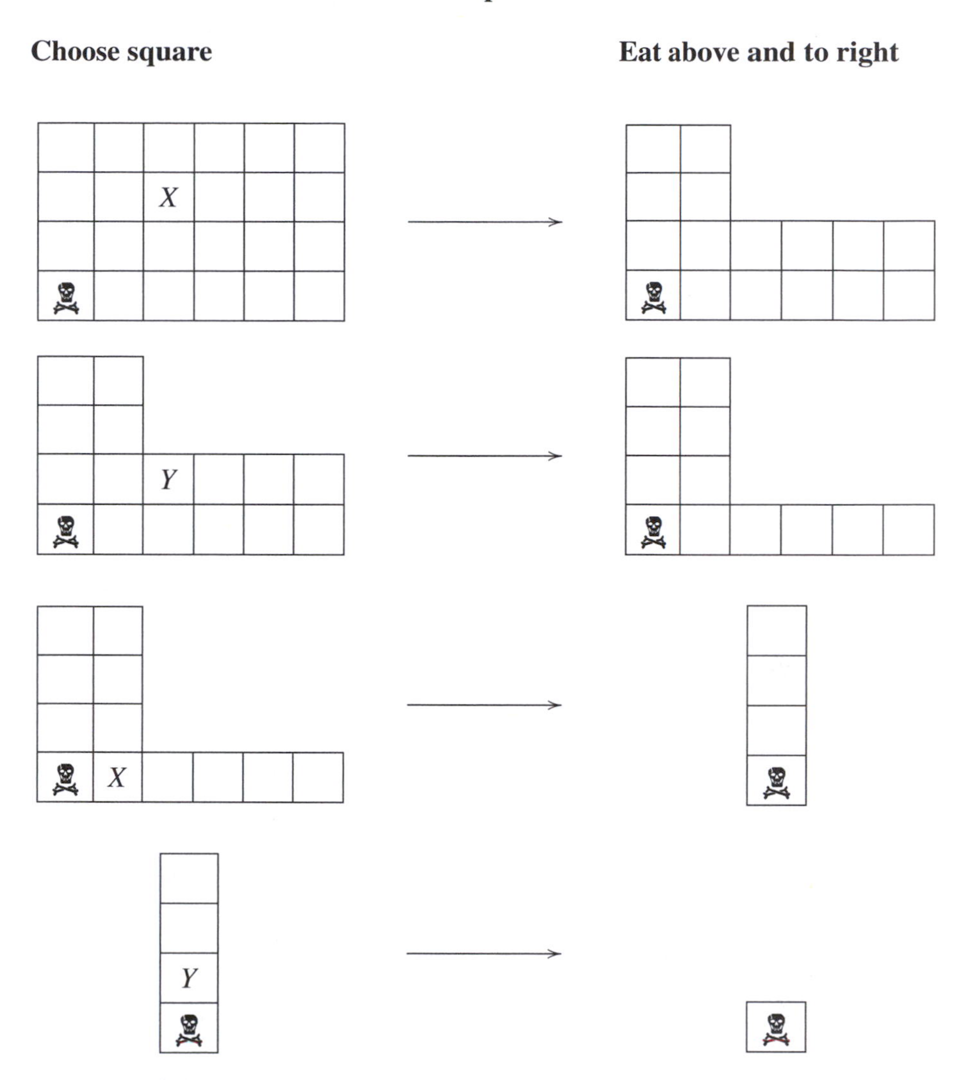

Player 1 is forced to eat the poison square and loses in this round.

As with any combinatorial game, Chomp presents a direct mathematical challenge: who wins the game in rational play? What is the winning strategy? We can begin by looking at the binary labels for some small subgames.

Binary Labels for Some Chomp Positions

Chomp position	Binary label	Chomp position	Binary label
	0	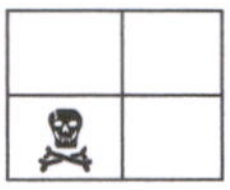	1
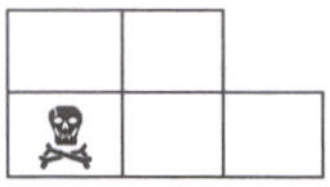	0	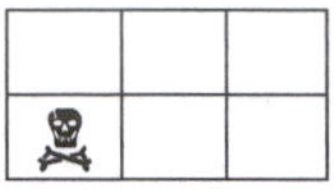	1
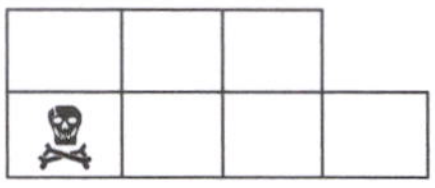	0	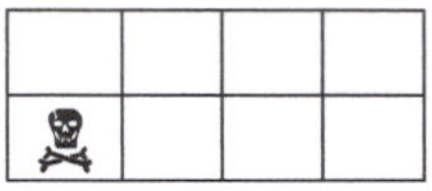	1

We have hardly scratched surface of the many possible positions in Chomp. Nevertheless, we can see a pattern developing. The full rectangles all have binary label 1. Notice that these are the starting positions for a particular game of Chomp. The proof of this fact, in general, introduces an important notion called *strategy stealing*.

Theorem 20.7 *Player 1 wins Chomp in rational play starting with any rectangular* $m \times n$ *block with either* $m > 1$ *or* $n > 1$.

Proof By Zermelo's Theorem, either Player 1 or Player 2 has a winning strategy in Chomp. We suppose, for a contradiction, that Player 2 has a winning strategy. In fact, suppose that Player 2 is so certain of his or her infallible strategy that he or she has made the algorithm openly available. Player 2's winning strategy contains, in particular, a response to every first move by Player 1. We imagine two different openings to the game.

First, suppose that Player 1 chooses as his or her first move to eat the upper-rightmost square, as shown

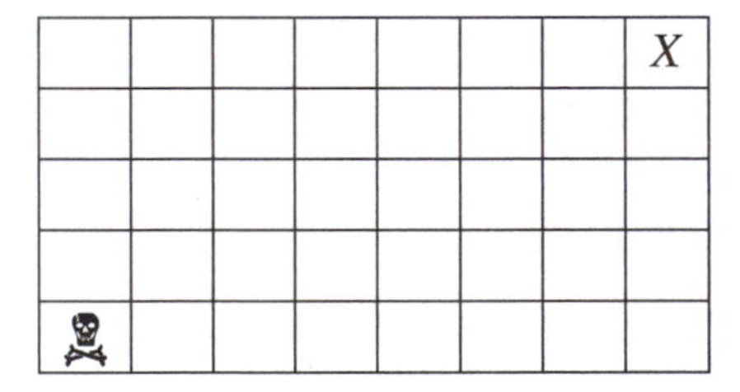

Player 2's winning strategy then provides a countermove, say, C, in response. Further, Player 2's algorithm provides a response to every subsequent move by Player 1 following this move C.

Now start the game over. This time Player 1 steals Player 2's move C for his or her first turn. The Chomp board then looks exactly the same for Player 2 on his or her first turn as it did originally for Player 1 on his or her second turn in the first version of the game. The point is that the upper-rightmost box marked X is eaten by every first move in the game. In the second version of the game, Player 1 has effectively stolen Player 2's winning position; he or she can now follow Player 2's algorithm to respond to every move starting with Player 2's response to move C until he or she has won the game. ■

The argument used in this proof deserves special mention.

efinition 20.8 A **strategy-stealing argument** is a proof that the first player in a combinatorial game is the winner in rational play. The argument depends on two properties of the game:

1. The game must be **symmetric**. That is, the two players must have the same available moves and same outcomes from these moves.
2. The game must have a first move available with the property that Player 1 is no worse off taking this move than he or she would be in allowing Player 2 to move first.

The strategy-stealing argument reveals the winning player in rational play but not the winning strategy. We formalize the possibilities as follows:

efinition 20.9 A **complete solution** of a combinatorial game is the identification of the winning player and the winning strategy in rational play. A **weak solution** is an identification of the winning player only.

Bouton's Theorem is a complete solution for Nim, whereas Theorem 20.7 represents a weak solution to Chomp.

Example 20.10 Hex

Hex is played on a diamond-shaped $n \times n$ board of hexagonal spaces. The four sides of the Hex board are colored, alternately, white and black. Here is a picture of the board for $n = 7$:

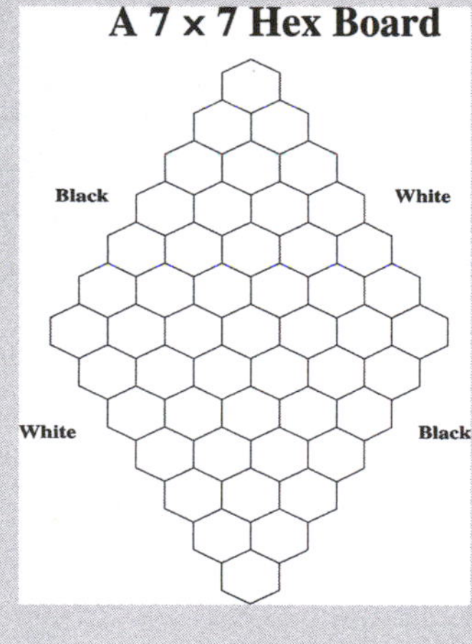

Player 1 has white hexagonal pieces, and Player 2 has black pieces. Each player, on his or her turn, places one of the pieces onto an open space on the board. The winner of the game is the player who creates a path of his or her pieces crossing from the one side of this color to the other. Thus Player 1 wins if he or she creates a path of white hexagons from one white side to the opposite white side. Player 2 wins if he or she creates a path of black hexagons from one black side to other black side.

We consider the following question:

Can Hex end in a tie?

To make the issue precise, suppose that all of n^2 positions have been covered with white and black hexagonal pieces. Is it possible that there is neither a path of white pieces connecting the white regions nor a path of black pieces connecting the black regions? The following elegant sketch is due to David Gale.

Theorem 20.11 *Hex is a win-lose game.*

Proof The idea is to imagine the white hexagons as a flow of water from two reservoirs represented by the white sides of the Hex board. We imagine the black hexagon pieces as blocks laid to form a dam aimed at keeping the water from flowing between the two reservoirs. Now it stands to reason that when the board is completely full, either the water flows between the reservoirs and there is a winning path of white pieces or, alternately, the dam has been successfully constructed. In the latter case, note that the black hexagonal pieces must form an unbroken path between the black sides. In this case, Player 2 wins. We see then that one player must win the game. ■

As a direct consequence, we deduce a weak solution to Hex.

Theorem 20.12 *Player 1 wins Hex in rational play.*

Proof The proof follows the same line as that of Theorem 20.7. We leave the details for you in Exercise 20.14. ■

Theorems 20.7 and 20.12 are examples of nonconstructive proofs. In both cases, we arrive at the conclusion that there exists a winning strategy for Player 1 without gaining any insight into how to find the strategy. Indeed, finding the complete solution to Chomp and Hex in the general cases is an open problem.

Exercises

20.1 Convert the following game trees to an equivalent Tree Game as in Definition 20.1. Apply the Ordinal Labeling Rule to the equivalent Tree Game to compute the Sprague–Grundy number of the combinatorial game.

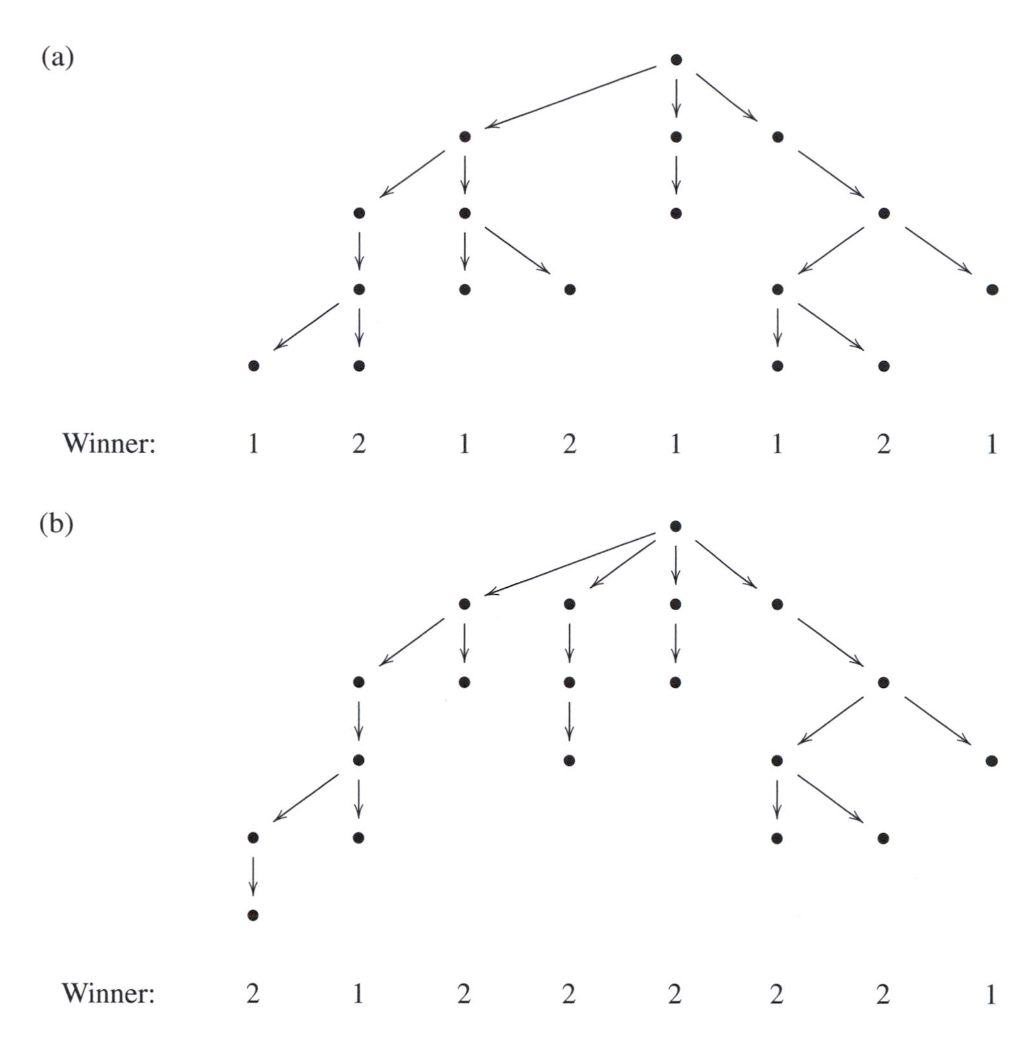

20.2 Determine the winner in rational play of the following versions of Northcott's game:

(a)

⊗		⊗		⊗		⊙	⊙
			⊙				
	⊙						
				⊙			
⊙					⊗		
			⊗			⊗	
							⊗
	⊗	⊙			⊙		

(b)

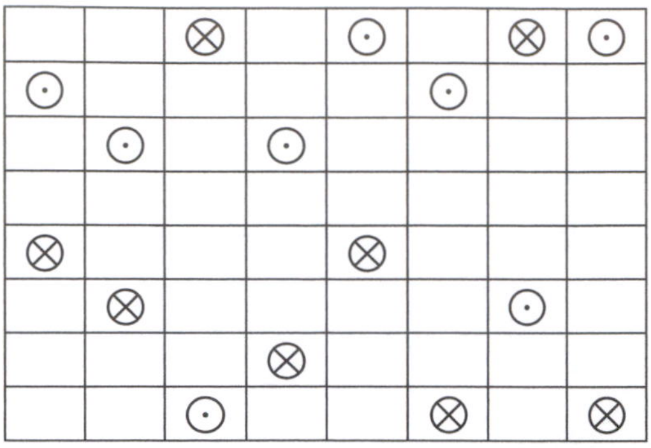

		⊗		⊙		⊗	⊙
⊙					⊙		
	⊙		⊙				
⊗				⊗			
	⊗					⊙	
			⊗				
		⊙			⊗		⊗

20.3 Consider the following variation on Northcott's Game. The game is played in two stages. In the first stage, the players alternate placing the checkers in the columns beginning with the leftmost and proceeding to the rightmost. Players must place their checkers on one of the four squares closest to their side of the board. Once the checkers have been placed in all eight columns in this column-by-column fashion, Player 1 begins the resulting version of Northcott's Game. Determine the winning player and winning strategy for this variation on Northcott's game.

20.4 (Nimble) The game Nimble is played on a single row of square numbered 1 to n. To begin, some coins are placed in various squares with, perhaps, more than one coin on a square, as follows:

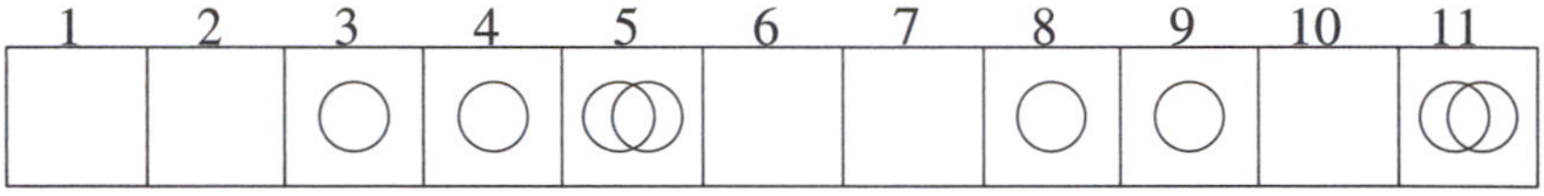

1	2	3	4	5	6	7	8	9	10	11
		○	○	○○			○	○		○○

Each player, on his or her turn, moves one coin not in the first square to the left any number of squares (but at least one). The winner is the last player to move. Prove that this game is simply Nim in disguise. Determine the winner of the game shown in rational play.

20.5 (Turning Turtles) This is a Nim-like game under a somewhat better disguise than Nimble. Ten turtles are lined up from left to right. Some are on their backs (X) and some right-side up (O). On his or her turn, each player must choose one right-side-up turtle and flip it onto its back (turn O into X). The player may, if he or she wishes, also change the status of one turtle to the left of this chosen one, either from O to X or from X to O. We translate a "Turtle configuration" (i.e., a list of X's and O's) to a Nim configuration as follows: view a right-side-up turtle in position n as a pile of n chips. This gives the Nim configuration for the game. For example,

Turtles	Nim Piles
XOXXOXOXXO	(2, 5, 7, 10)
OXXOOOXXOX	(1, 4, 5, 6, 9)

(a) Show that the Nim move of removing an entire pile of n chips corresponds to the Turning Turtles move of flipping an O in position n to an X.
(b) Consider the Nim move of reducing a pile of n chips to a pile of k chips. Since we have a pile of n chips, in Turning Turtles, there must be an O in position n. Flip this O to an X. We now want to add k chips back. If there is an X in position k, we may simply flip this to an O and thereby add a pile of k chips in the Nim configuration. Suppose that there is an O in position k. Prove that flipping this O to an X yields a Nim configuration that is strategically the same as to that obtained by simply reducing the pile of n chips to k chips. *Hint:* Prove that the Nim configuration with two piles of k is strategically the same as the configuration with both piles of k eliminated.
(c) Find the winners of the preceding Tuning Turtles game in rational play.

20.6 Consider the game Turning Turtles, but in this case a player, having turned an O to an X in position n, may now, if he or she wishes, flip the turtle in position $n-1$ (either from O to X or from X to O). Prove that this version of Turning Turtles is strategically the same as Two-Pile Nim.

20.7 (Wythoff's Game) Here is a game with a design similar to Two-Pile Nim. Start with two piles of chips. On his or her turn, each player can take a chip or chips from one of the two piles, or the player may take an equal number of chips from both piles at once. The winner is the last player to take chips. Prove that the smallest configurations with binary label 0 are: (0, 0), (1, 2), (3, 5), (4, 7), and (6, 10). Find the next four smallest configurations with binary label 0.

20.8 Recall the game Taking Chips for general n and k (Example 4.14). The game begins with a pile of n chips. Each player, on his or her turn, must take at least one chip and no more than k chips from the pile. The last player to move wins. Compute the Sprague–Grundy numbers for the following versions of Taking Chips Misère:

(a) $n=5$ and $k=2$
(b) $n=7$ and $k=3$
(c) $n=12$ and $k=3$
(d) $n=22$ and $k=6$

20.9 Let n and k be positive integers with $k<n$. Find a formula for the Sprague–Grundy number for the Misère version of Taking Chips with $n=$pile size and $k=$maximum taken on a turn. See Exercise 17.9.

20.10 Extend the table started in this chapter by computing the Sprague–Grundy numbers for both the standard and Misère versions of Nim$(4,m)$ for $m=0,\ldots,4$ and Nim$(5,m)$ for $m=0,\ldots,5$. Can you make a conjecture about the correspondence between Sprague–Grundy numbers for the original and Misère versions of Two-Pile Nim?

20.11 Determine the binary labels for the following positions in the game Chomp. For the positions with binary label 1, find a first move in rational play.

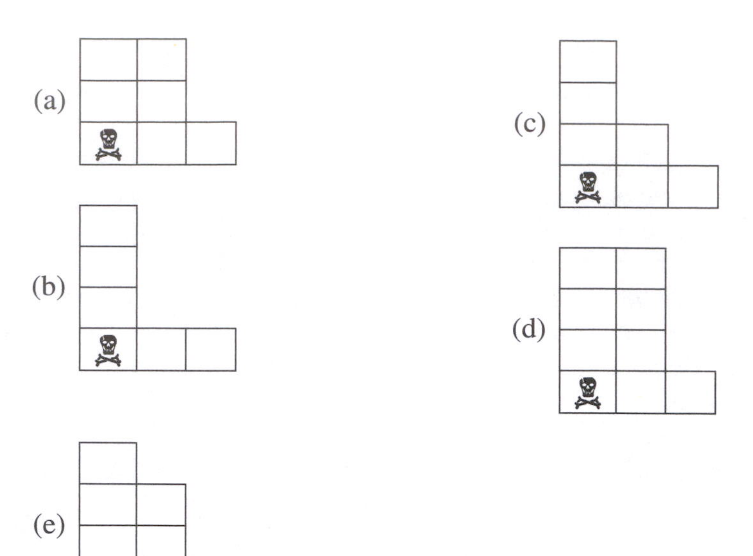

20.12 Compute the Sprague–Grundy number of the following games:

(a) Nim Misère$(1,2,2)$
(b) Nim Misère$(1,2,3)$
(c) Nim Misère$(1,2,4)$
(d) 3×2 Chomp
(e) 2×3 Chomp
(f) 3×3 Chomp

20.13 In the game Hex, explain why both white and black cannot simultaneously have a winning path.

20.14 Prove that Hex satisfies the requirements for a strategy-stealing argument as in Definition 20.8 and so prove Theorem 20.12.

20.15 Explain why the strategy-stealing argument (Definition 20.8) does not apply to the game Nim.

20.16 (Subset Takeaway Game) The **Subset Takeaway Game** was introduced by David Gale. Given a set A, each player must name a nonempty subset of A other than A itself, that is, a **proper subset** of A. There is one rule: No player may name a subset of A that contains a subset already mentioned during the game. The first player with no legal move loses. Thus Player 1 loses the game when $A = \{1\}$ because there are no proper subsets of A. Determine the winner of the Subset Takeaway Game in rational play for the following cases:

(a) $A = \{1,2\}$
(b) $A = \{1,2,3\}$
(c) $A = \{1,2,3,4\}$
(d) Can you make a conjecture for A any finite set?

21 Borda versus Condorcet

In this chapter, we explore the debate over voting methods that marked the birth of social choice theory as an academic field. Let us briefly recall the history and point of contention. On one side, we have the Chevalier Jean Charles de Borda, a mathematician and military scientist. Borda proposed his voting method in a lecture to the French Academy of Sciences urging the body to adopt his voting method for member elections. Borda's method, as we know, converts voter preferences into a single ranking, the social welfare. While ties do occur, they are not so common. In 1780, Borda published the first mathematical proof of fairness. He proved that the Borda Count can never elect a candidate ranked last by a majority of the voters (a Majority Loser). We proved this result ourselves in Theorem 12.15. In fact, Borda proved a stronger theorem. Answering the open question from Chapter 12, we state the following theorem:

Theorem 21.1 Borda's Theorem *The Borda Count satisfies the Condorcet Loser Criterion.* ■

On the other side of the debate is the Marquis Nicolas de Condorcet, a prominent intellectual of the Enlightenment period in France. Condorcet also had mathematical evidence supporting his political ideals for elections. Condorcet had a proof that the best chance for achieving the "correct" result in an election is to obey the majority opinion. Condorcet's Jury Theorem (Theorem 7.10) was stated in Chapter 7. For Condorcet, the majority opinion is expressed through head-to-head Simple Majority elections. As we have seen, the Borda Count does not always select the majority winner as the social choice and so violates a basic principle of fairness.

It was Condorcet himself who pointed out the principal obstacle to his own majoritarian ideals. When there are three or more candidates, the majority opinion, as expressed through head-to-head elections, can be fundamentally conflicted. We can have a Condorcet Paradox. Moving ahead almost 200 years to 1951, Kenneth Arrow used the Condorcet Paradox to prove his Impossibility Theorem, beginning the modern era of social choice theory.

In this chapter, we explore the mathematical landscape of the debate between Borda and Condorcet. We begin by proving a version of Condorcet's Jury Theorem. Our proof uses the idea of "introducing a conditional" discovered in our analysis of the game Craps in Chapter 14. We then consider two examples:

one a challenge for Borda and the other a challenge for Condorcet. These examples together reveal the complexity of social preferences and motivate our proof of a *realization theorem* for social preferences proved by James McGarvey. McGarvey's Theorem (Theorem 21.8) establishes that, indeed, anything is possible as far as conflicts in social preferences go. Next, we prove Borda's Theorem (Theorem 21.1). Finally, we introduce a compromise voting method proposed by Duncan Black that combines the virtues of the Borda Count with Condorcet's ideals of majoritarian rule.

A Jury Theorem

Our analysis of the game of Craps depended on introducing a conditional to compute a probability. We apply this technique again to prove the following version of the Jury Theorem:

Theorem 21.2 *Suppose that we are given a jury pool in which each juror will vote correctly with a fixed probability p with* $1/2 < p < 1$. *Suppose that the jurors vote without consulting. Let M and N be be any odd integers with* $M < N$. *Then*

$$P(\text{“Correct verdict with } M \text{ jurors”}) < P(\text{“Correct verdict with } N \text{ jurors”})$$

Proof Although the statement is certainly believable, to compute the probabilities is forbidding. Fortunately, we only need to prove an inequality. Further, we may assume that $N = M + 2$. The general result then follows by an induction argument (see Exercise 21.10).

Begin with the M jurors. Since M is odd, when these jurors vote, there is a verdict; the verdict is either incorrect or correct. We are interested in the situation where adding the two new jurors changes the verdict. We define our condition then to be this event: " Two new jurors change verdict." We prove that the two new jurors change the verdict from incorrect to correct with higher probability than the reverse. That is, we prove that

$$P(\text{“Incorrect to correct”} \mid \text{“Two new jurors change verdict”}) >$$
$$P(\text{“Correct to incorrect”} \mid \text{“Two new jurors change verdict”})$$

We focus on the given event "Two new jurors change verdict." This event requires that the margin of victory among the M jurors be equal to one vote; otherwise, the two new jurors cannot change the verdict. Writing $M = 2k + 1$ for some k, we may assume that the vote is either k to $k + 1$ for the correct verdict or $k + 1$ to k for the correct verdict. We then have the following branching:

Probability Tree when M Jurors Reach a Verdict by a Margin of 1

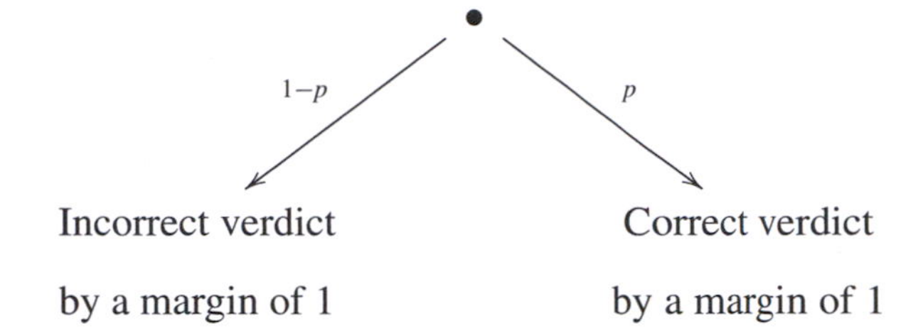

The branch probabilities correspond to the probability that a "swing" voter votes either incorrectly with probability $1-p$ or correctly with probability p. We leave the careful proof of this fact to Exercise 21.11. Adding the two new jurors gives the following tree:

Probability Tree when Two New Jurors Change the Verdict

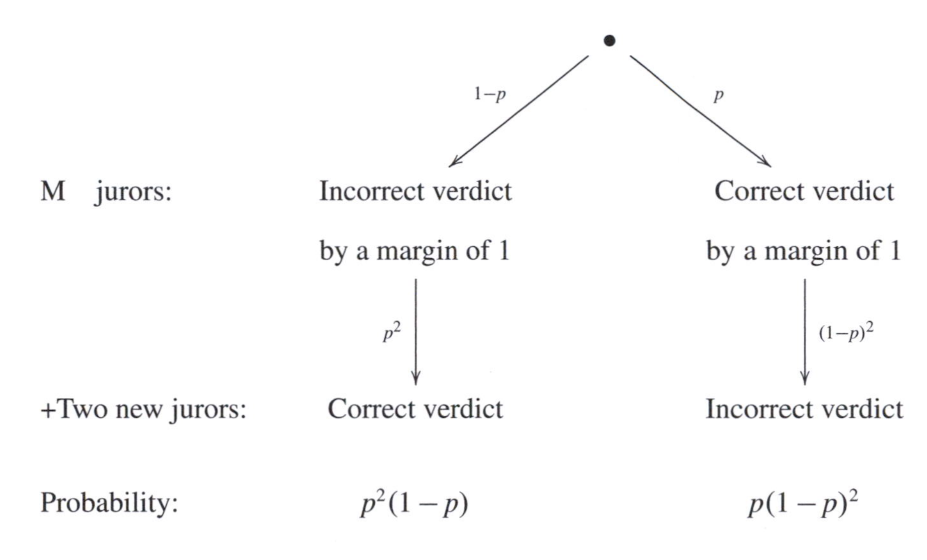

The left branching is the case where the two new jurors correct an incorrect verdict by the original M jurors. The right branching is where they reverse a correct verdict by the M jurors. Let $D(p)$ be the difference between the two probabilities. That is,

$$\begin{aligned} D(p) &= p^2(1-p) - p(1-p)^2 \\ &= p(1-p)(p-(1-p)) \\ &= p(1-p)(2p-1) \end{aligned}$$

Since p and $1-p$ are positive, the sign of $D(p)$ depends on the term $2p-1$. This term is positive because $p > 1/2$. We conclude that the probability of the two new jurors correcting an incorrect verdict is larger than that of them reversing a correct verdict. ■

We remark that Theorem 21.2 does not imply the Jury Theorem (Theorem 7.10). The Jury Theorem asserts that the probability of a correct verdict approaches 1 as the number of jurors increases indefinitely. We have only proved that the probabilities increase with the number of jurors. Nevertheless, our version does support Condorcet's belief in the value of obeying the majority opinion of a well-informed populous.

The Conflict

We delve into the debate between Borda and Condorcet by considering two examples. We imagine the first example as Condorcet's challenge to Borda:

Example 21.3 Condorcet's Challenge

Consider the following preference table of 195 voters choosing between four candidates A, B, C, and D:

Preference Table for Condorcet's Challenge

Number of voters	50	49	48	48
1st choice	A	A	B	C
2nd choice	B	C	C	D
3rd choice	C	D	D	A
4th choice	D	B	A	B

Applying the Borda Count to this preference table gives the following ranking:

Borda Count Social Welfare for Condorcet's Challenge

Candidate	Borda points
C	388
A	345
B	244
D	193

Condorcet's critique of the Borda Count begins with the observation that candidate A has a majority of the first-place votes and should be at the top of the ranking. Digging deeper, recall that we introduced the **social preference graph** in Chapter 5 to summarize the various relations of social preference between candidates. Specifically, the social preference graph is the complete directed graph with arrows $X \longrightarrow Y$ indicating that X beats Y in a head-to-head election. We have the following:

Social Preference Graph for Condorcet's Challenge

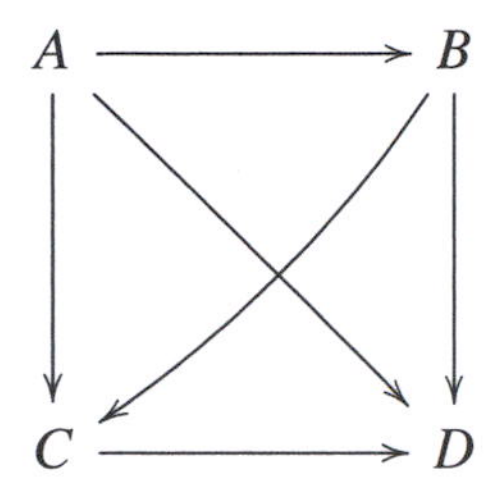

If we count the number of head-to-head wins for each candidate, we obtain the following ranking, which we call the *Condorcet ranking*:

Condorcet Ranking for Condorcet's Challenge

Candidate	Head-to-head wins
A	3
B	2
C	1
D	0

We make this official with the following definition:

efinition 21.4 The **Condorcet ranking** of candidates in an election is the ranking obtained by counting the number of wins for each candidate when all candidates are run head to head in Simple Majority elections.

This example supports Condorcet's basic critique of the Borda Count. Not only does the majority candidate A lose the Borda Count, but the winner C only beats one other candidate in a head-to-head election. The Borda Count fails to recognize a clear mandate for the ranking $A \quad B \quad C \quad D$ as determined by the Condorcet ranking.

We next consider a possible response to Condorcet's Challenge:

Example 21.5 Borda's Response

Consider the following preference table of 47 voters choosing between five candidates A, B, C, D, and E:

Preference Table for Borda's Response

Number of voters	10	12	11	14
1st choice	A	E	E	C
2nd choice	B	A	B	D
3rd choice	C	B	D	E
4th choice	D	C	A	A
5th choice	E	D	C	B

As usual, the Borda Count renders a definitive social welfare:

Borda Count Social Welfare for Borda's Response

Candidate	Borda points
E	120
A	101
C	88
B	87
D	74

However, running each of the 10 pairs of head-to-head elections gives the following deadlocked Condorcet ranking:

Condorcet Ranking for Borda's Response

Candidate	Head-to-head wins
A	2
B	2
C	2
D	2
E	2

This example highlights the issue with the Condorcet ranking. How do we determine a winner of this election or, more problematically, a social welfare using the Simple Majority Method? We next explore two mathematical problems motivated by these examples.

On the Possible Social Preference Graphs

A preference table provides a comparison of every pair of candidates in an election. The resulting social welfare graph records the results of these head-to-head contests. Of course, given any list of candidates, we can arbitrarily assign winners and losers to head-to-head contests without a preference table but just according to our whim. The result is called a *tournament*.

efinition 21.6 Given a collection of candidates A, B, C, and so on, a **tournament** is an assignment of winner and loser to every pair of candidates.

Mathematically, a *tournament* corresponds to a true round-robin in sports in which every pair of teams meets one time. Of course, the difficulty in determining a champion from a round-robin is precisely the issue raised by Borda's Response. We continue to speak the language of social choice theory and use the term "candidates" instead of "players" or "teams." Given candidates X and Y, we indicate that X beats Y, as usual, with an arrow $X \longrightarrow Y$. A tournament is then seen to be a complete directed graph with the candidates as nodes, as in Definition 5.15. Here is an example:

(21.1) **A Tournament with Five Candidates**

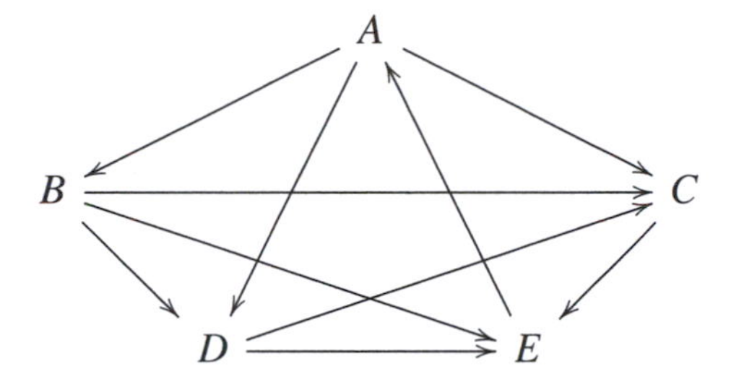

It is easy to count the number of different tournaments possible for a given set of candidates, as the following theorem shows:

Theorem 21.7

$$n(\textit{distinct tournaments with n candidates}) = 2^{\binom{n}{2}}$$

Proof If there are n candidates total, then the number of pairs of candidates is given by "choosing" two candidates at a time: $\binom{n}{2}$. For each pair of candidates X, Y, we have two choices of arrows: $X \longrightarrow Y$ or $X \longleftarrow Y$. Now apply the Principle of Counting. ■

As mentioned earlier, if we eliminate the possibility of ties, then a preference table gives rise to a tournament, namely, the social preference graph. For an illustration, we have Example 21.5:

Borda's Response **Social Preference Graph**

Number	10	12	11	14
1st	*A*	*E*	*E*	*C*
2nd	*B*	*A*	*B*	*D*
3rd	*C*	*B*	*D*	*E*
4th	*D*	*C*	*A*	*A*
5th	*E*	*D*	*C*	*B*

$\Longrightarrow$

Given an arbitrary tournament, we can ask the converse question: is there a preference table having this exact tournament as a social preference graph? This type of question is called a **realization problem** in mathematics, that is, a problem asking for a general construction of objects of a certain type. If a preference table can be produced, we say that the given tournament is **realized** as a social preference graph. We consider this problem for the Tournament with Five Candidates (21.1). Here is a summary of the realization problem:

A Realization Problem

Preference Table? **Tournament with Five Candidates**

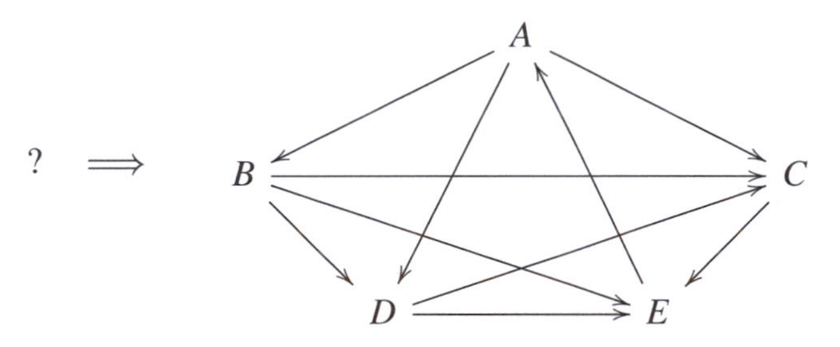

We solve this particular problem now. We introduce 10 voters $p_1,\ldots,p_{10}$ whose rankings of the candidates will create this tournament via head-to-head elections.

Our first step is to pick out one candidate to remove. We take out *E*. Now we notice that the other four candidates are actually transitively ordered by the tournament: *A* *B* *D* *C*. That is, *A* beats *B* , *C*, and *D*; *B* beats *D* and *C*, while *D* beats *C*. We define the preferences of voters p_1 and p_2 so as to achieve two things: (1) p_1 and p_2 both rank the first four candidates in the correct order: *A* *B* *D* *C*; and (2) p_1 and p_2 disagree regarding all head-to-head contests involving *E*. The following assignments achieve both goals:

p_1	p_2
A	E
B	A
D	B
C	D
E	C

We say a collection of voters $\mathcal{V}$ with preferences for a slate of candidates is **neutral** about two candidates X and Y if these voters reach a tie when voting on X versus Y by Simple Majority. Condition (2) ensures that $\mathcal{V} = \{p_1, p_2\}$ is neutral about E versus any other candidate A, B, C, or D.

The tournament specifies that E should beat A head to head. We achieve this with our next two voters, p_3 and p_4. We assign preferences to these voters so that (1) both voters prefer E to A and (2) the voters disagree about every other pair of candidates. The needed assignments are as follows:

p_3	p_4
E	D
A	C
B	B
C	E
D	A

Notice the trick: we have placed the relation "E over A" for (1) at the top of the ranking $B \quad C \quad D$ for the first voter, and then we have put "E over A" at the bottom of the reverse ranking $D \quad C \quad B$ for the second voter. Combined, the voters p_1, p_2, p_3, and p_4 now have social preferences for the transitive ranking $A \quad B \quad D \quad C$ and also for A over E. Furthermore, these four voters are neutral about E versus B, C, and D.

We proceed in the same manner to define preferences for p_5 and p_6. These voters will both prefer B to E, as specified by the tournament. They will have opposite preferences for all other pairs:

p_5	p_6
B	D
E	C
A	A
C	B
D	E

Continuing in this manner, we define the preferences for p_7 and p_8 to achieve the needed relation C over E and p_9 and p_{10} to achieve the relation D over E. The result is the following solution to the realization problem:

Preference Table Realizing Tournament with Five Candidates

Number	p_1	p_2	p_3	p_4	p_5	p_6	p_7	p_8	p_9	p_{10}
1st	A	E	E	D	B	D	C	D	D	C
2nd	B	A	A	C	E	C	E	B	E	B
3rd	D	B	B	B	A	A	A	A	A	A
4th	C	D	C	E	C	B	B	C	B	D
5th	E	C	D	A	D	E	D	E	C	E

The same ideas can be applied to prove the following general result:

Theorem 21.8 McGarvey's Theorem *Every tournament can be realized as the social preference graph for some preference table.*

Proof We prove this by induction on the number n of candidates (nodes) in the tournament.

Base Case: When $n = 1$, there is nothing to prove, and when $n = 2$, the result is easy: if the tournament is $A \to B$, we simply introduce one voter who prefers A to B.

Induction Hypothesis: We assume that any tournament with n candidates can be realized as the social preference graph for some preference table.

Induction Step: Start with a tournament with $n+1$ candidates. For definiteness, let us refer to this tournament as $\mathcal{T}_{n+1}$. We must build a preference table $\mathcal{P}_{n+1}$ realizing this tournament. Choose a candidate, say, X, from among the $n+1$ candidates in the tournament $\mathcal{T}_{n+1}$. Removing X and all edges either to or from X gives a tournament with only n candidates. Let us call this tournament $\mathcal{T}_n$. By the Induction Hypothesis, there is some preference table $\mathcal{P}_n$ whose social preferences for pairs of candidates not involving X are as specified by the arrows in $\mathcal{T}_n$.

The preference table $\mathcal{P}_n$ consists of a collection of voters, say, $\mathcal{V}$, each ranking the n candidates A, B, C, D, and so on but with no mention of candidate X. We construct a collection of voters $\mathcal{V}'$ who also rank X. We begin by building a preliminary set of such voters $\overline{\mathcal{V}}$.

Given a voter p in the collection $\mathcal{V}$, we define two clones of p. Both of these clones will rank all $n+1$ candidates. We call the first clone p'. He or she will rank candidate X first and then have the same social preferences for A, B, C, D, and so on below X as p does. The second clone p'' will rank X last and copy p's ranking for B, C, D, and so on above A. The idea is shown here:

Original Voter in $\mathcal{V}$	**Clone 1**	**Clone 2**
p	p'	p''
	X	A
A	A	C
C	C	B
B	B	D
D	D	X

Let $\overline{\mathcal{V}}$ consist of all first and second clones of our original voters in $\mathcal{V}$. With these preferences, the collection $\overline{\mathcal{V}}$ of voters is neutral about X versus A, B, C, D, and so on, and yet the voters $\overline{\mathcal{V}}$ have the same social preferences for pairs of candidates that do not involve X as the original voters in $\mathcal{V}$.

We now add $2n$ more voters to introduce the correct social preferences for the n pairings A and X, B and X, C and X, and so on. For A and X, we introduce two new voters r and q . These voters agree on the correct preference for A and X and disagree on all other pairs. Recall the following trick:

Arrow in $\mathcal{T}_{n+1}$

$$A \longleftarrow X$$

r	q
X	D
A	C
B	B
C	X
D	A

We continue with X and B, X and C, and so on adding pairs of voters to $\overline{\mathcal{V}}$ to introduce the correct social preference as specified by the Arrow in $\mathcal{T}_{n+1}$. The result is a new collection $\mathcal{V}'$ of voters and a corresponding preference table $\mathcal{P}_{n+1}$. By construction, $\mathcal{P}_{n+1}$ realizes the tournament $\mathcal{T}_{n+1}$, completing the induction step and the proof. ■

How do we interpret Theorem 21.8 as regards the debate between Borda and Condorcet? For Borda supporters, the result reinforces the point that social preferences are too complicated to directly produce a social welfare. Since anything is possible, we need an efficient and decisive method for interpreting the social will, namely, the Borda Count.

Condorcet supporters might counter that Theorem 21.8 merely sketches the difficult territory we face when making a social choice. It does not remove the requirement that we obey majority opinion when possible. For the example in Condorcet's Challenge, there is a clear mandate for the social welfare based on the Condorcet Ranking. The Borda Count ignores this mandate. In a more conflicted situation such as Borda's Response, the Borda Count delivers a social welfare, but for these voters, it is hard to say whether this ranking is most appropriate.

Condorcet Losers and the Borda Count

Condorcet's critique of the Borda Count focused on the violation of the Condorcet Winner Criterion. We now prove Borda's Theorem (Theorem 21.1) to the effect that a Condorcet Loser (such as candidate D in Condorcet's Challenge) can never win the Borda Count. As was the case with the Borda Count and the Majority Loser Criterion (Theorem 12.15), the proof of Borda's Theorem depends on some basic arithmetic.

Suppose that k voters rank m candidates for a Borda Count election where, as usual, we award 0 Borda points for a last-place vote, 1 Borda point for a next-to-last-place vote up to $m-1$ Borda points for a first-place vote. By Corollary 12.14, we have the formula

Total Borda Points with *m* Candidates and *k* Voters

$$B_{\text{total}} = \frac{km(m-1)}{2}$$

We deduce a formula for the average number of Borda points per candidate, written B_{avg}. That is,

Average Borda Points with *m* Candidates and *k* Voters

$$B_{\text{avg}} = \frac{B_{\text{total}}}{m} = \frac{k(m-1)}{2}$$

We use this expression to prove the following theorem:

Theorem 21.1 Borda's Theorem *The Borda Count satisfies the Condorcet Loser Criterion.*

Proof We prove that a Condorcet Loser always receives strictly less than the average Borda points B_{avg} in a Borda Count election. Assume, as earlier, that a preference table is given with m candidates $A_1, \ldots, A_m$ and k voters $p_1, \ldots, p_k$. Let

$$B_j(A_i) = \text{total Borda points awarded to candidate } A_i \text{ by voter } p_j$$

and

$$B(A_i) = \text{total Borda points awarded to candidate } A_i$$

The total Borda points $B(A_i)$ for candidate A_i is just the sum of the points awarded by each voter. Thus

$$B(A_i) = B_1(A_i) + B_2(A_i) + \cdots + B_k(A_i)$$

To compute $B_j(A_i)$, we count how many candidates voter p_j ranks above A_i. We then subtract this number from $m - 1$. That is,

$$B_j(A_i) = (m-1) - \text{number of candidates } p_j \text{ ranks above } A_i$$

We can rewrite this, of course, as

$$\text{Number of candidates } p_j \text{ ranks above } A_i = (m-1) - B_j(A_i)$$

If we sum the left-hand side over all voters, we get the total number of times that a candidate is ranked above A_i in some voter's individual ranking. Summing the right-hand side of this equation over all voters, we get $k(m - 1) - B(A_i)$. Combining, we derive the following counting formula:

Counting Formula for General Candidate

$$\text{Number of times a candidate is ranked above } A_i \text{ in some voter's individual ranking} = k(m-1) - B(A_i)$$

Now suppose that A_i is a Condorcet Loser. Then each other candidate A_j beats A_i head to head. This means that more than half the k voters rank A_j over A_i. That is,

$$\text{Number of times } A_j \text{ is ranked above } A_i \geq \frac{k+1}{2}$$

Summing both sides of this inequality over all $m - 1$ candidates A_j other than A_i, we get

Inequality for Condorcet Loser

$$\text{Number of times a candidate is ranked above Condorcet Loser } A_i \text{ in some voter's individual ranking} \geq \frac{(m-1)(k+1)}{2}$$

We now combine the counting formula with the inequality formula for the Condorcet Loser A_i. We obtain the following inequality:

$$k(m-1) - B(A_i) \geq \frac{(m-1)(k+1)}{2}$$

Solving for $B(A_i)$ and applying some algebra, we obtain the following:

$$\begin{aligned} B(A_i) &\leq k(m-1) - \frac{(m-1)(k+1)}{2} \\ &= k(m-1) - \frac{k(m-1)}{2} - \frac{m-1}{2} \\ &= \frac{k(m-1)}{2} - \frac{m-1}{2} \\ &< \frac{k(m-1)}{2} \\ &= B_{\text{avg}} \end{aligned}$$

We conclude that a Condorcet Loser A_i cannot possibly win the Borda Count. Otherwise, all m candidates have strictly less then the average Borda points, an obvious impossibility. ■

We highlight the key step in this proof as the following theorem:

Theorem 21.9 *A Condorcet Loser always receives less than the average number of Borda points in a Borda Count election.* ■

A Compromise

We briefly analyze a compromise voting method introduced by Duncan Black in 1958. We have the following definition:

Definition 21.10 The **Black Method** is the voting method defined as follows: given a preference table, if some candidate X is the Condorcet Winner, then X is the Black social choice. Otherwise, the Black social choice is the Borda Count social choice for the given preference table.

We can quickly deduce the following positive result:

Theorem 21.11 *The Black Method satisfies the Condorcet Winner and the Condorcet Loser criteria.*

Proof The Black Method always elects a Condorcet Winner by definition. By Theorem 21.1, the Black Method never elects a Condorcet Loser. ■

As with nondictatorial voting methods, the Black Method is manipulable (see Exercise 21.4). We prove the following theorem:

Theorem 21.12 *The Black Method exhibits the No-Show Paradox.*

Proof Recall from Definition 6.11, the No-Show Paradox occurs when a voter (or voting bloc) achieves a better outcome for the election according to his or her true preferences by not submitting a ballot than when he or she participates and votes sincerely. Consider the following preference table:

True Preference Table

Number of voters	**1**	**10**	**10**	**1**	**1**
1st choice	C	A	B	A	B
2nd choice	A	C	C	B	A
3rd choice	B	D	D	C	C
4th choice	D	B	A	D	D

We see here that candidate A is a Condorcet Winner and so is the Black social choice. Now suppose that the first voter is a no-show. The preference table then becomes

Preference Table with First Voter Removed

Number of voters	**10**	**10**	**1**	**1**
1st choice	A	B	A	B
2nd choice	C	C	B	A
3rd choice	D	D	C	C
4th choice	B	A	D	D

In this case, there is no Condorcet Winner. Thus we run the Borda Count and find that C is the new Black social choice. Since the first voter prefers C to A, the proof is complete. ■

One historical question that we have not addressed in this section is

What voting method did Condorcet propose as an alternative to the Borda Count?

Condorcet did briefly describe an alternate voting method (see Exercise 21.6), but the details are somewhat ambiguous. We describe two methods inspired by Condorcet and a method inspired by the Borda Count in the exercises. It is worth emphasizing that the debate begun by Borda and Condorcet over the fairest voting method continues to be actively pursued in academic journals and on websites to this day.

Exercises

21.1 Consider the following tournament with four candidates A, B, C, and D:

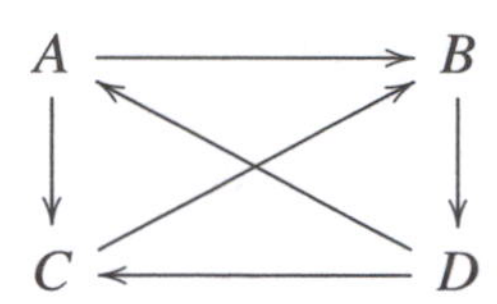

(a) Find a preference table realizing this tournament such that the Borda Count social choice is A.

(b) Find another preference table realizing this tournament such that Borda Count social choice is B.

21.2 For each of the following tournaments, follow the procedure of the proof of Theorem 21.8 to find a preference table solving the realization problem:

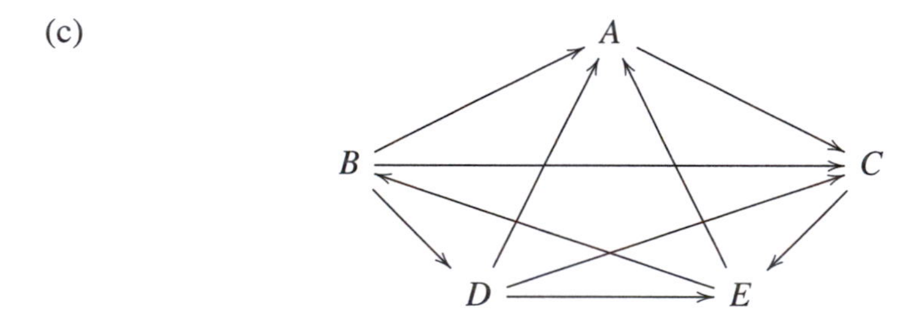

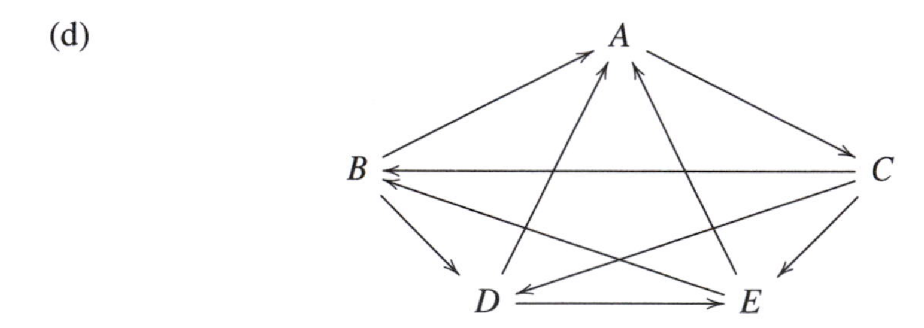

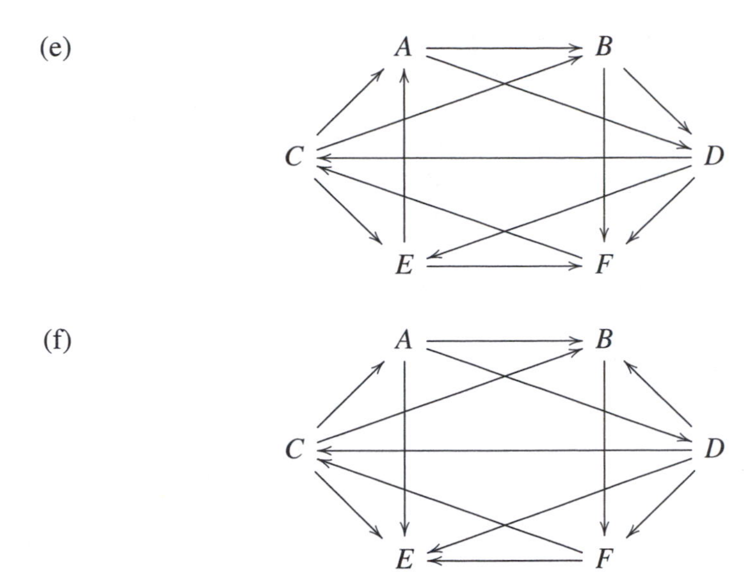

21.3 Follow the proof of Borda's Theorem (Theorem 21.1) to prove that a Condorcet Winner always receives more than the average Borda points B_{avg} in a Borda Count.

21.4 Give an example to show that the Black Method is manipulable.

21.5 (Nanson Method) The **Nanson Method** is the social choice method defined as follows: given a preference table, run the Borda count. All candidates receiving less than B_{avg} Borda points are eliminated. The Borda count is run again without these candidates, and again, all candidates with less than the average Borda points in the new election are eliminated. Repeat until one candidate remains. This candidate is the Nanson social choice.

(a) Determine the Nanson social choice for the following preference table:

Number	**11**	**13**	**8**	**14**	**10**	**16**
1st choice	A	B	C	D	E	C
2nd choice	B	C	D	E	A	B
3rd choice	C	D	B	A	B	A
4th choice	D	E	A	B	C	E
5th choice	E	A	E	C	D	D

(b) Prove that the Nanson Method satisfies the Condorcet Winner Criterion.
(c) Prove that the Nanson Method satisfies the Condorcet Loser Criterion.
(d) Give an example to show that the Nanson Method is manipulable.

21.6 (Condorcet Method) Condorcet described the following method for producing a social choice using pairwise comparisons called the **Condorcet Method**: given a preference table, construct the social preference graph, and label each arrow with the margin of victory. For example, if A beats B head to head by 30 votes, we write $A \xrightarrow{30} B$. If there is a Condorcet Winner, that candidate is the Condorcet social choice. Otherwise, remove arrows from the social preference graph one by one in increasing order until some candidate is the head of every arrow in the social preference graph. The unbeaten candidate is the Condorcet social choice.

(a) Determine the Condorcet social choice for the following preference table:

Number	10	20	16	15	24	18
1st choice	A	B	C	D	E	A
2nd choice	B	C	E	B	D	B
3rd choice	C	E	D	A	C	E
4th choice	D	D	A	E	A	D
5th choice	E	A	B	C	B	C

(b) Prove that the Condorcet Method satisfies the Condorcet Winner Criterion.
(c) Prove that the Condorcet Method satisfies the Condorcet Loser Criterion.
(d) Give an example to show that the Condorcet Method is manipulable.

21.7 (Kemeny Method) John Kemeny described a method for producing a social welfare in the spirit of Condorcet. Given a preference table with m candidates, each possible ordering of the m candidates is given points by each voter according to how many comparisons the given order shares with the voter. For example, suppose that voter p prefers A to B and B to C (and so p prefers A to C). We write this as ABC. Then the ranking ABC receives 3 points from p because it agrees with all three rankings of voter p. The ranking A over C over B receives 2 points from voter p because it agrees with voter p on the ranking A over B and A over C. Here is the preference table for voter p:

Points Awarded by Voter *p* with Ranking ABC

Ranking	Points
ABC	3
ACB	2
BAC	2
BCA	1
CAB	1
CBA	0

The points for each ranking are totaled for all voters, and the ranking receiving the most points is the **Kemeny social welfare**. Determine the Kemeny social welfare for the following preference tables:

(a)

Number	7	5	4
1st choice	A	B	C
2nd choice	B	C	B
3rd choice	C	A	A

(b)

Number	3	2	4	1
1st choice	A	C	B	A
2nd choice	B	B	A	C
3rd choice	C	A	C	B

(c)

Number	7	5	4
1st choice	A	D	C
2nd choice	B	B	D
3rd choice	C	C	A
4th choice	D	A	B

Note that there are $4! = 24$ orders to consider in this case!

21.8 We introduce a "fairness" criterion for social choice methods that we call the **Invariance Criterion**, defined as follows: given a preference table, construct the social preference graph that is a tournament $\mathcal{T}$. We say that a voting method satisfies Invariance if every preference table with the same candidates resulting in the same preference table $\mathcal{T}$ produces the same social choice by the voting method. Prove by example that each of the voting methods Plurality, Borda Count, Hare, Coombs, and Dictatorship violate the Invariance Criterion. Can you think of a voting method that satisfies the Invariance Criterion?

21.9 Prove that any tournament with n candidates can be realized by a preference table with $2^{\binom{n}{2}}$ voters.

21.10 In the proof of Theorem 21.2, we showed that for any odd M,

$$P(\text{“Correct verdict with } M+2 \text{ voters”}) > P(\text{“Correct verdict with } M \text{ voters”})$$

Use this result to give an induction proof for the stated result. That is, prove that for given any $N > M$ both odd,

$$P(\text{“Correct verdict with } N \text{ voters”}) > P(\text{“Correct verdict with } M \text{ voters”})$$

Hint: We may write $N = M + 2k$ for some $k = 1, 2, 3, \ldots$.

21.11 Prove the following statement, used in the proof of Theorem 21.2: let $M = 2k+1$, and suppose that M jurors vote independently for either the correct verdict with probability p or the incorrect verdict with probability $1-p$. Prove that

$$P(\text{"Correct verdict"} \mid \text{"Margin = 1"}) = p$$

and

$$P(\text{"Incorrect verdict"} \mid \text{"Margin = 1"}) = 1-p$$

22 The Sprague–Grundy Theorem

Given two sequential-move strategic games, we create a new sequential-move game by allowing moves to be taken in either game. This idea provides a notion of "addition" for sequential-move games, a quick way to turn simple games into more complex ones. It ultimately leads to a new notion of addition called the *Nim sum* for ordinal numbers and a beautiful theorem proved, independently, by Roland Sprague and Patrick Grundy that generalizes Bouton's Theorem (Theorem 17.11). In this chapter, we formulate and prove the Sprague–Grundy Theorem (Theorem 22.4). As an application, we introduce the game *Hackenbush* and show how the Sprague–Grundy Theorem can be used to solve this takeaway game.

Game Sums

We motivate the idea of a game sum with the following example:

Example 22.1 Checkers and Chips

Each player starts with three checkers. There is a pile of seven chips between them. On his or her turn, a player can place some of the checkers (at least one) in the center, or he or she may take either one, two, or three chips from the pile. The game is over when both of the following two conditions are met: (1) all six checkers have been placed in the center, and (2) all the chips have been removed from the center. The last player to move wins the game

The game Checkers and Chips is actually two separate games run in parallel. We call the first game Stacking Checkers and denote it by $\mathcal{G}_1$. In Stacking Checkers, each player has three checkers. The players alternate placing checkers in the center until each player has run out of checkers. The winner is the player to move last. The second game is a version of Taking Chips with $n=7$ and $k=3$ (see Example 4.14). We denote this game by $\mathcal{G}_2$. The game Checkers and Chips will be called the *game sum* of the games $\mathcal{G}_1$ and $\mathcal{G}_2$ and written

$$\text{Checkers and Chips} = \mathcal{G}_1 \bigoplus \mathcal{G}_2$$

Notice that both $\mathcal{G}_1$ and $\mathcal{G}_2$ are takeaway games in the sense that the last player to move wins. We will work exclusively with takeaway games in this chapter.

Definition 22.2 Suppose that we are given takeaway games $\mathcal{G}_1$ and $\mathcal{G}_2$. The **game sum** $\mathcal{G}_1 \bigoplus \mathcal{G}_2$ is then the takeaway game in which each player, on his or her turn, chooses an available move in either game. The game sum ends when there are no available moves in either game. The last player to move wins.

We remark that while the definition is only made for two games, the idea directly extends to the case of n takeaway games $\mathcal{G}_1, \mathcal{G}_2, \ldots, \mathcal{G}_n$, giving a game sum written

A Game Sum of *n* Games

$$\mathcal{G}_1 \bigoplus \mathcal{G}_2 \bigoplus \cdots \bigoplus \mathcal{G}_n.$$

While the individual games Stacking Checkers and Taking Chips are quite simple, the game sum Checkers and Chips is considerably more complicated. Can you determine which player can guarantee victory in Checkers and Chips?

We can visualize the process of summing games and the added complexity by considering the effect of summing two Tree Games as in Definition 10.10. Recall that a Tree Game is the takeaway game based on a given rooted tree in which players alternately move down the tree toward the end nodes. When an end node is reached, the player whose turn it is loses the game. We consider two very simple trees. We have labeled the nodes for convenience:

Two Rooted Trees

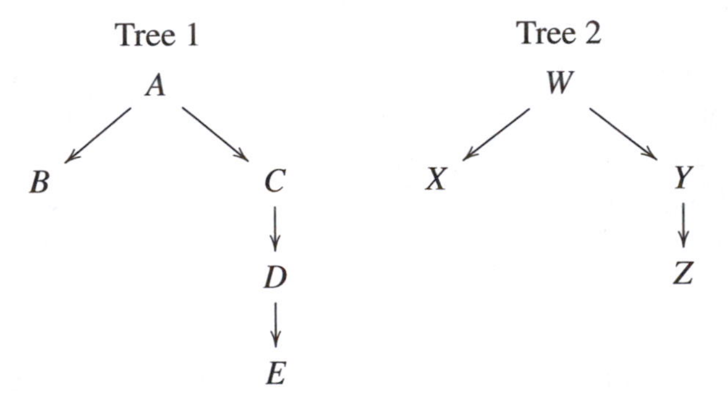

Analyzing rational play for these games is trivial: in both games, Player 1 can win on the first move.

Let $\mathcal{T}$ denote the Tree Game corresponding to the game sum of these two games. The nodes of the game sum are then ordered pairs of nodes in the original two trees. For example, the root node is now (A, W). From this node, Player 1 may move in Tree 1 to (B, W) or (C, W), he or she may move in Tree 2 to (A, X) or (A, Y). Here is the full tree:

Rooted Tree for the Game Sum

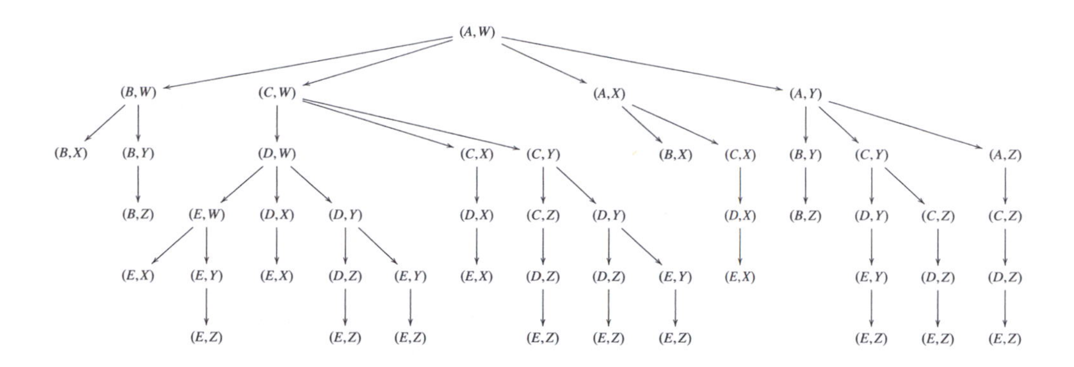

Which player wins this Tree Game in rational play?

The canonical example of a game obtained by summing simpler games is the game of Nim. Recall that we write $\mathrm{Nim}(m_1,\ldots,m_k)$ for the game Nim with k piles such that the ith pile has m_i chips. For each integer m_i, the game $\mathrm{Nim}(m_i)$ is just a game of One-Pile Nim. Of course, the strategy to win One-Pile Nim is simple: just take all the chips. We have the following identification:

Nim as a Game Sum

$$\mathrm{Nim}(m_1,\ldots,m_k) = \mathrm{Nim}(m_1) \bigoplus \mathrm{Nim}(m_2) \bigoplus \cdots \bigoplus \mathrm{Nim}(m_k)$$

That is, the game $\mathrm{Nim}(m_1,\ldots,m_k)$ is the game sum of the One-Pile Nim games. This observation again illustrates how complex games can be built out of very simple individual games.

Binary Numbers and Sprague–Grundy Numbers

Bouton's Theorem (Theorem 17.11) determines the winning player and winning strategy for the game Nim. The Sprague–Grundy Theorem (Theorem 22.4) represents a generalization of Bouton's Theorem. The key ingredients for the Sprague–Grundy Theorem are binary numbers (as introduced in Chapter 17) and the Ordinal Labeling Rule (Definition 17.1). We briefly recall these two notions here.

First, recall that the **binary representation** of an integer n is the sequence of 0s and 1s obtained by writing n as a sum of powers of 2. For example, $27 = 16+8+2+1$, so we write $27 =_2 11011$. Similarly, $22 = 16+4+2$, so $22 =_2 10110$. As in Chapter 17, we picture the binary form of a number in **binary column form** as follows:

Binary Column Form

Power of 2	27
2^4	1
2^3	1
2^2	0
2^1	1
2^0	1

Power of 2	22
2^4	1
2^3	0
2^2	1
2^1	1
2^0	0

Recall the Ordinal Labeling Rule introduced in Chapter 17. Given a rooted tree $\mathcal{T}$, the **ordinal labels** of the nodes are determined by working backward from the end nodes to the root. The end nodes are first labeled 0. In general, a node with all branching nodes already labeled is itself labeled with the smallest ordinal that is not represented by an adjacent node. For example, if there is no node adjacent to the given node labeled 0, then the given node is labeled 0. More generally, a node is labeled n provided that the labels 0 through $n-1$ all occur on adjacent nodes but the ordinal n does not.

The label on the root is called the *Sprague–Grundy number* (Definition 17.3) of the tree. Given a rooted tree $\mathcal{T}$, we write

$$SG(\mathcal{T}) = \text{the Sprague–Grundy number of } \mathcal{T}$$

Given a takeaway game $\mathcal{G}$ with game tree $\mathcal{T}$, we write

$$SG(\mathcal{G}) = SG(\mathcal{T})$$

and call this number the **Sprague–Grundy number** of the game $\mathcal{G}$.

Nim Sums

We define a new operation on ordinal numbers using the binary column format. The **Nim sum** of ordinals m and n, written $m \oplus n$, is obtained by first adding the binary digits in each row of the binary column format of m and n. If the sum is even (in this case, 0 or 2), then we put a 0 in this position in the binary column form for $m \oplus n$. If the sum is odd (1), we put a 1 in this position. The resulting number is the Nim sum $m \oplus n$ in binary column form. For example,

A Nim Sum of Two Numbers

Power of 2	27	22	$27 \oplus 22 = 13$
2^4	1	1	0
2^3	1	0	1
2^2	0	1	1
2^1	1	1	0
2^0	1	0	1

From here on, we will omit the first column of powers of 2 for convenience.

The Nim sum has some strange properties. For instance, you can check that $8 \oplus 7 = 15$, whereas $7 \oplus 7 = 0$. For any ordinal a observe that $a \oplus 0 = 0 \oplus a = a$. Here are some other useful facts that are easy to prove.

Theorem 22.3 *Let a, b, c be ordinals. Then*

1. $a \oplus a = 0$.
2. $a \oplus b = b \oplus a$.
3. $(a \oplus b) \oplus c = a \oplus (b \oplus c)$.
4. *If $a \oplus b = a \oplus c$ then $b = c$.*

Proof The proofs of (1)–(3) are immediate from the definition. For (4), suppose that $a \oplus b = a \oplus c$. Then

$$b = 0 \oplus b = (a \oplus a) \oplus b = a \oplus (a \oplus b) = a \oplus (a \oplus c)$$
$$= (a \oplus a) \oplus c = 0 \oplus c = c \qquad \blacksquare$$

Theorem 22.3 (2) and (3) imply that we may compute the Nim sum of multiple ordinals $m_1, \ldots, m_k$ without worrying about the order. The description of this sum in terms of binary column form is similar to the case of two ordinals. Write each ordinal m_i in binary column form. Add the entries in each row. If the sum is even, we put a 0 in the corresponding row for $m_1 \oplus \cdots \oplus m_n$. If the sum is odd, we put a 1 in this row. Here is an example of a Nim sum of four ordinals:

A Nim Sum of Four Ordinals

35	17	41	13	$35 \oplus 17 \oplus 41 \oplus 13 = 22$
1	0	1	0	0
0	1	0	0	1
0	0	1	1	0
0	0	0	1	1
1	0	0	0	1
1	1	1	1	0

The Sprague–Grundy Theorem

We have now assembled the ingredients needed to state the Sprague–Grundy Theorem. Consider a game sum $\mathcal{G}_1 \bigoplus \mathcal{G}_2 \bigoplus \cdots \bigoplus \mathcal{G}_k$ of takeway games. We seek a formula for the Sprague–Grundy number

$$SG\left(\mathcal{G}_1 \bigoplus \mathcal{G}_2 \bigoplus \cdots \bigoplus \mathcal{G}_k\right)$$

of the game sum directly in terms of the individual Sprague–Grundy numbers $SG(\mathcal{G}_1), \ldots, SG(\mathcal{G}_k)$. The formula is the content of the following famous theorem:

Theorem 22.4 Sprague–Grundy Theorem *Given takeaway games $\mathcal{G}_1, \mathcal{G}_2, \ldots, \mathcal{G}_k$, we have*

$$SG\left(\mathcal{G}_1 \bigoplus \mathcal{G}_2 \bigoplus \cdots \bigoplus \mathcal{G}_k\right) = SG(\mathcal{G}_1) \oplus SG(\mathcal{G}_2) \oplus \cdots \oplus SG(\mathcal{G}_k). \qquad \blacksquare$$

Let us interpret this theorem for the game Nim. First, we observe that

$$SG(\text{Nim}(n)) = n.$$

(See Theorem 17.5.) Thus we have the following direct consequence of Theorem 22.4, which explains the terminology of "Nim sum":

Corollary 22.5 *Let $m_1, \ldots, m_k$ be positive integers. Then*

$$SG(Nim(m_1, \ldots, m_k)) = m_1 \oplus \cdots \oplus m_k$$

Proof We have $\text{Nim}(m_1, \ldots, m_k) = \text{Nim }(m_1) \bigoplus \text{Nim }(m_2) \bigoplus \cdots \bigoplus \text{Nim }(m_k)$ and $\text{Nim}(m_i) = m_i$. The result now follows from Theorem 22.4 ■

We note that Corollary 22.5 extends Bouton's Theorem. For example, we have that

$$SG(\text{Nim}(35, 17, 41, 13)) = 35 \oplus 17 \oplus 41 \oplus 13 = 22$$

Since the Sprague–Grundy number is, in particular, nonzero, we conclude that Player 1 wins Nim(35, 17, 41, 13) in rational play, as specified by Theorem 17.11.

We now give the proof of our main result:

Proof of Theorem 22.4 We will prove this result for two games $\mathcal{G}_1$ and $\mathcal{G}_2$. The general result for k games follows from an easy induction argument (see Exercise 22.8).

Suppose that $SG(\mathcal{G}_1) = a$ and $SG(\mathcal{G}_2) = b$. We must prove that $SG(\mathcal{G}_1 \bigoplus \mathcal{G}_2) = a \oplus b$. Let $\mathcal{T}_1, \mathcal{T}_2$, and $\mathcal{T}$ denote the rooted trees corresponding to $\mathcal{G}_1, \mathcal{G}_2$, and $\mathcal{G}_1 \bigoplus \mathcal{G}_2$, respectively. Note that

$$\text{Depth}(\mathcal{T}) = \text{Depth}(\mathcal{T}_1) + \text{Depth}(\mathcal{T}_2)$$

We proceed by induction on the depth of $\mathcal{T}$:

Base Case: When depth $(\mathcal{T}) = 1$ then one of the games $\mathcal{G}_1$ or $\mathcal{G}_2$ has no moves and so has Sprague–Grundy number = 0. The other game then has depth 1 and so has Sprague–Grundy number = 1. Thus

$$SG(\mathcal{G}) = 1 = 1 \oplus 0 = SG(\mathcal{G}_1) \oplus SG(\mathcal{G}_2).$$

As usual, we formulate our induction hypothesis in terms of A-rooted subtrees of $\mathcal{T}$ as in Definition 10.12. Since $\mathcal{T}$ corresponds to a game sum, each node of $\mathcal{T}$ corresponds to a pair of nodes (X,Y), with X a node of $\mathcal{T}_1$ and Y a node of $\mathcal{T}_2$. Suppose that the root node of $\mathcal{T}$ is (A,B). Then a node (X,Y) is the root of a proper subtree of $\mathcal{T}$ provided that either $X \neq A$ or $B \neq Y$. We introduce the following notation: we write $\mathcal{G}_1(X)$ for the game corresponding to an X-rooted subtree of $\mathcal{T}_1$ and set $x = SG(\mathcal{G}_1(X))$. Similarly, $\mathcal{G}_2(Y)$ will denote the game corresponding to the Y-rooted subtree of $\mathcal{T}_2$ and $y = SG(\mathcal{G}_2(Y))$. Finally, we write $\mathcal{G}(X,Y)$ for the game corresponding to the (X,Y)-rooted subtree of $\mathcal{T}$. Thus $\mathcal{G}(A,B) = \mathcal{G}_1 \bigoplus \mathcal{G}_2$ With this notation in place, we have the following:

Induction Hypothesis: Assume that for all nodes (X,Y) of $\mathcal{T}$ with either $X \neq A$ or $Y \neq B$, we have $SG(\mathcal{G}(X,Y)) = x \oplus y$.

Induction Step: We prove that $SG(\mathcal{G}(A,B)) = a \oplus b$. By the definition of the Sprague–Grundy number of the game $\mathcal{G}(A,B)$, this amounts to showing that the root node (A,B) of $\mathcal{T}$ has adjacent nodes labeled $0,1,2,\ldots,(a \oplus b) - 1$ and that there is no adjacent node labeled $a \oplus b$. Now the nodes adjacent to the root node of $\mathcal{T}$ are of two types: They are either of the form (X,B), where X is adjacent to the root node A in $\mathcal{T}_1$ or they are of the form (A,Y), where Y is adjacent to the root node B in $\mathcal{T}_2$. By the Induction Hypothesis, we know that the label on a node (X,B) is $x \oplus b$ and that the label on a node (A,Y) is $a \oplus y$. We prove the two needed facts separately. First, we have the following:

Claim 1: *If X is adjacent to the root node A of $\mathcal{T}_1$, then $x \oplus b \neq a \oplus b$. Similarly, if Y is adjacent to the root node B of $\mathcal{T}_2$, then $a \oplus y \neq a \oplus b$.*

Proof of Claim 1 Suppose that X is adjacent to the root node A of $\mathcal{T}_1$. If $x \oplus b = a \oplus b$, then by Theorem 22.3 (4) we have $a = x$. This contradicts the fact that $SG(\mathcal{G}_1) = a$; there can be no node adjacent to A in $\mathcal{T}_1$ with ordinal label a. The second statement is proved the same way. ■

Claim 2: *If c is any ordinal with $c < a \oplus b$, then either (1) there exists a node X of $\mathcal{T}_1$ adjacent to the root node A of this tree with $x \oplus b = c$ or (2) there exists a node Y of $\mathcal{T}_2$ adjacent to B such that $a \oplus y = c$.*

Proof of Claim 2 For definiteness, let us assume that 2^m is the highest binary digit in $a \oplus b$. We are given that $c < a \oplus b$. This means we can find a row in the binary column formats for these two ordinals such that c has a 0 and $a \oplus b$ has 1 in this position. Let k denote the highest such row where this difference occurs. Then c and $a \oplus b$ have identical binary digits in the rows corresponding to $2^{k+1}, 2^{k+2}, \ldots, 2^m$. Since c and $a \oplus b$ have the same binary digit in these rows, when we take the Nim sum of c and $a \oplus b$, we obtain an ordinal with 0 as the binary digit in these rows. We picture this as follows: here we use an asterisk to indicate that row entries are the same (0 or 1) for the rows above the 2^k row. We use a question mark to indicate that we do not know anything about these entries. With these conventions, we have the following table:

Nim Sum $c \oplus (a \oplus b)$

Power of 2	c	$a \oplus b$	$c \oplus (a \oplus b)$
2^m	*	*	0
⋮	⋮	⋮	⋮
2^{k+2}	*	*	0
2^{k+1}	*	*	0
2^k	0	1	1
2^{k-1}	?	?	?
⋮	⋮	⋮	⋮
2^2	?	?	?
2^1	?	?	?
2^0	?	?	?

Next, since $a \oplus b$ has a 1 in the 2^kth row, either a or b (but not both) must have a 1 in this row also. Let us suppose that it is a with a 1 in the row corresponding to 2^k. The Nim sum $a \oplus (c \oplus (a \oplus b))$ is then as follows:

Nim Sum $a \oplus (c \oplus (a \oplus b))$

Power of 2	a	$c \oplus (a \oplus b)$	$a \oplus (c \oplus (a \oplus b))$
2^m	*	0	*
⋮	⋮	⋮	⋮
2^{k+2}	*	0	*
2^{k+1}	*	0	*
2^k	1	1	0
2^{k-1}	?	?	?
⋮	⋮	⋮	⋮
2^2	?	?	?
2^1	?	?	?
2^0	?	?	?

Here we are observing that $a \oplus (c \oplus (a \oplus b))$ has the same binary digits as a in the rows $2^{k+1}, 2^{k+2}, \ldots, 2^m$ and has a 0 in the row 2^k. But a has a 1 in the 2^k position.

It follows easily that $a > a \oplus (c \oplus (a \oplus b))$ (see Exercise 22.9). Since $a = SG(\mathcal{G}_1)$, there must be a node X adjacent to the root node A in the tree $\mathcal{T}_1$ with ordinal label

$$x = a \oplus (c \oplus (a \oplus b)) = c \oplus (a \oplus a) \oplus b = c \oplus b$$

Note that we have used Theorem 22.3 to simplify. Now, since $SG(\mathcal{G}(X)) = x$, our Induction Hypothesis gives

$$SG(\mathcal{G}(X,B)) = x \oplus b = (c \oplus b) \oplus b = c$$

as needed. ■

Claims 1 and 2 imply that $SG(\mathcal{G}) = a \oplus b$, which completes the induction and the proof. ■

Hackenbush

We conclude with an illustration of how the Sprague–Grundy Theorem can be applied to solve a takeaway game.

Example 22.6 Hackenbush

The game Hackenbush is played with a rooted tree that is customarily drawn opening upward (so as to resemble an actual tree). The players take turns selecting branches of the tree and removing this branch and also all branches above this branch that are now no longer connected to the root. The last player to move wins.

We analyze the game of Hackenbush corresponding to the following tree:

A Tree for a Game of Hackenbush

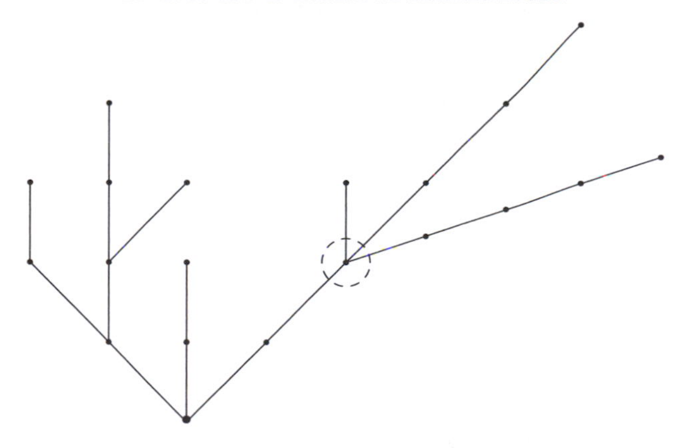

The key to solving this game is the recognition that Hackenbush contains several games of Nim. Consider the circled node. The three branchings that emerge from this node may be viewed as Nim piles of sizes 1, 3, and 4. We thus may replace these branches with a single Nim pile (or branching) of size $1 \oplus 3 \oplus 4 = 6$. That is,

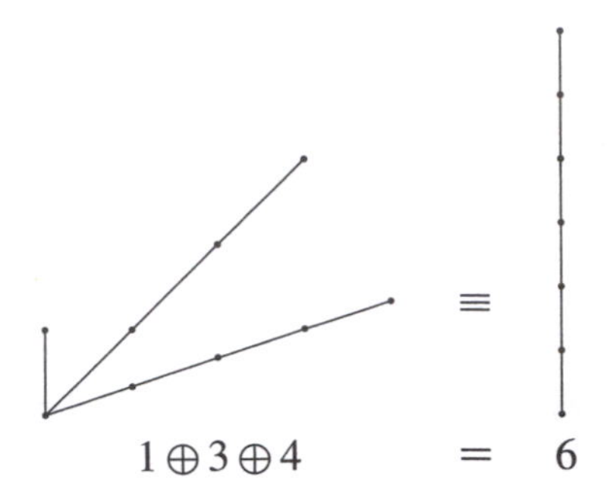

The original tree now may be rewritten with the three branchings from the circled node replaced with a single branching of length 6, as follows:

An Equivalent Tree for a Game of Hackenbush

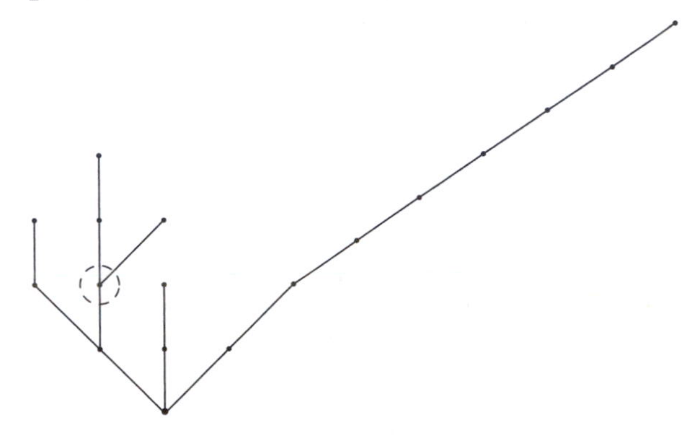

Next, focus on the two branchings emerging from the circled node in this new tree. Here we have $1 \oplus 2 = 3$, so we may replace these two branchings with a single branching of length 3, as follows:

Another Equivalent Tree for a Game of Hackenbush

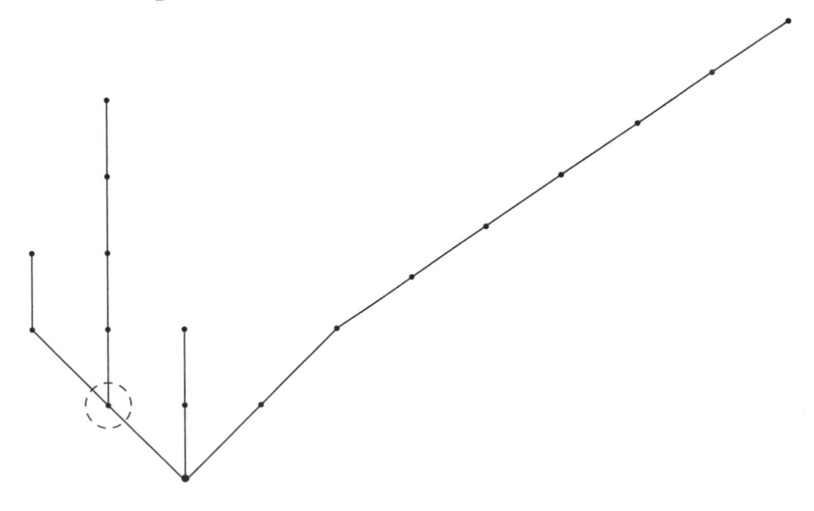

Next, observe that the two branchings from the circled node in this version of the tree have Nim sum $2 \oplus 4 = 6$. Making this replacement yields a tree with three branchings from the root of lengths 7, 2, and 8. Thus we see that the Sprague–Grundy number for the game is

$$SG(\text{“A game of Hackenbush”}) = 7 \oplus 2 \oplus 8 = 13$$

We remark that while our method effectively determines the Sprague–Grundy number for the game, some thought is required to determine the winning strategy, in this case for Player 1. We leave this challenge for you as an exercise.

Exercises

22.1 Write down the rooted tree corresponding to the game sum of the following two Tree Games:

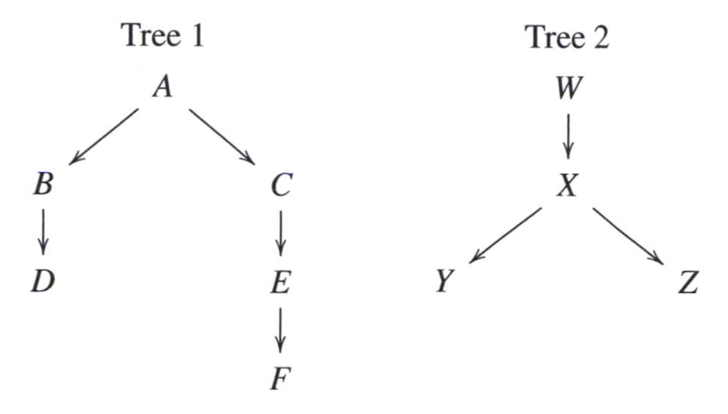

Compute the Sprague–Grundy number of the game sum.

22.2 Use Theorem 22.4 to compute the Sprague–Grundy numbers of the following games:

(a) Nim(7, 8, 9, 10)
(b) Nim(12, 14, 15, 11)
(c) Nim(22, 40, 38, 16, 25)
(d) The game sum of two versions of Taking Chips: $\mathcal{G}_1$ with $n=22$ and $k=5$ and $\mathcal{G}_2$ with $n = 33$ and $k = 6$ (see Example 4.14 and Exercise 17.9).
(e) The following version of Northcott's Game (see Example 20.3):

Player 1 ($\odot$)

		$\otimes$				$\otimes$	
$\odot$			$\odot$	$\odot$			
	$\odot$						
							$\odot$
$\otimes$					$\odot$		
						$\odot$	
	$\otimes$		$\otimes$		$\otimes$		$\otimes$
		$\odot$		$\otimes$			

Player 2 ($\otimes$)

(f) The game sum of the games described in parts (a) through (e).

22.3 Suppose that $43 \oplus x = 27$. Find x.

22.4 Consider the following sequential-move game. Each player has a large pile of chips. Player 1 makes two nonempty, nonequal piles of any size in the center. Player 2 then makes a third nonempty pile of any size. The players then play the game of Nim with the three piles. Player 1 goes first. Who wins this game in rational play? Justify your answer.

22.5 Compute the Sprague–Grundy number for the game Checkers and Chips (Example 22.1).

22.6 Compute the Sprague–Grundy numbers for the following games of Hackenbush:

(a)

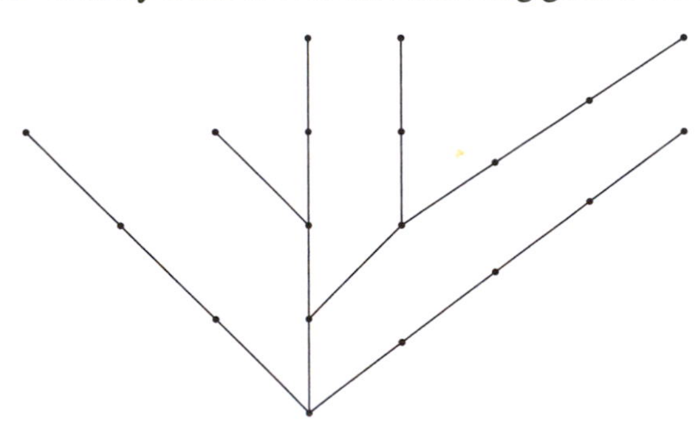

(b)

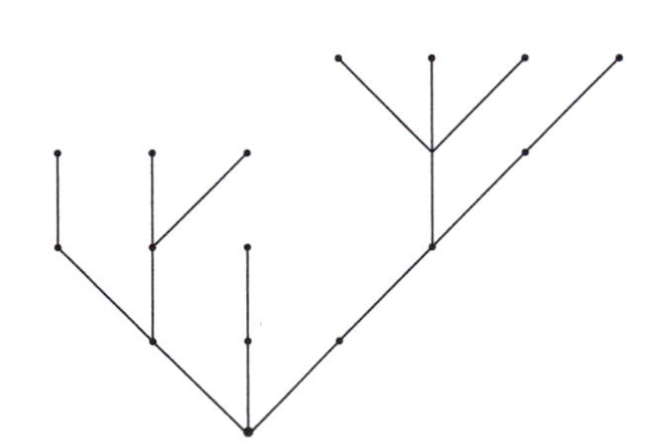

(c)

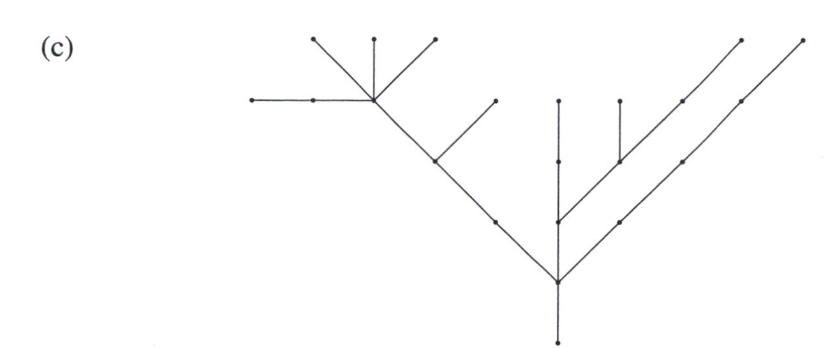

22.7 Define two win-lose sequential-move strategic games $\mathcal{G}_1$ and $\mathcal{G}_2$ to be **equivalent**, and write $\mathcal{G}_1 \cong \mathcal{G}_2$ if $SG(\mathcal{G}_1 \bigoplus \mathcal{G}_2) = 0$ (compare Definition 20.1). Use Theorem 22.4 to prove the following properties of this relation:

(a) *Reflexive Property*: $\mathcal{G} \cong \mathcal{G}$ for all $\mathcal{G}$.
(b) *Symmetric Property*: $\mathcal{G}_1 \cong \mathcal{G}_2$ implies $\mathcal{G}_2 \cong \mathcal{G}_1$ for all $\mathcal{G}_1, \mathcal{G}_2$.
(c) *Transitive Property*: $\mathcal{G}_1 \cong \mathcal{G}_2$ and $\mathcal{G}_2 \cong \mathcal{G}_3$ implies $\mathcal{G}_1 \cong \mathcal{G}_3$ for all $\mathcal{G}_1, \mathcal{G}_2, \mathcal{G}_3$.

A relation $\cong$ on a set $\mathcal{S}$ satisfying the reflexive, symmetric, and transitive properties is called an **equivalence relation** on $\mathcal{S}$. Prove that the preceding equivalence relation on the set of win-lose sequential-move strategic games satisfies the following further property: if $\mathcal{G}_1 \cong \mathcal{G}_1'$ and $\mathcal{G}_2 \cong \mathcal{G}_2'$, then $\mathcal{G}_1 \bigoplus \mathcal{G}_2 \cong \mathcal{G}_1' \bigoplus \mathcal{G}_2'$.

22.8 Use induction to complete the proof of Theorem 22.4. That is, prove that

$$SG(\mathcal{G}_1 \bigoplus \mathcal{G}_2) = SG(\mathcal{G}_1) \oplus SG(\mathcal{G}_2)$$

for any two win-lose sequential-move strategic games $\mathcal{G}_1$ and $\mathcal{G}_2$ implies that

$$SG\left(\mathcal{G}_1 \bigoplus \mathcal{G}_2 \bigoplus \cdots \bigoplus \mathcal{G}_n\right) = SG(\mathcal{G}_1) \oplus SG(\mathcal{G}_2) \oplus \cdots \oplus SG(\mathcal{G}_n)$$

for any n such games.

22.9 Prove the following fact used in the proof of Theorem 22.4: let x and y be integers such that, in binary column format, (1) x has a 1 in the row corresponding to the power 2^k, while y has a 0 in this row, and (2) both x and y have the same value in the rows above 2^k (i.e., in rows $2^{k+1}, 2^{k+2}, \ldots, 2^m$). Then $x > y$.

22.10 Determine a first move in rational play for the game Hackenbush introduced after Example 22.6. Determine the first move for the games introduced in Exercise 22.6.

22.11 Recall Nim Misère is played like Nim except that the last player to move is the winner (see Chapter 20). Prove that when either $m > 1$ or $n > 1$,

$$SG(\text{Nim Misère}(m, n)) = m \oplus n$$

23 Arrow's Impossibility Theorem

Kenneth Arrow's 1951 doctoral thesis launched a new wave of research on the mathematics of social choice. As with Nash's Equilibrium Theorem in game theory, Arrow's Impossibility Theorem has become the centerpiece of modern social choice theory, inspiring countless extensions, variants, alternate proofs, and responses.

We prove Arrow's Theorem (Theorem 18.19) in this chapter following the elegant argument given by Amayarta Sen. We begin by reviewing the basic ingredients of the statement (social welfare methods and their fairness criterion). Next, we introduce the idea of a *dictating set* and characterize the Pareto Criterion in terms of this notion. We then state *Arrow's Principle*, which asserts that dictating power passes from a set of voters to some proper subset. The Impossibility Theorem is then deduced as a direct consequence of Arrow's Principle. Our main work consists of proving Arrow's Principle. We prove this result using two further principles called the *Power Contraction* and *Power Expansion Principles*.

We conclude with some developments in social choice theory appearing after Arrow's thesis. We prove a theorem of Sen, called the *Sen Coherence Theorem*, that gives conditions on a preference table that ensure that the social preference relations are transitive. When the voters have Sen-coherent preferences, Condorcet's ideal of majority rule can be realized. We also introduce two voting methods that have attracted recent interest, *Approval Voting* and *Majority Judgment*. These methods do not fit neatly into our framework for voting methods. We briefly discuss the fairness properties of these methods.

Social Welfare and Fairness

Recall that a social welfare method is a mechanism that takes a preference table and produces a ranking of the candidates called the *social welfare*. We assume that ties are broken (see Convention 5.9) so that the ranking is strict, first to last. Recall that each of our social choice methods extend to a social welfare methods. For instance, we have the following result of Example 18.13:

Social Welfare Methods

Preference Table

Number	7	5	8	2
1st choice	A	C	D	B
2nd choice	B	A	C	D
3rd choice	C	B	A	C
4th choice	D	D	B	A

$\Longrightarrow$

Social Welfare

Plurality	Borda	Hare	Coombs
D	C	A	C
A	A	D	A
C	D	C	B
B	B	B	D

Recall that the Plurality Method and the Borda Count extend to social welfare methods by ranking candidates according to the number of first-place votes and the number of Borda points, respectively. The Hare and Coombs Methods both extend by ranking candidates according to the round of elimination.

For Arrow's Theorem, we treat the social welfare method as the unknown quantity. For example, we might have a social welfare method, Method Z, that happens to produce the following:

Social Welfare Method *Z*

Peference Table

Number	7	5	8	2
1st	A	C	D	B
2nd	B	A	C	D
3rd	C	B	A	C
4th	D	D	B	A

$\Longrightarrow$

Social Welfare

Method Z
B
D
A
C

We certainly cannot judge Method Z from this one election. To get more information, we could apply Method Z to other preference tables. In fact, Method Z can be applied to any preference table involving any collection of candidates, voters, and voter rankings of candidates.

We next recall the two fairness criteria appearing in the statement of Arrow's Theorem. First, let us establish some general notation. Given an election, we write $\mathcal{V}$ for the set of all voters. Subsets of $\mathcal{V}$ will be denoted $\mathcal{P}, \mathcal{Q}$, and so on and individual voters by p, q, and so on. The letters A, B, C, D, and so on will be used for candidates. Given a set of voters $\mathcal{P}$, we write

$$\mathcal{P}' = \text{all voters not in } \mathcal{P}$$

Recall that a collection of voters $\mathcal{P}$ is called a *voting bloc* if all the voters in $\mathcal{P}$ have the exact same rankings of the candidates. We say that $\mathcal{P}$ is a **partial voting bloc** if all voters in $\mathcal{P}$ agree on their comparison of (at least) one pair of candidates.

For example, $\mathcal{P}$ is a partial voting bloc if all voters in $\mathcal{P}$ have candidate A ranked above B. The voters in $\mathcal{P}$ may disagree about other candidates, and there may be other voters who rank A over B. Further, we do not require all voters in $\mathcal{P}$ to have A and B ranked in the same position. We indicate this simple piece of information in a **partial preference table**, by which we mean a preference table with only partial information about the voters preferences. For example, we might write the following:

A Partial Preference Table

$\mathcal{P}$	$\mathcal{P}'$
A over B	No information

The set of voters $\mathcal{P}'$ need not be a partial voting bloc in this scenario. Here is another example:

Another Partial Preference Table

$\mathcal{P}$	$\mathcal{P}'$
A over B	B over C

In this case, both $\mathcal{P}$ and $\mathcal{P}'$ are partial voting blocs.

Recall that the Pareto Criterion for social welfare methods (Defintion 18.16) requires that in every election in which every voter ranks candidate A over B, candidate A is ranked above B in the social welfare ranking of the candidates. We can picture a social welfare method satisfying Pareto using the following partial preference table:

The Pareto Fairness Criterion

Partial Preference Table		
$\mathcal{V}=$ all voters		Social Welfare
A over B	$\Longrightarrow$	A over B

We recall that the Borda Count social welfare method satisfies the Pareto Criterion, whereas the Plurality, Hare and Coombs social welfare methods violate this criterion because of technicalities with ties (see Theorem 18.17).

The interpretation of Arrow's Independence Criterion for social welfare methods is more complicated. However, the idea of a partial preference table is useful here as well. Recall from Definition 18.10 that the Independence Criterion requires that if candidate A is ranked above candidate B in the social welfare, then candidate A remains ranked above candidate B when voters change their preferences for other candidates as long as they retain their preferences for A versus B.

We can recast the Independence Criterion as follows: given a preference table, let $\mathcal{P}$ be the set of all voters who rank A over B so that $\mathcal{P}'$ consists of all voters who rank B over A. The sets $\mathcal{P}$ and $\mathcal{P}'$ are then partial voting blocs (in each set, all voters agree on their rankings of A versus B). Suppose that the outcome of the election has A ranked over B in the social welfare as follows:

First Election

Partial Preference Table

$\mathcal{P}$	$\mathcal{P}'$		Social Welfare
A over B	B over A	$\Longrightarrow$	A over B

The Independence Criterion requires that we can rerun the election making changes to the voters preferences. As long as $\mathcal{P}$ and $\mathcal{P}'$ are still partial voting blocs with respect to A and B in the sense that every one in $\mathcal{P}$ prefers A to B and everyone in $\mathcal{P}'$ prefers B to A, the social welfare will have A ranked above B again as follows:

Another Election

Partial Preference Table

$\mathcal{P}$	$\mathcal{P}'$		Social welfare
A over B	B over A	$\Longrightarrow$	A over B

We will make use of partial voting blocs and preference tables repeatedly in what follows.

Dictating Sets

Arrow's Theorem raises the specter of a Dictatorship in the quest for fairness in elections. A Dictatorship does have the virtue of simplicity, if few other virtues. The official definition of this method as a social welfare method is as follows:

Definition 23.1 A **Dictatorship** is the social welfare method satisfying the following condition: for each collection of voters $\mathcal{V}$, we can identify one voter p such that in any election held by the voters $\mathcal{V}$, the ranking that p submits for the candidates is always the social welfare ranking for the candidates.

Here is an illustration with four candidates:

A Dictatorship

Partial Preference Table / **Social Welfare**

p	Other voters		Dictatorship (p dictator)
A	Any ranking		A
C	of A, B,	$\Longrightarrow$	C
D	C,D		D
B			B

Given a collection of voters $\mathcal{V}$, we say the voter p is the **dictator** for the collection $\mathcal{V}$. Notice that the dictator p can change when the voting collection $\mathcal{V}$ changes. The definition of a dictatorship thus requires a choice of p for each collection $\mathcal{V}$. When $\mathcal{V}$ is fixed, a Dictatorship is quite simple: p's ranking of the candidates is the social welfare.

Now suppose that $\mathcal{V}$ is fixed and we are given a social welfare method, say, Method Z. The idea of a Dictatorship with dictator p extends to subsets $\mathcal{P}$ of $\mathcal{V}$ in the presence of a social welfare method.

Definition 23.2 We say that a subset $\mathcal{P}$ of $\mathcal{V}$ is a **dictating set** for $\mathcal{V}$ and a social welfare method if, whenever $\mathcal{P}$ votes as a bloc, the common ranking of the candidates chosen by $\mathcal{P}$ is always the social welfare.

An illustration is as follows:

A Dictating Set

Partial Preference Table

$\mathcal{P}$	Voters not in $\mathcal{P}$		Method Z social welfare
A	Any ranking		A
C	of $A, B,$	$\Longrightarrow$	C
D	C, D		D
B			B

The notion of a dictator p for $\mathcal{V}$ is recovered by the following equivalence result:

Theorem 23.3 *A social welfare method for a collection of voters $\mathcal{V}$ is a dictatorship with Dictator p if and only if the set $\{p\}$ is a dictating set.*

Proof The result is a direct consequence of our definitions. ■

As a sample result, we investigate the possibilities for dictating sets for the Borda Count. Not surprisingly, they are quite limited.

Theorem 23.4 *Let $\mathcal{V}$ be a collection of voters. A subset $\mathcal{P}$ of $\mathcal{V}$ is a dictating set for the Borda Count social welfare method if and only if $\mathcal{P} = \mathcal{V}$.*

Proof It is easy to see that $\mathcal{P} = \mathcal{V}$ is a dictating set for the Borda Count. We focus on the converse statement. Suppose that $\mathcal{P}$ is a proper subset of $\mathcal{V}$. Let $\mathcal{Q}$ denote the complement of $\mathcal{P}$ in $\mathcal{V}$. Let p, q, and n denote the number of voters in $\mathcal{P}$, $\mathcal{Q}$, and $\mathcal{V}$, respectively. We consider an election with $n+1$ candidates $A_1, A_2, \ldots, A_{n+1}$. We suppose the voters $\mathcal{P}$ and $\mathcal{Q}$ vote as blocs as follows:

Preference Table

$\mathcal{P}$	$\mathcal{Q}$
A_1	A_{n+1}
A_2	A_1
A_3	A_2
$\vdots$	$\vdots$
A_n	A_{n-1}
A_{n+1}	A_n

Notice that candidate A_n receives a total of p Borda points (all from $\mathcal{P}$) and that candidate A_{n+1} receives $q \cdot n$ Borda points (all from $\mathcal{Q}$). Since $\mathcal{P}$ is a proper subset of $\mathcal{V}$, $q \geq 1$ and $p < n \leq q \cdot n$. Thus A_n is ranked below A_{n+1} in the Borda Count social welfare, and $\mathcal{P}$ is not a dictating set. ■

The Pareto Criterion can be characterized in terms of dictating sets via the following important equivalence result:

Theorem 23.5 *A social welfare method satisfies the Pareto Criterion if and only if the full collection of voters $\mathcal{V}$ is always a dictating set for $\mathcal{V}$ and the given method.*

Proof The proof is just a matter of unraveling definitions. Suppose that the voting method satisfies the Pareto Criterion so that any ranking of two candidates that is held by all voters is preserved in the social welfare. We check that $\mathcal{V}$ is a dictating set for $\mathcal{V}$, so we assume that $\mathcal{V}$ votes as a bloc. In this case, all the rankings of candidates are unanimous, so the social welfare is just the ranking of the candidates chosen by the bloc $\mathcal{V}$.

Conversely, suppose the full set of voters $\mathcal{V}$ is a dictating set. This means that when all voters in $\mathcal{V}$ rank candidate A over B, this ranking occurs in the social welfare. Thus the social welfare method satisfies the Pareto Criterion by definition. ■

Inheriting Power

The proof of Arrow's Theorem depends on a surprising fact about voting power in a social welfare system satisfying Independence and Pareto fairness. The precise statement is Arrow's Principle, given next.

The idea can be viewed as a transfer of power from parent to child using the following terminology: let $\mathcal{P}$ be a subset of the collection of all voters $\mathcal{V}$. Let $\mathcal{Q}$ be a subset of $\mathcal{P}$. If $\mathcal{Q}$ is not empty and not all of $\mathcal{P}$ ($\mathcal{Q}$ is a proper subset of $\mathcal{P}$), then we say that $\mathcal{Q}$ is a **child** of $\mathcal{P}$ and $\mathcal{P}$ is a **parent** of $\mathcal{Q}$. Let $\mathcal{Q}'$ denote the set of all voters in $\mathcal{P}$ but not in $\mathcal{Q}$. We will call the sets $\mathcal{P}, \mathcal{Q}$ and $\mathcal{Q}'$ a **family**. We picture this as follows:

A Family of Voters

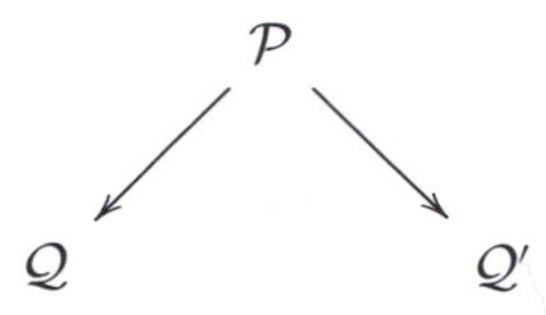

Now suppose that we are given a social welfare method, Method Z, and $\mathcal{P}$ is a dictating set for Method Z elections with voting set $\mathcal{V}$. Recall that this means that if all voters in $\mathcal{P}$ agree about a comparison of any two candidates A and B, then this ranking of A and B occurs in the Method Z social welfare ranking. The Power Contraction Principle asserts that if Method Z satisfies the Pareto and Independence Criteria, then one of $\mathcal{P}$'s children, $\mathcal{Q}$ or $\mathcal{Q}'$, inherits the power of dictating set from their parent $\mathcal{P}$.

Theorem 23.6 Arrow's Principle *Suppose that we are given a social welfare method for at least three candidates satisfying the Pareto and Independence fairness criteria. Let $\mathcal{V}$ be a collection of voters and $\mathcal{P}$ a subset of $\mathcal{V}$. Suppose that $\mathcal{P}$ is a dictating set for the given social welfare system and the voters $\mathcal{V}$. Let $\mathcal{Q}$ be a proper subset of $\mathcal{P}$ and $\mathcal{Q}'$ the set of voters in $\mathcal{P}$ but not in $\mathcal{Q}$. Then one of the sets $\mathcal{Q}$ or $\mathcal{Q}'$ is a dictating set.* ■

In our terminology, Theorem 23.6 asserts that the family tree of a dictating set $\mathcal{P}$ takes the following form:

Family Tree of a Dictating Set

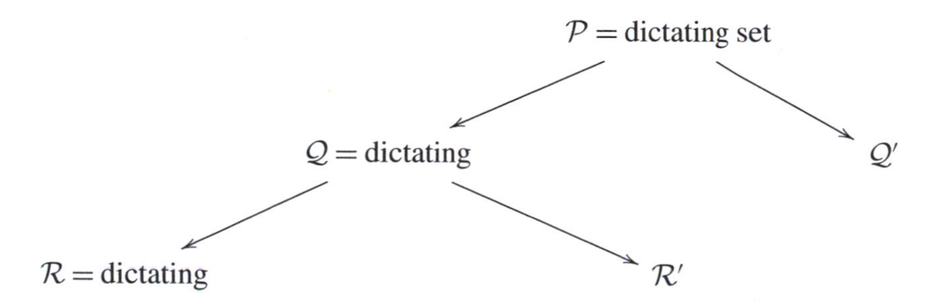

The tree can be extended as long as there are more children. In fact, notice that a dictating set $\mathcal{P}$ has no children precisely when $\mathcal{P} = \{p\}$ contains just one voter. However, in this case, the voter p is a dictator by Theorem 23.3. Arrow's Principle thus allows us to identify a dictator in a dictating set and leads to a proof of Arrow's Impossibility Theorem.

Theorem 23.7 Arrow's Impossibility Theorem *The only social welfare method with three or more candidates that satisfies both the Independence and Pareto Criteria is a Dictatorship.*

Proof Suppose that we are given a social welfare method that satisfies the Pareto and Independence Criteria. Let $\mathcal{V}$ be a collection of voters. Since the social welfare method satisfies Pareto fairness, $\mathcal{V}$ is a dictating set by Theorem 23.5.

Write $\mathcal{V} = \{p_1, p_2, \ldots, p_n\}$, where the p_i are the individual voters. Let $\mathcal{P}_i = \{p_1, p_2, \ldots, p_i\}$ so that $\mathcal{P}_i$ is a proper subset of $\mathcal{P}_{i+1}$ and $\mathcal{P}_n = \mathcal{V}$. We have the following family tree for $\mathcal{V}$:

Family Tree for $\mathcal{V}$

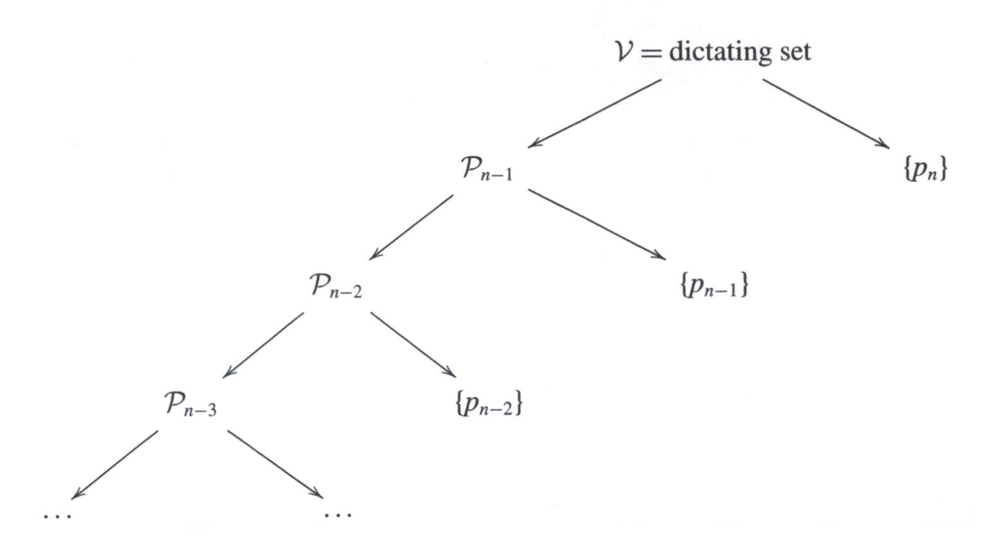

By Arrow's Principle (Theorem 23.6), working down the tree from the root, at each stage, one of the two sets is a dictating set. If it is ever the singleton set $\{p_i\}$, then p_i is the needed dictator. If dictating power is always passed to the larger set $\mathcal{P}_i$, we wait until the last stage when the family tree is

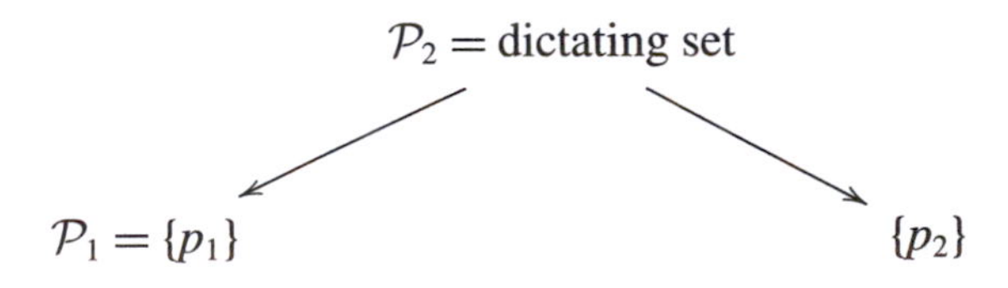

We conclude, in this case, that either p_1 or p_2 is the dictator. ■

Proof of Arrow's Principle

It remains to prove Arrow's Principle. Our goal is to identify dictating sets $\mathcal{P}$ for a collection of voters $\mathcal{V}$. In order to do so, we introduce a seemingly much weaker notion of voting power.

Definition 23.8 Given a social welfare method, a collection of voters $\mathcal{V}$ and two candidates A and B in an election, we say that a subset $\mathcal{P}$ of $\mathcal{V}$ is **decisive for A over B** if, whenever all voters in $\mathcal{P}$ rank A over B, then A is ranked over B in the social welfare.

Here is the picture with partial preference tables. As usual, $\mathcal{P}'$ denotes voters in $\mathcal{V}$ that are not in $\mathcal{P}$.

$\mathcal{P}$ is decisive for A over B

$\mathcal{P}$	$\mathcal{P}'$		Social welfare
A over B	No constraint on A versus B	$\Longrightarrow$	A over B

Now suppose that the social welfare method in question satisfies the Independence Criterion. Then, to show $\mathcal{P}$ is decisive for A over B, we may impose conditions on how voters rank other pairs of candidates. These rankings do not affect the social welfare ranking of A versus B. As long as we impose no constraints on how the voters in $\mathcal{P}'$ rank A and B, the preceding scenario implies that $\mathcal{P}$ is decisive for A over B. We will use this observation repeatedly in what follows.

As a first step toward the proof of Arrow's Principle, we show that the power of decisiveness is passed, in some form, from parent to one child.

Theorem 23.9 Power Contraction Principle *Suppose that we are given a collection of voters $\mathcal{V}$ and a social welfare method for at least three candidates satisfying the Pareto and Independence Criteria. Suppose that a subset of votes $\mathcal{P}$ is decisive for candidate A over B. Let $\mathcal{Q}$ be a proper subset of $\mathcal{P}$ and $\mathcal{Q}'$ the set of voters in $\mathcal{P}$ but not in $\mathcal{Q}$. Then either $\mathcal{Q}$ or $\mathcal{Q}'$ is decisive for some pair of candidates.*

Proof Let C be a third candidate. Let $\mathcal{P}'$ denote the voters in $\mathcal{V}$ not in $\mathcal{P}$. Consider the following election in which the voters in $\mathcal{P}'$ are not constrained:

Partial Preference Table

$\mathcal{P}$		$\mathcal{P}'$
$\mathcal{Q}$	$\mathcal{Q}'$	
A over B	A over B	No constraints
A over C	C over B	
No constraint on B versus C	No constraint on A versus C	

Since every voter in $\mathcal{P}$ has A ranked above B, we must have A ranked above B in the social welfare. There are three possible social welfare rankings in which A is ranked over B:

Social welfare 1	Social welfare 2	Social welfare 3
A over B	A over C	C over A
B over C	C over B	A over B

In the first two rankings, A is above C. The voters in $\mathcal{Q}$ all have A ranked above C, and all the other voters (those in $\mathcal{Q}'$ and $\mathcal{P}'$) have no constraint on their rankings of A versus C. Thus, if social welfare 1 or 2 is the result, we can see that $\mathcal{Q}$ is decisive for A over C. In social welfare 3, C is ranked above B. The voters in $\mathcal{Q}'$ have C ranked above B, and all the other voters are free to rank C and B as they choose. Thus $\mathcal{Q}'$ is decisive for C over B. ■

Theorem 23.9 shows that some form of power is passed in the sense of Arrow's Principle. We next prove a truly remarkable fact that may be paraphrased as "a little power goes a long way."

Theorem 23.10 Power Expansion Principle *Suppose that we are given a social welfare method for at least three candidates satisfying the Pareto and Independence Criteria with voters $\mathcal{V}$. Suppose that a subset $\mathcal{P}$ of voters is decisive for some candidate A over another candidate B. Then $\mathcal{P}$ is a dictating set for $\mathcal{V}$.*

Proof Let X and Y be an arbitrary pair of candidates. We prove that $\mathcal{P}$ is decisive for X over Y. We first handle the case where X and Y are distinct from A and B. Consider the following partial preference table:

$\mathcal{P}$	$\mathcal{P}'$
X	X over A
A	B over Y
B	No constraint on X versus Y
Y	

What is the social welfare for this election? Since $\mathcal{P}$ is decisive for A over B, we must have A over B in the social welfare. Since every voter has X ranked above A, by the Pareto Criterion, X must be ranked above A. Similarly, every voter has B ranked above Y, so B must be ranked above Y in the social welfare. The only possibility for the social welfare, then, is

Social welfare
X over A
A over B
B over Y

In particular, X is ranked above Y in the social welfare. Since $\mathcal{P}'$ was not constrained in its ranking of X versus Y this means that $\mathcal{P}$ is decisive for X over Y.

It remains to consider the various cases in which one of X and Y is allowed to be either A or B. We treat the most interesting case (where $X = B$ and $Y = A$) and leave the remaining for the exercises.

We prove that if $\mathcal{P}$ is decisive for A over B, then $\mathcal{P}$ is decisive for B over A. We introduce a third candidate C. We first show that $\mathcal{P}$ is decisive for C over B.

Consider the following partial preference table:

$\mathcal{P}$	$\mathcal{P}'$
C	B over C
A	No other constraints
B	

Since every voter has C ranked above A, we must have this ranking in the social welfare by the Pareto Criterion. Since $\mathcal{P}$ is decisive for A over B, this ranking holds in the social welfare also. The only possibility for the social welfare ranking is

Social welfare
C over A
A over B

Since no constraint was placed on C versus B among the voters in $\mathcal{P}'$, we conclude that $\mathcal{P}$ is decisive for C over B.

Next, we show that $\mathcal{P}$ is decisive for C over A. Consider the following partial preference table:

$\mathcal{P}$	$\mathcal{P}'$
C	B over A
B	No other constraints
A	

Since we now know that $\mathcal{P}$ is decisive for C over B and every voter ranks B over A, the social welfare ranking is:

Social welfare
C over B
B over A

We conclude that $\mathcal{P}$ is decisive for C over A.

Finally, to show that $\mathcal{P}$ is decisive for B over A, we use the following partial preference table:

$\mathcal{P}$	$\mathcal{P}'$
B	B over C
C	No other constraints
A	

Since $\mathcal{P}$ is decisive for C over A, and B is ranked above C by all voters, we conclude that the social welfare is as follows:

Social welfare
B over C
C over A

We conclude that $\mathcal{P}$ is decisive for B over A, as needed. ■

We apply the preceding two results to prove Arrow's Principle:

Proof of Theorem 23.6 Suppose that we are given a social welfare method with at least three candidates satisfying Independence and Pareto fairness, a collection of voters $\mathcal{V}$, and a dictating set $\mathcal{P}$. Let $\mathcal{Q}$ be a proper subset of $\mathcal{P}$ and $\mathcal{Q}'$ the set of voters in $\mathcal{P}$ but not in $\mathcal{Q}$. Since $\mathcal{P}$ is a dictating set, by definition, $\mathcal{P}$ is decisive for A over B for any pair of candidates A and B. By Theorem 23.9, either $\mathcal{Q}$ and $\mathcal{Q}'$ is decisive for some pair of candidates. By Theorem 23.10, this set ($\mathcal{Q}$ or $\mathcal{Q}'$) is then a dictating set, as needed. ■

The proof again reveals the Condorcet paradox as the basic obstruction to producing a "fair" social welfare.

After Arrow

Arrow's thesis marked the start of a renaissance in social choice theory. We have seen some of the theorems in the same vein that appeared after the Impossibility Theorem, including Sen's Impossibility Theorem (Theorem 18.22), Mackay's Theorem (Theorem 18.12), and notably, the Gibbard-Satterthwaite Theorem (Theorem 6.4). We conclude with two examples of other types of developments in the field. The first concerns conditions on voter preferences that ensure that a fair social welfare can be achieved. We then introduce two voting methods that seem to offer an escape from Arrow's Impossibility Theorem.

Sen Coherence

Recall from Definition 5.13 that the relation of social preference is determined by the Simple Majority Method. We say that candidate A is socially preferred to candidate B if a majority of voters prefer A to B. Transitivity is then the property that, for any three candidates A, B, and C, if A is socially preferred to B and B is socially preferred to C, then A is socially preferred to C. A Condorcet Paradox occurs when social preferences are not transitive (Definition 6.2).

If social preferences are transitive, then the social welfare can be taken to be the social preference ranking. In this case, the social welfare truly fulfills Condorcet's ideal of majority rule: each comparison of candidates in the social welfare obeys the majority preference of the voters. It is natural then to ask

What conditions on a preference table will ensure
that social preferences are transitive?

An answer to this question was given by Sen in the terms of a *coherence* condition on the preference table. We introduce this condition now.

finition 23.11 A preference table is said to be **Sen coherent** if, for each set of three candidates A, B, and C, at least one of the three candidates, say, A, satisfies at least one of the following three conditions:

1. No voter ranks A above both B and C.
2. No voter ranks A between B and C.
3. No voter ranks A below both B and C.

We note that Sen coherence requires that for each choice of three candidates in the election, one statement of nine possible is true. Specifically, if A, B, and C are the three candidates, then candidate A can satisfy any one of conditions (1), (2), or (3), or candidate B can satisfy one of the three conditions, or candidate C can satisfy one of the conditions. We prove the following theorem:

Theorem 23.12 Sen Coherence Theorem *Suppose that we are given a preference table for an odd number of voters and at least three candidates. If the preference table is Sen coherent, then the relation of social preference is transitive.*

Proof Let A, B and C be any three candidates. Suppose that A is socially preferred to B and B is socially preferred to C. We must show that A is socially preferred to C. Since the preference table is Sen coherent, we know that one of the candidates A, B, or C satisfies one of the three preceding conditions (1), (2), or (3). We are thus faced with nine possible cases, which we may denote as A_1, A_2, A_3, B_1, B_2, B_3, C_1, C_2, and C_3. As we will see, certain of theses cases are impossible:

Case A_1: We assume that no voter ranks A above both B and C. Every voter then either ranks A below both B and C or between B and C. We group voters into blocs according to their preferences for these three candidates. We can then write

Number of voters	w	x	y	z
1st choice	B	C	B	C
2nd choice	C	B	A	A
3rd choice	A	A	C	B

We know that A is socially preferred to B, so $z > w+x+y$. Since B is socially preferred to C, we have $w+y > x+z$. By substitution, we obtain

$$z > w+x+y = (w+y)+x > (x+z)+z = 2z+x$$

This inequality is clearly impossible, so we can eliminate this case.

Case A_2: In this case, we assume that no voter ranks A between B and C. Here we have the following partial preference table:

Number of voters	w	x	y	z
1st choice	A	A	B	C
2nd choice	B	C	C	B
3rd choice	C	B	A	A

Since A is socially preferred to B, we see $w+x > y+z$. This inequality implies that A is socially preferred to C as well, so this case leads to the desired result.

Case A_3: We assume that no voter ranks A below both B and C. The partial preference table takes the following form:

Number of voters	w	x	y	z
1st choice	A	A	B	C
2nd choice	B	C	A	A
3rd choice	C	B	C	B

Since B is socially preferred to C, we obtain $w+y > x+z$. Now note that

$$w+x+y = (w+y)+x > (x+z)+x \geq z$$

We conclude that A is socially preferred to C, as needed.

The remaining cases proceed similarly. We handle two more and leave the rest to you as Exercise 23.7

Case B_3: We assume that no voter ranks B below both A and C. Here we have

Number of voters	w	x	y	z
1st choice	A	C	B	B
2nd choice	B	B	A	C
3rd choice	C	A	C	A

Since A is socially preferred to B, we have $w > x+y+z$. Since B is socially preferred to C, we obtain $w+y+z > x$. Now note that

$$w+y > (x+y+z)+y = x+z+2y \geq x+z$$

We conclude that A is socially preferred to C, as needed.

Case C_2: We assume that no voter ranks C between A and B. The preference table is thus:

Number of voters	w	x	y	z
1st choice	A	B	C	C
2nd choice	B	A	A	B
3rd choice	C	C	B	A

Since B is socially preferred to C, we obtain $w+x > y+z$. This inequality already implies that A is socially preferred to C. ■

We remark that Duncan Black gave the first general condition on the preference table to ensure transitivity. The notion is called *single-peakedness*.

Nonpreferential Voting

Sen's Coherence Theorem gives conditions on the preferences of voters that ensure the transitivity of social preferences. We conclude with two methods that eliminate preference ballots in the strict sense, allowing voters to submit intransitive preferences. Since these methods do not fulfill the requirements for a voting method as in Definition 2.7, we refer to them as **nonpreferential voting methods** to avoid confusion.

Approval Voting

The idea behind this nonpreferential method is simple. Instead of having voters rank the candidates, voters decide whether or not they "approve" of each possible candidate. The candidate with the most voter approval is then the social choice.

finition 23.13 **Approval Voting** is the nonpreferential social choice method whereby each voter submits a list of the candidates of whom he or she approves. The candidate with the most approval votes is the **Approval Voting social choice**.

Recall the preference table for Family Dinner Decision (Example 5.3) from Chapter 5, in which seven voters (family members) are trying to decide between four dining options: Thai (T), Mexican (M), Greek (G), and a diner (D).

Preference Table for Family Dinner Decision

Family member	**Mom**	**Dad**	**Big bro.**	**Sister 1**	**Sister 2**	**Sister 3**	**Baby bro.**
1st choice	T	T	M	G	G	G	D
2nd choice	M	M	T	M	M	M	T
3rd choice	D	D	D	D	D	D	M
4th choice	G	G	G	T	T	T	G

We can imagine the approval votes in this election as follows:

Possible Approval Votes for Family Dinner Decision

Family member	**Mom**	**Dad**	**Big bro.**	**Sister 1**	**Sister 2**	**Sister 3**	**Baby bro.**
Thai	✓	✓	×	×	×	×	×
Mexican	✓	✓	✓	✓	✓	✓	×
Greek	×	×	×	✓	✓	✓	×
Diner	×	×	×	×	×	×	✓

In this scenario, Mexican is the Approval Voting social choice with six approval votes.

By doing away with voter preference rankings and focusing instead on approval of candidates, Approval Voting appears to avoid the problems afflicting our preferential voting methods. For example, one might expect that there is little incentive to vote insincerely because voters are only listing the candidates that are acceptable options. Indeed, our fairness criteria do not directly apply to Approval Voting because it is not, strictly speaking, a voting method. However, if we perform the translation of preferences to approval votes, as earlier, we can then analyze this method using our framework of fairness criteria. In practice, this analysis is easy because we have a good deal of freedom in deciding which candidates are approved for a given preference table. Here is an illustration:

Theorem 23.14 *Approval Voting violates the Independence Criterion.*

Proof We consider a scenario with three candidates A, B, and C and three voting blocs with two, three, and four voters, respectively. The blocs of two and four voters approve their top two choices, whereas the bloc of three voters approves only its top choice. Originally, we have:

Original Preference Table

Number	2	3	4
1st choice	A	B	A
2nd choice	B	A	C
3rd choice	C	C	B

$\Longrightarrow$

Original Approval Votes

Number	2	3	4
A	✓	×	✓
B	✓	✓	×
C	×	×	✓

The Approval Voting social choice is candidate A. We now imagine that all voters retain their preferences for A versus B but some change their preferences for C. Specifically, we suppose that the bloc of four voters switches from ranking C over B to ranking B over C. We then have

New Preference Table

Number	2	3	4
1st choice	A	B	A
2nd choice	B	A	B
3rd choice	C	C	C

$\Longrightarrow$

New Approval Votes

Number	2	3	4
A	✓	×	✓
B	✓	✓	✓
C	✓	×	×

The winner of the second election is candidate B, a violation of Independence. ■

We remark that while Approval Voting would appear to satisfy the Pareto Criterion, the issue of ties in the social choice arises as it did with the Plurality, Hare and Coombs Methods and leads to a violation (see Exercise 23.10).

Majority Judgment

We introduce a voting system designed by Michel Balinski and Rida Laraki representing an extension of Approval Voting. In this system, voters judge each candidate's fitness according to a common scale. We will use the letter grades A, B, C, D, and F for this purpose. The candidates are ranked by virtue of their *median* grades. Ties are broken by computing *rank 1 medians*. We explain this terminology now.

Given a list of grades chosen from A, B, C, D, and F, the **median** grade is the middle grade obtained when the list is ordered from highest (A) to lowest (F). Here is an example with an odd number of grades:

Median with odd number: A A A B B B (C) C C D D F F

When there are an even number of grades, we define the median to be the smaller of the two middle grades. Here is an example with an even number of grades:

Median with even number: A A B C C (D) D D F F

Given a list of grades as earlier, the **rank 1 median**, written **median$_1$**, is obtained by removing the median grade and then determining the median of the remaining grades. Here is an example

Original scores: A A A B B B (C) C C D D F F

Median$_1$: A A A B B B (C) C D D F F

Definition 23.15 The **Majority Judgment** social welfare method is the voting system in which voters judge candidates according to a common grading system. The rankings of the candidates are determined by their median grades, with ties between candidates broken by the rank 1 median scores.

We illustrate the method using the Family Dinner Decision. We imagine that the family members judge the dinner options as follows:

Judgments for Family Dinner Decision

Family member	Mom	Dad	Big bro.	Sister 1	Sister 2	Sister 3	Baby bro.
Thai	A	A	C	F	F	F	F
Mexican	B	B	A	B	B	B	F
Greek	C	B	F	A	A	A	F
Diner	B	B	C	C	C	C	A

We then compute the medians:

Medians for Family Dinner Decision

	Median						
Thai	A	A	C	(F)	F	F	F
Mexican	A	B	B	(B)	B	B	F
Greek	A	A	A	(B)	C	F	F
Diner	A	B	B	(C)	C	C	C

Finally, we break the tie for first place using the rank 1 medians:

Tie Breaker: Mexican and Greek

	Median$_1$					
Mexican	A	B	B	(B)	B	F
Greek	A	A	A	(C)	F	F

The Majority Judgment social welfare is Mexican   Greek   Diner   Thai.

Notice that if voters only assign the grades A or F, the Majority Judgment method reduces to Approval Voting. Thus any violation of a fairness criterion by Approval Voting is a violation by Majority Judgment. We prove the following result:

Theorem 23.16 *The Majority Judgment method violates the Majority Winner Criterion and exhibits the No-Show Paradox.*

Proof We prove both statements using the same example involving seven voters ($p_1,p_2,p_3,p_4,p_5,p_6,p_7$) voting on two candidates (X and Y). Here are the judgments of the voters for the two candidates:

Voters	p_1	p_2	p_3	p_4	p_5	p_6	p_7
X	A	A	A	C	C	C	C
Y	B	B	B	B	F	F	F

We observe that X has median grade C, and Y has a median grade B. Thus Y is the Majority Judgment social choice even though a majority of the voters prefer X to Y.

Now suppose that p_1 does not vote in the election. Then the judgments are

Voters	p_2	p_3	p_4	p_5	p_6	p_7
X	A	A	C	C	C	C
Y	B	B	B	F	F	F

In this case, X is the Majority Judgment social choice, which is a better outcome for p_1. This is an example of the No-Show Paradox (Definition 6.11). ■

The preceding examples highlight the fact that despite the limitations sketched by Arrow's Impossibility Theorem and the Gibbard-Satterthwaite Theorem, the pursuit of fair and democratic voting methods remains an important and actively studied problem.

Exercises

23.1 Given a social choice method, say, Method Z, we can create a social welfare method called the **recursive** social welfare as follows: Given a preference table, run Method Z to find the social choice, say, X. Candidate X is then ranked first in the social welfare and removed from the preference table. The Method Z social choice for the new preference table, say Y, becomes the second candidate in the recursive Method Z social welfare and so on.

(a) Find the recursive social welfare for the following preference table using the Plurality, Borda Count, Hare, and Coombs Methods.

Number of voters	**5**	**8**	**6**	**10**	**7**
1st choice	*A*	*C*	*E*	*D*	*B*
2nd choice	*B*	*E*	*D*	*B*	*A*
3rd choice	*C*	*A*	*C*	*A*	*D*
4th choice	*D*	*B*	*A*	*E*	*E*
5th choice	*E*	*D*	*B*	*C*	*C*

(b) Give an example to show that the Recursive Borda Count social welfare need not be the same as the Borda Count social welfare.

(c) Repeat part (b) for the Hare Method.

23.2 Here is a social welfare method representing a compromise between two extremes. First, define a social choice method called the **Liberal Dictator**. If the preference table has a Condorcet Winner, that candidate is the social choice. Otherwise, a fixed dictator p chooses the social choice. The **Liberal Dictator** social welfare method is then defined to by applying recursion to this social choice method, as in Exercise 23.1.

(a) Find the Liberal Dictator social welfare for the preference table in Exercise 23.1, part (a). Take the dictator to be one of the voters in the first bloc of five voters.

(b) Decide whether the Liberal Dictator social welfare method satisfies the Pareto Criterion. Prove your answer.

(c) Decide whether the Liberal Dictator social welfare method satisfies the Independence Criterion. Prove your answer.

23.3 The definition of a dictating set can be formulated for a social choice method as follows: a subset $\mathcal{P}$ of all voters $\mathcal{V}$ is a dictating set for a social choice method if, whenever all the voters in $\mathcal{P}$ rank a candidate A at the top of their individual rankings, then A is the social choice.

(a) Prove that a voting method satisfies the Majority Winner Criterion if and only if any group of voters $\mathcal{P}$ numbering more than half the total in $\mathcal{V}$ is a dictating set.

(b) Prove that a set $\mathcal{P}$ of voters is a dictating set for the Plurality Method if and only if $\mathcal{P}$ has a majority of the voters.

(c) Consider a Borda Count election with 21 voters and 5 candidates. Find the smallest number k such that any collection $\mathcal{P}$ of k voters is a dictating set for this social choice.

23.4 Prove that no social welfare method can have two distinct dictators. Explain why this implies that, in the preceding proof of Arrow's Principle (Theorem 23.6), social welfare 3 is impossible.

23.5 Complete the proof of Arrow's Power Expansion Principle (Theorem 23.10) by proving the following statements. In each case, the proof is similar to that given in the text. Assume that the social welfare method satisfies the Pareto and Independence Criteria. Assume that $\mathcal{P}$ is decisive for A over B.

(a) Prove that $\mathcal{P}$ is decisive for A over Y for some third candidate Y.
(b) Prove that $\mathcal{P}$ is decisive for B over Y for some third candidate Y.
(c) Prove that $\mathcal{P}$ is decisive for X over A for some third candidate X.
(d) Prove that $\mathcal{P}$ is decisive for X over B for some third candidate X.

23.6 Prove that the following preference table is Sen Coherent. Determine the transitive social preference order.

Number	**12**	**7**	**14**	**22**	**8**	**9**	**10**
1st choice	A	C	B	B	A	C	B
2nd choice	B	B	A	D	D	B	A
3rd choice	C	D	D	C	B	D	C
4th choice	D	A	C	A	C	A	D

23.7 Complete the proof of Theorem 23.12 by proving that in cases B_1, B_2, C_1, and C_3 either candidate A beats C head to head or the given scenario is impossible.

23.8 Give an example of a preference table that is not Sen Coherent but does produce transitive social preferences.

23.9 Determine the Majority Judgment social welfare for the election with 31 voters and 4 candidates (W, X, Y, and Z) with grades A, B, C, D, F.

Number	**7**	**4**	**2**	**6**	**5**	**3**	**4**
W	A	B	C	D	B	C	F
X	B	A	B	C	C	B	A
Y	C	B	D	A	A	A	B
Z	C	F	A	D	C	C	C

23.10 Give an example proving that Approval Voting violates the Pareto Criterion because of tie-breaking considerations.

23.11 Give an example proving that Approval Voting violates the Condorcet Loser Criterion.

23.12 Give an example proving that Approval Voting is manipulable.

23.13 **Range Voting** is a nonpreferential voting system in which each voter assigns a number from 0 to 9 to each candidate. The candidate with the highest average total score (points awarded divided by number of voters) is the Range Voting social choice. Prove that

(a) Range Voting violates the Majority Winner Criterion.
(b) Range Voting violates the Majority Loser Criterion.
(c) Range Voting is manipulable.

In each case, you should start with a preference table and assign scores to candidates based on the rankings.

Suggestions for Further Reading

Chapter 19

A great source for paradoxes in probability theory is the text by Frederick Mosteller [27]. For a more philosophical discussion, see William Eckhardt [13]. The Sleeping Beauty Paradox (Example 19.2) has been written about extensively. Adam Elga's paper [15] is a good starting point for delving deeper into this literature.

Chapter 20

The definitive book on combinatorial games is the multivolume text by Berlekamp, Conway, and Guy [5]. David Gale's paper on Hex [18] gives a rigorous and accessible proof that Hex is a win-lose game (Theorem 20.11).

Chapter 21

Modern analysis of the debate begun with the papers of Borda [11] and Condorcet [12] is extensive. H. P. Young [53] carefully interprets Condorcet's work to uncover a plausible candidate for Condorcet's social welfare method. Young argues that Condorcet's social welfare corresponds to the Kemeny Method social welfare introduced in Exercise 21.7. John Kemeny's original paper describing his method is quite readable [21]. Donald Saari has been a vocal proponent of the Borda Count (see, for example, ref. 39). A recent article of Mathias Risse argues the side of Condorcet [38]. David McGarvey's paper [26] on realizing social preference graphs (Theorem 21.8) is quite accessible. The strong version of Condorcet's Jury Theorem (Theorem 7.10) can be proved as an application of the *Central Limit Theorem* for binomial probability, as found in the standard statistics text.

Chapter 22

Our proof of the Sprague–Grundy Theorem (Theorem 22.4) follows the argument given by Thomas Ferguson in his excellent online lecture notes [16]. These notes also contain extensions of the game Hackenbush (Example 22.6), as well as many other applications of the Sprague–Grundy Theorem in combinatorial game theory.

Chapter 23

Our proof of Arrow's Impossibility Theorem (Theorem 18.19) follows Sen [45]. Sen's article also contains helpful commentary on Arrow's Theorem. Sen's Coherence Theorem (Theorem 23.12) is proved in ref. 43. The books by Jerry Kelly [20] and Hannu Nurmi [31] give an overview of social choice theory after Arrow. The method of Approval Voting is analyzed by Steven Brams and Peter Fishburn [10]. The Majority Judgment Method is introduced by Michel Balinski and Rida Laraki in their paper [2]. Finally, Michael Philips [36] gives a philosophical discussion of the question of whether preference ballots are appropriate for social welfare problems.

Bibliography

1. Arrow, Kenneth (1951/1963), *Social Choice and Individual Values*, Wiley, New York.
2. Balinski, Michel, and Rida Laraki (2007), "A theory of measuring, electing and ranking," *Proceedings of the National Academy of Sciences, USA* **104**:8720–5.
3. Banzhaf, John (1968), "One man, *n* votes?: Mathematical analysis of voting power and effective representation," *George Washington Law Review* **36**:808–23.
4. Bouton, Charles (1901), "Nim, a game with a complete mathematical theory," *Annals of Mathematics* **14**:35–9.
5. Berlekamp, Elwyn, John Conway, and Richard Guy (1982), *Winning Ways for Your Mathematical Plays*, Academic Press, London.
6. Black, Duncan (1958), *Theory of Committees and Elections*, Cambridge University Press, Cambridge.
7. Blau, Julian (1975), "Liberal values and independence," *The Review of Economic Studies* **42**:395–401.
8. Brams, Steven (1994), *Theory of Moves*, Cambridge University Press, Cambridge.
9. Brams, Steven (2011), *Game Theory and the Humanities: Bridging Two Worlds*, MIT Press, Cambridge, MA.
10. Brams, Steven, and Peter Fishburn (1978), "Approval voting," *American Political Science Review* **72**:831–47.
11. Borda, Jean-Charles de (1781), "Mémoire sur les élections au Scrutin," *Histoire de l'Académie Royale des Sciences Paris*.
12. Condorcet, The Marquis de (1785), *Essai sur l'Application de l'Analyse à la Probabilité des Décisions Rendues à la Pluralité des Voix*, L'Imprimerie Royale, Paris.
13. Eckhardt, William (2012), *Paradoxes in Probability Theory*, Springer-Verlag, New York.
14. Felsenthal, Dan (2010), "Review of paradoxes afflicting various voting procedures where one out of *m* candidates ($m \geq 2$) must be elected," in *Assessing Alternative Voting Procedures*, London School of Economics.
15. Elga, Adam (2000), "Self-locating belief and the sleeping beauty problem," *Analysis* **60**:143–7.
16. Ferguson, Thomas, "Game theory"; available online at www.math.ucla.edu/~tom/Game_Theory/comb.pdf.
17. Fishburn, Peter (1977), "Condorcet social choice functions," *SIAM Journal on Applied Mathematics* **33**:469–89.
18. Gale, David (1979), "The game of Hex and the Brouwer fixed-point theorem," *American Mathemaitcal Monthly* **86**:818–27.
19. Gibbard, Allan (1973), "Manipulation of voting systems: A general result," *Econometrica* **41**:587–601.
20. Kelly, Jerry (1978), *Arrow Impossibility Theorems*, Academic Press, New York.
21. Kemeny, John (1959), "Mathematics without numbers," *Daedalus* **88**:577–91.

22. Kuhn, Steven (2009), "Prisoner's dilemma," *The Stanford Encyclopedia of Philosophy*; available online at plato.stanford.edu/archives/spr2009/entries/prisoner-dilemma/.
23. Luce, Duncan, and Howard Raiffa (1957/1989), *Games and Decisions*, Dover, New York.
24. Mackay, Alfred (1973), "A simplified proof of an impossibility theorem," *Philosophy of Science* **40**:175–7.
25. May, Kenneth (1952), "A set of independent, necessary and sufficient conditions for simple majority decision," *Econometrica* **20**:680–4.
26. McGarvey, David (1953), "A theorem on the construction of voting paradoxes," *Econometrica* **21**:608–10.
27. Mosteller, Frederick (1987), *Fifty Challenging Problems in Probability with Solutions,* Dover, New York.
28. Nanson, Edward (1883), "Methods of elections," *Transactions and Proceedings of the Royal Society of Victoria* **19**:197–240.
29. Nash, John (1950), "Non-cooperative games," Ph.D. thesis, Princeton University, Princeton, NJ.
30. Nurmi, Hannu (1983), "Voting procedures: A summary analysis," *British Journal of Political Science* **13**:181–208.
31. Nurmi, Hannu (1999), *Voting Paradoxes and How to Deal with Them*, Springer-Verlag, New York.
32. O' Neill, Barry (1986), "International escalation and the dollar auction," *Journal of Conflict Resolution* **30**:33–50.
33. Osborne, Martin (1995), "Spatial models of political competition under plurality rule: A survey of some explanations of the number of candidates and the positions they take," *Canadian Journal of Economics* **28**:261–301.
34. Osborne, Martin, and Ariel Rubinstein (1994), *A Course in Game Theory*, MIT Press, Cambridge, MA.
35. Packel, Edward (1981), *The Mathematics of Gambling and Gaming,* Mathematics Association of America, Washington, DC.
36. Philips, Michael (1989), "Must rational preferences be transitive?," *Philosophical Quarterly* **39**:477–83.
37. Press, William, and Freeman Dyson (2012), "Iterated prisoner's dilemma contains strategies that dominate any evolutionary opponent," *Proceedings of the National Academy of Science USA* **109**:10409–13.
38. Risse, Mathias (2005), "Why the Count de Borda cannot beat the Marquis de Condorcet," *Social Choice and Welfare* **25**:95–113.
39. Saari, Donald (1990), "The Borda dictionary," *Social Choice and Welfare* **7**:279–317.
40. Saari, Donald (1997), "Are individual rights possible?," *Mathematics Magazine* **70**:83–92.
41. Satterthwaite, Mark (1975), "Strategy-proofness and Arrow's conditions: Existence and correspondence theorems for voting procedures and social welfare functions," *Journal of Economic Theory* **10**:187–217.
42. Selten, Reinhard (1978), "The chain store paradox," *Theory and Decision* **9**:127–59.
43. Sen, Amayarta (1966), "A possibility theorem on majority decisions," *Econometrica* **34**:491–9.
44. Sen, Amayarta (1970), "The impossibility of a Paretian liberal," *Journal of Political Economy* **78**:152–7.
45. Sen, Amayarta (1995), "Rationality and social choice," *American Economic Review* **85**:1–25.
46. Sen, Amayarta (2009), "From social choice to development: The influence of Nicholas de Condorcet and Jean-Charles de Borda," *Revue Tiers Monde* **53**:263–7.

47. Straffin, Philip (1988), "The Shapley-Shubik and Banzhaf power indices as probabilities," in *The Shapley Value: Essays in Honor of Lloyd Shapley*, ed. Alvin Roth, Cambridge University Press, Cambridge.
48. Straffin, Philip (1993), *Game Theory and Strategy*, Mathematics Association of America, Washington, DC.
49. Taylor, Alan, and Allison Pacelli (2009), *Mathematics and Politics*, Springer-Verlag, New York.
50. Taylor, Alan, and William Zwicker (1992), "A characterization of weighted voting," *Proceedings of the American Mathematical Society* **115**:1089–94.
51. von Neumann, John, and Oscar Morgenstern (1944), *The Theory of Games and Economic Behavior*, Princeton University Press, Princeton, NJ.
52. Williams, J. D. (1954/1982), *The Compleat Strategyst*, Dover, New York.
53. Young, H. P. (1988), "Condorcet's theory of voting," *American Political Science Review* **82**:1231–44.

Index

1989–90 NBA MVP Election, 57
1990 Irish presidential election, 58
1992 U.S. presidential election, 21
2000 Florida presidential election, 18
2012 U.S. presidential election, 89

Arrow's Impossibility Theorem
  proof, 356
  statement, 272
Arrow's Principle, 355
Arrow, Kenneth, 7
auction, 39
  dollar, 241
  sealed bid, 40
  Vickrey, 249

backward induction solution, 44, 237
Banzhaf, John, 93
Battle of the Sexes, 15
Bayes' Law, 212
binary labeling rule, 47
binary number, 255
  binary expansion, 288
  column form, 256
Birthday Problem, 289
Blackjack, 32
Borda, Jean-Charles de, 7
Borel, Émile, 5
Bouton's Theorem, 259
Brams, Steven, 246
Bridge, 215

Chain-Store Game, 238
Chicken, 120
Chomp, 306
combinatorial identity, 198
Condorcet ranking, 319
Condorcet, Nicolas de, 7, 315
Craps, 203

dictating set, 353
directed graph, 67
  complete, 67
dominated-strategy solution, 42

election, 17
expected value, 29

fairness criterion, 175
  Anonymity, 266
  Condorcet Loser, 182
  Condorcet Winner, 177
  Independence
    social choice, 268
    social welfare, 271
  Majority Loser, 179
  Majority Winner, 176
  Minimal Liberalism, 274
  Monotonicity
    social choice, 264
    social welfare, 276
  Neutrality, 266
  Nonperversity, 276
  Pareto
    social choice, 275
    social welfare, 271

game, 11
  combinatorial, 46
  equivalence, 347
  fairness of, 33
  gambling, 23
  iterated, 236
  Misère, 303
  mixed-strategy, 111
  outcome, 11
  partial-conflict, 37
  payoff, 11
  pure-chance, 12
  pure-strategy, 12, 108
  repeated, 106
  sequential-move, 12
  simultaneous-move, 14

game (*cont.*)
  subgame, 31
  sum, 336
  symmetric, 234
  takeaway, 47
  total-conflict, 45
  win-lose, 12
  zero-sum, 15
Gibbard-Satterthwaite Theorem, 79

Hackenbush, 343
Hex, 309
Hotelling-Downs Electoral Game, 130
Hunting Stag, 247

impossibility theorem, 268

Jury Theorem, 100, 316

Law of Combinations, 164
Law of Conditional Probability, 168
Law of Conjunctions and Negation, 161
Law of the Probability Tree, 26, 170
Linearity of Expectation, 32, 170

Mackay's Theorem, 269
manipulable, 79
mathematical invariant, 193
mathematical model, 233
mathematical statement, 139
  biconditional, 142
  conditional, 140
  conjecture, 140
  contrapositive, 141
  converse, 141
  counterexample, 140
  existential, 144
  universal, 143
maximin strategy, 46, 125
May's Theorem, 267
McGarvey's Theorem, 324
minimax strategy, 125
Minimax Theorem, 223
  2×2 case, 225
  $2 \times n$ case, 228
  graphical representation, 226
  mixed-strategy equilibrium, 223
  pure-strategy equilibrium, 224
minimum excluded number, 251
mixed strategy, 109
Monte Hall Problem, 281
myopia, 237

Nash equilibrium, 120
Nash, John, 5
Nim, 5, 258, 340
  binary representation, 257
  configuration, 258
  definition, 49
  Misère, 304
  Nim sum, 338
  Two-Pile, 49
Nimble, 312
nonmyopic equilibrium, 245
Northcott's Game, 302

O'Neill's Theorem, 243
ordinal labeling rule, 251

paradox
  Bertrand's Box, 294
  Betrand's, 291
  Condorcet, 79
  Liar, 152
  No-Show, 84
  of the Chair, 87
  Prisoner's, 293
  Simpson's, 290
  St. Petersburg, 295
Pascal's triangle, 197
payoff matrix, 14
perfect information, 43
permutations, 164
Poker
  Betting on Diamonds, 105, 128
  Bluffing, Calling, and Raising, 112, 227
  Five-Card Stud, 4, 203
  hierarchy of hands, 206
  Straight Poker, 163
Power Contraction Principle, 357
Power Expansion Principle, 358
preference table, 60
  partial, 351
  Sen coherent, 361
Pricing Game, 38
Principle of Counting, 151
Principle of Induction, 146
  for rooted trees, 149
Prisoner's Dilemma, 38, 119
  generalized, 250
  iterated, 236
probability, 23, 159
  conditional, 167
  empirical, 157
  event, 23, 158

independent, 169
mutually exclusive, 161
experiment, 23, 158
sample space, 158
simple outcome, 158
tree, 26

rational play, 51
Rock-Paper-Scissors, 19
Roulette, 33

saddle point, 125
Sen Coherence Theorem, 361
Sen's Impossibility Theorem, 274
Sen, Amayarta, 272
Sleeping Beauty Problem, 283
Smith set, 74
social preference, 66
transitive, 78
social preference graph, 67, 324
realization, 324
social welfare method, 69, 270
Sprague–Grundy number, 253, 310, 337
Sprague–Grundy Theorem, 340
strategy stealing, 309
Subset Takeaway Game, 314

Taking Chips, 47, 254
Taylor-Zwicker Theorem, 195
Theory of Moves, 244
third-party candidate, 21
tournament, 321
transfer votes, 62
tree
A-rooted subtree, 149
abridged, 31
binary, 152
branch, 13
branch probability, 26
depth, 148
end node, 13
game, 13
path, 148
probability, 26
root, 13
Tree Game, 147
Turning Turtles, 312
Two-Envelopes Problem, 294

U.N. Security Council, 188
modified, 192
U.S. Electoral College, 89
U.S. Senate, 8, 196

value bet, 212
Venn diagram, 172
von Neumann, John, 5
voting bloc, 18
partial, 350
voting method, 16
Approval Voting, 363
Black Method, 328
Borda Count, 61
Bucklin Method, 184
Condorcet Method, 331
Coombs Method, 65
Copeland Method, 74
Dictatorship, 63
Double Plurality Method, 184
Hare Method, 62
Kemeny Method, 332
Majority Judgment, 366
Nanson Method, 331
nonpreferential, 363
Pairwise Comparisons, 68
Plurality Method, 60
Range Voting, 370
recursive, 367
Simple Majority Method, 66
Single Transferable Vote, 74

weak solution, 309
weakly dominated strategy, 41
weighted voting system, 92
Wythoff's Game, 313

Yahtzee, 173
yes-no voting system, 92
Banzhaf power index, 94
critical player, 93
equivalence, 191
Shapley-Shubik power index, 103
trade robust, 193
weighted voting system, 92
winning coalition, 93

Zermelo's Theorem, 49
for combinatorial games, 301
for takeaway games, 146